Competitive Strategy Analysis in the Food System

Competitive Strategy Analysis in the Food System

EDITED BY

Ronald W. Cotterill

Westview Press

BOULDER • SAN FRANCISCO • OXFORD

Copyright © 1993 by Westview Press, Inc., except for chapters 10 and 13, which are works of the United States Government

Published in 1993 in the United States of America by Westview Press, Inc., 5500 Central Avenue, Boulder, Colorado 80301-2877, and in the United Kingdom by Westview Press, 36 Lonsdale Road, Summertown, Oxford OX2 7EW

Library of Congress Cataloging-in-Publication Data
Competitive strategy analysis in the food system / edited by Ronald W.
  Cotterill.
    p.   cm.
  Includes bibliographical references.
  ISBN 0-8133-8638-1
  1. Food industry and trade—United States.   2. Grocery trade—
United States.   3. Food—United States—Marketing.   I. Cotterill,
Ronald.
HD9005.C589   1993
381'.456413'00973—dc20   92-34588
                                                               CIP

Printed and bound in the United States of America

The paper used in this publication meets the requirements
of the American National Standard for Permanence of Paper
for Printed Library Materials Z39.48-1984.

10   9   8   7   6   5   4   3   2   1

Contents

v

PART FOUR
Market Structure-Strategy Studies

PART FIVE
The Food Retailing Debate

PART SIX
Strategies in International Markets

PART SEVEN
Private and Public Decision Making

Tables and Figures

Figures

Acknowledgments

The "Competitive Strategy Analysis in the Food System" conference, on which this book is based, was organized by Regional Research Project NE-165: Private Strategies, Public Policies, and Food System Performance. Members of the conference organizing committee included Ronald Cotterill, Olan Forker, Jeffrey Perloff, Terry Roe, Richard Rogers, Randall Torgerson, and Randall Westgren. Financial support for the conference was provided by the Food Marketing Policy Center at the University of Connecticut; Cooperative State Research Service, USDA; The Agricultural Cooperative Service, USDA; and The Farm Foundation, Chicago, Illinois.

The editor wishes to thank the organizing committee for reviewing all papers and posters presented at the conference and selecting the particular chapters that appear in this book. A special thank you goes to the superb staff of the Food Marketing Policy Center. Andrew Franklin, production supervisor, provided liaison with authors and the publisher and produced camera-ready copy for all the graphs, figures, charts, and equations in this book. Irene Dionne, secretary, did extensive word processing. Finally the editor wishes to especially acknowledge and thank Julie Caswell and Richard Rogers, Department of Resource Economics, University of Massachusetts at Amherst; and NE-165/Food Marketing Policy Center colleagues for their council, advice, and support in this endeavor.

Ronald W. Cotterill
Executive Director, NE-165
Director, Food Marketing Policy Center

PART ONE

Introduction and Overview

1

Introduction and Methodological Overview

Ronald W. Cotterill

This book analyzes how firms compete in food markets. It is timely for several reasons. The organization of the farm-to-retail food system continues to evolve towards fewer, larger firms, and those firms tend to occupy leading positions in several food industries or markets. During the 1980s seller concentration in food manufacturing industries and retail food markets has continued to increase. The food system has also become global. Thus, the performance of large firms in oligopolistic markets is more important than ever in any assessment of the overall performance of the food sector. Concerning performance, between 1979 and 1989 fifteen of the top twenty supermarket chains were merged or highly leveraged with significant price premiums being paid out to acquired firm stockholders (Cotterill, 1991). Due to similar events in food manufacturing the stock prices of food firms in the Standard and Poor's 500 Index far outpaced the general market, increasing 900% between 1980 and 1989.[1]

Competitive strategy analysis seeks to identify the source of this increase in profitability for leading firms in the food system. It became a distinct field of business management during the 1980s. The field grew with the advent of the Strategic Planning Institute, its profit impact of market strategy (PIMS) project (Schoeffler *et al.*, 1974, Buzzell *et al.*, 1975, and Buzzell and Gale 1987), and the appearance of several texts on the subject, including the seminal work by Michael Porter, *Competitive Strategy Analysis* (1980). Porter recognized that research on industrial organization that is oriented towards antitrust and market regulation issues can also be presented to corporate executives as strategies for obtaining competitive advantage. Managers who strive to obtain and sustain long run profitability in excess of competitive rates of return have a vested interest in the sources of market power and efficiency. Consequently, competitive strategy analysis focuses upon many factors including market share, concentration, cost conditions, relative efficiency of different sized firms, advertising and product differentiation, pricing strategies, strategic groups, barriers to entry, mobility barriers, research and development, strategic planning processes, and public policy. Except for research and development, this book addresses each of these.

The following section describes organization and contents. A brief synopsis of each chapter is presented. The final section of this chapter describes how different empirical tests for market power and efficiency are related. Since eleven chapters address the issue and four different empirical methods are used, a methodological introduction seems appropriate. This moreover is consistent with a basic objective of the NE-165 conference where these papers first appeared. In addition to presenting diverse empirical approaches that conference sought to compare and contrast alternative methods of analysis.

Overview of Chapters

The chapters in this book are organized into six sections: structure performance studies, new empirical industrial organization (NEIO) studies, structure-strategy studies, the debate on the influence of market structure on prices and profits in the food retailing industry, strategies in international markets, and finally private and public decision making. In the first section Cotterill and Iton estimate a structure profit relationship for food manufacturing. They (1) test alternative models at the business unit level, (2) demonstrate that the relationship between concentration and profits depends upon model specification, (3) conclude that the most appropriate model provides evidence for a positive concentration profit relationship at the business unit level and (4) reject the Demsetz efficiency hypothesis as an underlying explanation for the relationship. In the next paper Frank and Henderson develop comprehensive measures of vertical coordination and analyze the determinants of the degree of vertical coordination. They report that transaction costs related to uncertainty are the most important determinant of increased vertical coordination. Frank and Henderson then analyze the impact of vertical coordination and other elements of market structure upon food manufacturing industry profitability and export propensities. They report that vertical coordination, concentration, and advertising are related to industry profits. Frank and Henderson hypothesized that increased vertical coordination would increase an industry's export propensity; however, they found the opposite.

The first of three NEIO studies in section two is by Wann and Sexton. It is a generalization of the conjectural variation method. The authors estimate decomposed price cost margins for the California pear processing industry to determine whether pear processors exercise market power against growers, and whether they exercise market power against retail food distributors in two processed product markets: canned pears and fruit cocktail. Wann and Sexton report modest price enhancement in both processed product markets relative to a benchmark market (fresh pears). They also find that (1) processors are exercising market power against growers in the procurement market, (2) closed membership pear processing cooperatives play a major role in the industry, and (3) processing cooperative conduct does not seem to reduce the degree of oligopsony power facing growers. The advent and operation of an industry-wide bargaining cooperative does, however, affect processor power.

In the second NEIO paper Durham and Sexton examine the California tomato processing market. They develop the supply side analog to the NEIO residual demand analysis of oligopoly and use the resulting residual supply model to analyze oligopsony power in the grower-tomato processor market. Since raw tomatoes are bulky and difficult to transport long distances, they examine several local processor market areas in California. Durham and Sexton find that the potential for the exercise of market power against growers due to spatial location is limited: only two of six market areas provide evidence for potentially significant noncompetitive pricing. They explain their results by noting there is substantial market overlap which generally occurs for two reasons. Firms from one area procure supply from other areas before or after their own region is in its most productive period, and to ensure against local shortfalls in planned delivery.

The final paper in the section, by Karp and Perloff, generalizes the conjectural variation model from a single to a multiperiod model. They analyze alternate dynamic pricing strategies in the international rice export markets. In the open loop equilibrium, firms choose their trajectory of output levels at the initial time period. In the feedback equilibrium, the industry chooses rules that allow them to adjust outputs in all time periods. Within each solution concept, one can observe competitive or noncompetitive conduct. Karp and Perloff also present Bayesian as well as classical estimates of the conjectural variation parameter for each solution concept, so there is an array of results. Using classical estimation for both solution concepts, they reject fully collusive conduct but not Cournot or competitive conduct in the rice industry.

Section three contains three papers that analyze advertising or price conduct in food markets. Forker and Lenz describe the extent of consumer targeted advertising by producers to increase demand for farm commodities. To date the research on generic advertising has focused upon estimating segments of the advertising sales response function and returns to producers due to advertising induced increased demand. Forker and Lenz conclude that generic advertising programs have generally been beneficial to the producers that have paid for them. However, several key research issues remain. Most notably, they maintain that research has yet to identify the total shape of the nonlinear advertising sales response function. Also, the impact of generic advertising upon brand advertising and product differentiation, and the magnitude of cross product advertising elasticities (e.g. the impact of almond advertising on the consumption of walnuts) need more extensive analysis.

In the next paper Haller presents a market structure-price analysis for a differentiated product oligopoly: the cottage cheese industry. Using scanner data of retail sales of individual brands of cottage cheese across 47 market areas in the fourth quarter of 1988, he is able to analyze the impact of a manufacturer's local market share, retailer concentration, and other variables upon retail price. Haller especially focuses upon the conduct of farmer cooperative versus investor owned (IOF) cottage cheese manufacturers. He reports that IOF prices are positively related to market share and cooperative prices are not. Also, the presence of cooperatives lowers the prices charged by IOFs. Thus, dairy processing cooperatives seem most interested in maximizing cottage cheese sales rather than profits. Retail grocery concentration is positively related to cottage cheese prices, but only marginally significant in one specification. Since cottage cheese is only one of several thousand products sold at retail, single product models of this type do not provide a powerful test of a retail concentration-price hypothesis.

In the third paper of the structure strategy section, Marion, Heimforth and Bailey provide a direct test of the retail structure-price hypothesis. They analyze changes in the Bureau of Labor Statistics food at home price index using a panel data set of 25 metropolitan statistical areas for 1977-1987. They are particularly interested in the penetration of warehouse stores, a new strategic group, and report that in markets with a significant amount of warehouse store penetration, established grocery store operators posted lower food price increases than operators in markets with no warehouse activity. They also report that changes in retail prices are positively related to changes in seller concentration.

The next section of this book presents four papers that continue a long-standing debate on the merits of structure, price and profit studies in food retailing. In 1977 the Joint Economic Committee of the U.S. Congress (U.S. Congress 1977a) sponsored research at the University of Wisconsin and subsequently held hearings on this topic (Marion, *et al.* 1977, 1979; U.S. Congress, 1977a, 1977b). The JEC study and hearings had two major impacts on research. First, its study of the concentration price relationship was one of the first empirical answers to the Demsetz critique (ie. the concentration-profit relationship is due to lower costs and not higher prices). Consequently, it encouraged Leonard Weiss and others to conduct concentration-price studies in a wide array of industries including food retailing (Weiss, 1989). Second, food retailing industry representatives strongly objected to the JEC study methods and conclusions. They requested that the USDA do additional research on the topic. A large scale USDA research effort commenced soon thereafter, and ultimately produced a report in 1989 (Kaufman and Handy).

In the first retail debate paper, Anderson critiques the JEC and eight subsequent studies of food retailing. He concludes that all studies completed to date are flawed and provide little reliable guidance for private or public policy formulation. Cotterill responds to Anderson in the second paper of the section. Next Geithman and Marion present evidence that the empirical methods used in the recent Kaufman and Handy study are flawed and are probably the reason why its results diverge from the other recent major studies of this industry. In the final paper Kaufman and Handy respond to the Geithman-Marion critique. These four papers clearly identify the strengths and weaknesses of particular approaches and may be particularly useful for researchers who wish to do structure-price research in other industries as well as food retailing.

Section five titled "Strategies in International Markets" contains two papers. Sheldon and Henderson use non-cooperative game theory to analyze branded product licensing as a strategic alternative to exporting or entry into a foreign market by direct foreign investment. In their simplest model, licensing is a vehicle for extracting rents from imperfectly competitive markets overseas. In a more complex model, the game's solution is sensitive to strategic interaction amongst incumbent firms and imperfect information about payoffs.

Lopez and Pagoulatos explore the structural and political determinants of trade barriers in the U.S. food manufacturing industries. They develop a model that analyzes the 1972 level of tariff barriers in four digit SIC census industries for the food manufacturing sector. Explanatory variables include labor's share of industry value added (positive and significant effect), the capital labor ratio (negative and significant), average wage rates (negative and significant), employment growth (negative and significant), and industry concentration (positive and significant). In a simulated welfare analysis of trade barriers, Lopez and Pagoulatos report that the deadweight welfare loss may be extremely small, however, the welfare transfers from consumers to producers can be significant. Rent seeking behavior by firms or producer groups can improve incomes.

The final section contains two papers. In the first, Westgren and Sonka examine strategic management decision making in the California fruit and vegetable processing industry. In this stable industry they report that more comprehensive planning leads to superior performance. This occurs at least in part because more comprehensive planning enhances management's planning abilities.

In the second paper, Preston and Connor examine the level and distribution of Federal Trade Commission antitrust enforcement activity in the food sector during 1981-1989. Rather than count investigations and cases, they use data obtained under the Freedom of Information Act on professional hours expended by staff lawyers and economists on each matter. They document that, excluding the big three shared monopoly cases inherited from the previous administration (Kellogg, *et al.*; ITT-Continental Baking; and General Foods), the number of professional hours devoted to antitrust activity dropped in the 1984-1989 period to 25 to 30 percent of the level in the 1981-84 period. Merger enforcement accounted for more than half of effort in 1981-1989 primarily because federal antitrust agencies encouraged the numerous acquiring firms to request prior approval of merger plans, including plans to eliminate horizontal overlap. Exclusive dealing, tying and price fixing matters accounted for the bulk of remaining effort. Except for the big three monopolization cases, little effort was allocated to monopolization and price discrimination.

A Comparison of Alternative Empirical Approaches to Strategy Analysis

Industrial organization theory provides an analytical foundation for competitive strategy analysis. During the 1980s there was an explosion in theory and new approaches to empirical analysis of market power and efficiency. This book illustrates four different approaches: traditional market structure-profit analysis, conjectural variation analysis, market structure-price analysis, and residual demand analysis. The last three were developed primarily in response to extensive criticism of the first.

There are many, some would maintain too many, oligopoly models. Yet in many of them one can deduce the market power hypothesis: firms have higher profits when they have large market shares in more concentrated markets with entry barriers because they have power over prices.[2] Consider the generalized Cournot model (Cowling and Waterson, 1976; Clark and Davies, 1982; and Scherer and Ross, 1990 p. 230). In an industry with homogenous goods and inverse market demand given by $p = p(x_1...x_n)$ a firm with constant marginal cost, c_i, a firm's profits are

$$\pi_i = px_i - c_i x_i \qquad (1)$$

The generalized Nash-in-quantities solution to the profit maximization problem for the ith firm gives

$$p = c_i + (1 + \lambda_i) p' x_i \qquad (2)$$

where p' is the slope of the inverse demand function and $\lambda_i = \sum_{i \neq j} \dfrac{\partial x_j}{\partial x_i}$ is the firm's conjectural variation. The conjectural variation for firm i is that firm's perception of how other firms will change their output when it changes its output. Equation 2 can be rewritten in terms of the price cost margin, commonly known as the Lerner index of market power, as follows:

$$\frac{p - c_i}{p} = \frac{s_i}{-\eta_m}(1 + \lambda_i) \qquad (3)$$

where s_i is the firm's market share and $\eta_m < 0$ is the market elasticity of demand.

Industry and Business Unit Structure Profit Analysis

Equation 3, when aggregated to the industry level, predicts that an industry's price cost margin is positively related to market concentration as measured by the Herfindahl index, H, and a weighted average conjectural variation parameter, λ, as well as the market elasticity of demand (Cowling and Waterson, 1976).

$$\frac{p - \bar{c}}{p} = \frac{H}{-\eta_m}(1 + \lambda) \qquad (4)$$

If the industry follows a Cournot strategy, $\lambda = 0$, and industry profitability is a linear function of seller concentration. Alternatively, if the strategic conduct of firms in the industry is a function of concentration, i.e. $\lambda = \lambda(H)$, one obtains a nonlinear relationship. Testing this market structure-profit model proceeds by regressing industry profit margins on seller concentration and control variables. These include measures of barriers to entry, growth in demand, measures of minimum efficient scale, vertical integration, advertising intensity, and capital intensity. Business unit level market structure-profit tests are motivated by equation 3 and analyze the market share-profit relationship. These disaggregate studies seek to explain, for example, Phillip Morris/General Foods' profitability in the coffee market as opposed to the average profitability of the coffee industry. PIMS business unit models of profitability have explained over 70 percent of the observed variation in business unit profits (Buzzel and Gale, 1987, p. 51).

Market structure-profit tests for market power at both the industry and business unit level have been criticized for three shortcomings: the failure to control for efficiency as a source of profitability, measurement problems associated with the use of accounting profit rates, and endogeneity of market structure with attendant estimation problems. The problems due to use of accounting measures, and endogeneity will be discussed at the end of this section. Concerning efficiency, Demsetz (1973) first argued that the industry concentration-profit relationship is due to the superior efficiency (lower cost) of leading firms and not power over price. To discriminate between the market power and efficiency hypothesis one needs to analyze the intra industry variation in profitability among different size firms as well as industry effects. The oligopoly model at the business unit level can be expanded to test the two hypotheses jointly (Clark, Davies and Waterson, 1984; Scherer and Ross, 1990, Chapter 11; Cotterill and Iton, this book).

Conjectural Variation Analysis

The conjectural variation approach to testing for market power, does not analyze variations in market structure across a set of industries. Rather it employs time series or cross section data from a particular industry. Such studies can be completed at the business unit level, however, the data needed for disaggregate studies are often not available.[3]

Industry level conjectural variation studies estimate the slope of the inverse market demand curve, p', from a demand equation and estimate an aggregate industry marginal cost, $\bar{c}$, from one or more cost equations. They then use an aggregate industry version of equation 2 to estimate λ, the weighted average conjectural variation parameter for the industry. Some studies estimate the aggregate industry conjectural elasticity (Applebaum, 1982). It is the market share weighted sum of firm conjectural elasticities that are defined as: $\theta_i = s_i (1 + \lambda_i)$ for firms $i = 1 \ldots n$. Either the conjectural variation or elasticity can be used to estimate an industry price cost margin. Industry level studies conclude that market power exists if this estimated price cost margin is significantly greater than zero.

Industry level conjectural variation studies, in one important dimension, correspond to industry level structure-profit studies. They can not distinguish between market power and efficiency. To avoid aggregation bias industry conjectural variation studies use a Gorman polar cost function and thus assume that firm total cost curves have different intercepts but identical constant slopes, i.e. different fixed cost levels but identical and equal marginal costs (Applebaum, 1982). As Clark and Davies (1982) demonstrate, this cost specification formally requires that all firms have equal market shares and identical conjectural variations in equilibrium. Applebaum (1982) also concludes that in equilibrium this requires all firms to have identical conjectural elasticities, however, Breshnahan disagrees with the Clark and Davies and Applebaum interpretation. He writes:

> In many circumstances, the lack of single-firm data will prohibit estimation of supply relations for individual firms. Instead aggregate industry data must be used. . . . When there is market power, however, different firms will almost certainly have different marginal costs in equilibrium. Analysis like that of Cowling and Waterson (1976) shows that noncooperative oligopoly will tend

to have variation in price-cost margins across firms unless they have identical, constant *MC*. In these circumstances, a stable industry marginal cost curve need not exist, and interpretation of [estimated conjectural variations and price cost margins] may be clouded. . . . It is better to follow Cowling and Waterson (1976) and interpret the aggregate (conjectural variation) as industry average conduct and the (corresponding) price cost margin as the average industry markup (Breshnahan, 1989, p. 1030).

The relevant point for the current discussion is that under either interpretation aggregate industry level models, by construction, cannot distinguish whether larger firms have lower marginal costs than smaller firms. In other words, industry level analysis cannot measure what proportion of an estimated industry price cost margin is due to market power and what proportion is due to efficiency. In principle, the power efficiency question can be analyzed in a conjectural variation study at the firm or business unit level where one can model and estimate a cost function that is sensitive to firm size—i.e., let marginal cost, c_i , *for* $i = 1 \ldots n$, vary among firms. To date Gollop and Roberts (1979) is the primary example.[4]

Differentiated Oligopoly: Residual Demand and Structure Price Analysis

The residual demand and the market structure-price approaches avoid the cost efficiency critique when testing for market power. They are best illustrated within the context of a differentiated products oligopoly model. Residual demand analysts estimate the residual demand curves for an individual product (business unit) and ascertain whether the residual demand curve has nonzero (negative) slope. If it does, the firm has market power (Baker and Breshnahan, 1985, 1988). Structure-price analysts usually analyze whether the market price level is higher and some also analyze whether larger market share firms have higher prices.[5]

Harris (1988) generalizes a dominant firm model to incorporate product differentiation. The model is Nash in prices (Bertrand). A dominant firm's own demand curve is a function of market demand, rival supply, and pricing by the rest of the oligopoly group: $q_1 \equiv q[q_m(p_1,\ldots),q_r(p_1,\ldots),p_r(p_1,\ldots)]$ where q_m = market demand, p_1 = price, q_r = rivals' summed supply, and $p_r(p_1) =$ a rival (price) response function.[6] Differentiating q_1 with respect to p_1 and manipulating, one obtains an expression for the dominant firm's own-price elasticity of demand, η_1,

$$\eta_1 = \eta_m \frac{q_m}{q_1} - \theta_s \frac{q_r}{q_1} + \eta_c \eta_r \qquad (5)$$

where $\eta_m < 0$ is the price elasticity of market demand. $\theta_s \gtreqless 0$ is the conjectural own-price elasticity of rival supply. It is the perceived percent change in rival supply for a one percent change in dominant firm price. $\eta_c \geq 0$ is the cross-price elasticity of firm 1 demand, and $\eta_r \gtreqless 0$ is the conjectural rival price response elasticity. This latter parameter is the perceived percent change in rival price for a one percent change in dominant firm price. Recognizing that q_m / q_1 is the inverse of firm 1's market share $(1 / s_1)$ and that $q_r / q_m \equiv (1 - s_1)$, and taking the simplest case of identical marginal costs, c, across firms, the profit maximizing price-cost margin may be written

$$pcm_1 = \frac{(p_1 - c)}{p_1} = \frac{1}{\eta_1} = f_1 = \frac{s_1}{-\eta_m + \theta_s(1 - s_1) - \eta_c \eta_r s_1}. \tag{6}$$

The firm's price-cost margin (pcm_1) equals the inverse of its residual price elasticity of demand; and equals f_1, the price flexibility or markup parameter. These in turn are a complex function of market share, market elasticity of demand, and cross price elasticity of demand and the two conduct parameters.

    Rather than specify f_1 as an observable variable and correlate it to other variables, residual demand studies specify f_1 as a parameter to be estimated and interpret it as a measure of market power. This requires that one assume that the right hand side of (6) is constant (Baker and Breshnahan, 1985, p. 427, footnote 1).[7] They estimate the price flexibility coefficient, f_1, within an applied demand analysis framework. It is the coefficient on own quantity in a log linear inverse residual demand equation. The inverse residual demand equation gives the firm's price as a function of its own quantity. In the equation estimated, price is also a function of a set of exogenous demand and cost shift variables that are obtained by substituting the rival firms' supply and demand conditions into the residual demand equation (Baker and Breshnahan, 1988). In this quasi-reduced form demand equation the only remaining endogenous variable is own quantity. Provided there are exogenous cost shift variables that affect only firm 1, one can use them as instrumental variables in a two or three stage least squares estimation routine to identify and estimate, f_1, the price flexibility coefficient.

    The structure-price approach addresses the Demsetz critique by solving for price as a function of costs as well as the variables on the right side of 6. Solving for p_1 gives

$$p_1 = \frac{c}{1 - f(s_1, \eta_m, \theta_s, \eta_s, \eta_r)}. \tag{7}$$

If the firm's conjectural rival supply response elasticity, θ_s, and its conjectural rival price response elasticity, η_r, are zero then the price equation reduces to:

$$p_1 = \frac{c}{1 - \dfrac{s_1}{\eta_m}}. \tag{8}$$

This is a rigorous statement of the Harvard School hypothesis that product differentiation confers market power. Even in the constant cost case a firm's price can vary and is a positive function of its market share. Market shares differ because different firms have different levels of product differentiation. The higher prices that large profit maximizing firms charge are due to the market power that differentiation confers, not higher costs.

Harris demonstrates that for a wide range of plausible values of cross price elasticity, η_c, and the behavioral parameters (θ_s, η_r) the partial derivative of equation 6 with respect to share is positive,

$$\frac{\partial pcm_1}{\partial s_1} > 0. \tag{9}$$

Since costs are constant in equation 6 this implies that the partial derivative of price with respect to share is positive. Harris also demonstrates that pcm_1 also is positively related to share when marginal costs are not constant (Harris, p. 276). However, if costs are a function of market share, or as Demsetz (1973) and Clark and Davies (1982) hypothesize, share is a function of costs, then one would have an ambiguous sign for the price share derivative. Higher market share firms may have higher prices due to product differentiation and market power and/or higher costs, or they may have lower prices due to lower costs. Measuring these relationships is central to competitive strategy analysis.

The Fisher-McGowan Critique and Endogeneity

Each of the four approaches presented here have received criticism. Two general criticisms deserve mention here: the Fisher-McGowan critique of accounting profits and endogeneity in estimated models. Fisher and McGowan (1983) argue that observed accounting profits say nothing about economic profits because they do not measure capital costs and thus costs accurately. Their criticism is not that marginal costs may be different than reported average costs, thereby making observed average profit rates poor measures of the Lerner index (Fisher McGowan, p. 83, footnote 7). If it were then the econometric estimation of marginal costs within conjectural variation studies or more careful specification of structure-profit studies (Clark, Davies, and Waterson, 1984, Martin, 1988) would answer the critique. Fisher explains:

> The problem is as follows: The numerator of the accounting rate of return in question is current profits; those profits are the consequence of investment decisions made in the past. On the other hand, the denominator is total capitalization, but some of the firm's capital will generally have been put in place relatively recently in the expectation of a profit stream much of which is still in the future. While the economic rate of return is the magnitude that properly relates a stream of profits to the investments that produce it, the accounting rate of return does not. By relating *current* profits to *current* capitalization, the accounting rate of return fatally scrambles up the timing.
>
> Moreover, this defect is not something that can be corrected by averaging, nor is it merely a start-up problem. It persists even in the steady-state growth. Unless the firm values its assets in the particular way long ago pointed out by Harold Hotelling, its accounting rate of return will not equal the economic rate of return. Further, that particular valuation method is totally impractical. For firms to use it would sometimes require taking negative depreciation; for observers to do so would require knowing the economic rate of return, so that computing the accounting rate would be pointless (Fisher, 1984, p. 509).

This measurement problem seems to suggest shifting from analysis of observed price-cost margins to NEIO estimation of price cost margins. However, NEIO models use the same accounting cost data to estimate marginal costs and price-cost margins and do not use the Hotelling depreciation formula. Thus, the alleged measurement problem persists in this class of industrial organization models.

One can illustrate that the Fisher McGowan critique is superfluous for both methods using work completed by Martin. He explicitly derives a structure profit model from a dynamic profit maximization problem with investment and depreciation. Martin demonstrates that the appropriate measurement of capital stock is the perpetual inventory formula (Martin, 1984, p. 502, equation 1). This in fact is the method used by rigorously specified conjectural variation studies (e.g. Wann and Sexton, this book). Neither NEIO nor structure-profit studies compute Hotelling depreciation, and given Martin's analysis, they should not compute it. Neither method answers the Fisher McGowan critique. Thus, the critique does not suggest that one method is superior to the other.[8]

Martin concludes a second paper wherein he develops models for the analysis of accounting profits as follows:

> it is incorrect to argue that market power cannot be analyzed using accounting rates of return *because accounting rates of return are a poor proxy for the internal rate of return.* Using a relatively straightforward, although detailed,

model of firm behavior, we can predict the impact of market power on accounting rates of return. Since we can predict the impact of market power on accounting rates of return, we can analyze accounting rates of return for the presence or absence of market power. In the analysis of market power, the relationship—or lack of relationship—between accounting rates of return and the internal rate of return is simply a red herring (Martin, 1988, p. 319).

Ravenscraft and Long (1984) point out the Fisher-McGowan critique is even more pernicious for other areas of economic analysis. If observed profits do not reflect economic profits due to accounting cost measurement problems, then the most basic signal of microeconomic theory is unobservable and market economies are rudderless. Unobservable economic profits cannot perform their central role in resource allocation.[9]

Moving on to the issue of endogeneity, two fundamental criticisms are usually raised under this rubric. The first is that policy makers cannot directly control an endogenous variable. They must change exogenous variables to change the equilibrium of the system. Critics expand this insight to conclude that one can (should) ignore endogenous relationships and focus upon reduced form relationships between endogenous and exogenous variables.[10] However, if one only estimates reduced form equations in the residual demand model or the structure-profit model, one would not be able to test for market power. The residual demand model is a quasi-reduced form model with an endogenous variable, own quantity, remaining in the equation so that one can estimate the structural parameter f_1, the estimated markup that is the Lerner index of market power. Similarly, in the structure-profit approach, some elements of market structure including market share may be partially endogenous in some models. Nonetheless, we remain interested in, for example, the structural coefficient for the relationship between market share and price-cost margins. The Demsetz critique maintains that lower costs directly increase PCMs and directly increase market share, and that *there is no endogenous relationship between market share and PCMs (no market power)*. In other words, the commonly documented relationship between market share and profits using ordinary least squares is spurious. To evaluate the Demsetz hypothesis one needs a structural model rather than a reduced form model to test whether the structural coefficient for market share in the PCM equation is zero.[11]

The second endogeneity criticism is that most empirical work has ignored the econometric implications of endogeneity. It has been directed almost exclusively at SCP studies, however, some conjectural variation studies also use less than fully simultaneous econometric procedures to estimate complex conjectural variation models. See, for example, Wann and Sexton (this book), Gollop and Roberts, (1979); Schroeter and Azzam, (1990); Azzam and Pagoulatos, (1990). These studies estimate supply, demand, and firm behavioral equations separately or in blocks. The gain from this estimation procedure is the ability to estimate very large and complex models that decompose conduct estimated price cost margins into several constituent parts such as oligopolistic and oligopsonistic components.

If a structural equation model is recursive and the error terms are not correlated then ordinary least squares estimation gives consistent and asymptotically efficient estimates (Kmenta, 1971, p. 586). The price cost margin-market share model, for example, is recursive since share can be modeled as a function of exogenous variables. Alternatively, the presence of an endogenous variable in structural equation may not preclude using OLS to obtain an approximately consistent estimate of its coefficient if that particular relation is stable in the data. For example, if the demand for a product shifts slowly and the supply shifts frequently, then the observed price quantity data points approximate the demand curve. Schmalensee makes this point but calls for more care when estimating such models:

> Even if . . . studies in industrial organization generally can only describe relations among long run equilibrium values of endogenous variables, such studies can make a contribution. But they should be designed, executed, and interpreted with due regard for their limitations (Schmalensee, 1989, p. 956).

Schmalensee also concludes that more rigorous modeling to identify and estimate key structural coefficients may not be possible because of a lack of exogenous variables. Yet progress in industrial organization on this issue may still be possible if one uses theory to specify parsimonious models that provide guidance on which variables are endogenous and which are exogenous and then resort to empirics to test models to determine which model is valid. Indeed the NEIO residual demand approach specifies that one or more variables such as quantity may be endogenous but there are variables that are exogenous to the industry (cost and demand shifters) that enable one to identify parameters on endogenous variables.

Summary

The four methods estimate different components of homogenous or heterogeneous oligopoly models. The Fisher-McGowan critique is, as Martin concluded, "a red herring" and does not argue for one approach over others. The conjectural variation models are more inclusive than the structure-profit in that they estimate cost and demand conditions and more exclusive in that they focus upon a particular industry to provide an econometric estimate of that industries' price cost margin. However, to date, this shift has come at the expense of analysis of the relationship between traditional market structure and market power and efficiency. Future conjectural variation studies at the business unit level may address these questions.

The market structure and conjectural variation approaches to the analysis of profitability may, as Breshnahan (1988, p. 1051) and Connor (1990, p. 1224) suggest, converge or at least become more closely articulated, in the near future. This could happen in three ways. First, structure-profit models may be derived rigorously from oligopoly theory to test specific conduct hypotheses. Several recent research efforts demonstrate that the analysis of observed profit rates within more tightly constructed market structure-profit models can provide scientific evidence on competitive strategy issues [Domowitz, Hubbard and Petersen (1987), Mueller (1990), Connor and Peterson (1992), Martin (1988), Cotterill and Iton (this book), Cotterill and Haller 1993]. This is not a dead end street. Actual firm and industry profitability are too important for the operation of a market economy to be dropped from the domain of industrial organization analysis. Second, estimated price cost margins from a set of conjectural variation studies could be regressed on elements of market structure (Clarke, Davies, and Waterson, 1984; Breshnahan 1988, p. 1051). Alternatively, some conjectural variation studies have included elements of market structure as determinants of estimated price cost margins when market structure has shifted during the time period analyzed. Wann and Sexton (this volume) for example test whether estimated price cost margins are affected by the presence of closed membership and/or bargaining cooperatives.

Residual demand and structure-price analysis are also related. The first estimates a residual demand elasticity or markup parameter assuming that the underlying conduct in the industry is constant. Structure-price analysis on the other hand analyzes the relationship between price and conduct which is hypothesized to be a function of market share, costs, demand elasticity, rival supply response and rival price response in the most general model. In principle the two methods can be combined. Baker and Breshnahan (1985), for example, test for of changes in market structure when estimating residual demand elasticities for beer firms by including instruments for the secular increase in seller concentration and the introduction of Lite beer. Structure-price studies in cross sectional data from local market industries, similarly may be able to use more sophisticated econometric techniques to estimate additional parameters such as the market elasticity of demand and possibly behavioral response parameters. In conclusion, the availability of data, the structure of the data set, data quality, and the particular hypothesis that one seeks to test may be critical when choosing among these alternative approaches to competitive strategy analysis.

Notes

1. This computation is based upon data from Goldman Sachs (1991), p. 1 and the 1990 Economic Report of the President, p. 401. The general market roughly tripled in value during this period.

2. Alternatively, the contestable market hypothesis predicts that firms with large market shares, even with scale economies, do not enjoy higher profits because there are no entry barriers. In the models presented here the number of firms is fixed; i.e. blockaded entry is assumed. If markets are contestable, there would be no relationship between market share and firm profits (Gilbert, 1989, Cotterill and Haller, 1993), estimated conjectural variations would equal the competitive value, and residual demand curves would be infinitely elastic.

3. When business unit data are available these studies use cross-section or panel data. Gallop and Roberts (1979), for example, use cross-section data for fifty-two firms in the coffee roasting industry to estimate a business unit level model.

4. Richard Sexton and I have had a very useful and stimulating discussion of this issue. In his most recent letter to me he stresses that an even handed discourse of this topic should include the following points that he raises. He writes:

> I continue to disagree with your assertions regarding an inability of conjectural variations modes to distinguish market power from efficiency. Lets take an example where we have a time series of price and cost data for a single industry. Suppose initially that only normal profits were being earned. But then suppose that one firm in the industry develops innovations that cause its costs to be lower than its rivals. That firm would gain market share, but market price would not necessarily fall. Rather, price would remain constant and the efficient firm would earn efficiency rents. Average industry cost, however, would fall. Your point is that a conjectural variation model would falsely impute the deviation between price and marginal cost that occurs here to market power. I think you are correct to the extent the analyst takes no steps to incorporate this evolution of the industry into the study. Suppose, though that in his equation to explain price as a function of costs, elasticity, and conjectural variation, the analyst also included one or more variables to account for the structural change, e.g., market share of the innovating firm. This variable will account for the increasing deviation over time between price and average industry marginal cost. The deviation will not be erroneously imputed to market power and the conjectural variations parameters as would be the case if the market evolution was not accounted for. To the extent that studies fail to account for these effects when there is a basis to suspect they exist, I think you have a point. However, as I tried to show above, the conjectural variations framework ought to be able to accommodate through appropriate variable inclusion these types of efficiency effects. . . . Thus, I submit that your concern may pinpoint an empirical deficiency of specific conjectural variations studies, but it is not a deficiency of the methodology per se, and the criticism lacks the force against these models that it has against structure-profit studies (Sexton, 1992).

Actually, I believe we are in agreement on the fundamental point. My assertion is not that the conjectural variation approach cannot analyze efficiency and power questions jointly, it is that aggregate industry level studies cannot do this. Virtually all the studies completed to date are aggregate industry studies. Moreover, none of these studies recognize that they may be measuring efficiency rather than power. Perhaps one can include market share and other elements of market structure such as concentration and barriers to entry in a conjectural variation model. However, I have serious reservations about including market share as a measure of only efficiency. Most industrial organization economists hypothesize that share measures both efficiency and power and look to more carefully constructed empirical tests to ascertain the importance of each (Scherer and Ross 1990, Ch. 11, Cotterill and Iton, this book). Following Gollop and Roberts (1979) an estimated cost function that is flexible enough to allow for economies of scale in a disaggregate C.V. model may be able to identify differential firm efficiency.

5. The Demsetz critique can be extended in the differentiated oligopoly use to include the hypothesis that higher quality leads to higher shares and higher prices. See Anderson (this book) and Cotterill (this book) for a more detailed analysis of this version.

6. Harris narrows the rivalrous reaction problem of n(n-1) firm interactions by assuming that the n-1 rival firms adopt identical prices based on identical demand, cost, and rival reaction functions. This does not affect the general results and simplifies the algebra.

7. It is possible to relax this assumption by modelling the flexibility coefficient as a function of variables in the right hand side of equation 6. Baker and Breshnahan, in fact, to test for "structural change" due to increasing concentration and the introduction of Lite beer in the brewing industry (Baker and Breshnahan, 1985, p. 435).

8. Conjectural variation studies with their direct focus upon estimating cost functions provide an opportunity to refine accounting cost data or substitute economic engineering based estimates to develop more accurate measures of marginal cost. This avenue for future applied research may benefit from extensive previous work by agricultural economists (French, 1977).

9. See Dennis Mueller (1990) for a succinct analysis of accounting versus economic returns. Mueller demonstrates that they can be used for analysis, but concludes as Leonard Weiss did in 1974:

Accounting conventions can and probably do introduce substantial errors of observation when using accounting returns as measures of economic returns. These errors tend to obscure the true relationships between profits and other economic variables our theories predict. We can expect low R^2 for equations that use accounting profits as the dependent variables, and the economic relationships we uncover are likely to be stronger than our econometric tests might suggest (Mueller 1990, p. 13-14).

10. See, for example, Perloff (1991) p. 6.

11. See Martin (1988) for a somewhat different and more general test statistic. The analysis, however, still estimates and uses the structural coefficient on market share.

References

Appelbaum, E. 1982. The Estimation of the Degree of Oligopoly Power. *Journal of Econometrics* 19: 287-299.

Azzam, A. and E. Pagoulatos. 1990. Testing Oligopolistic and Oligopsonistic Behavior:

An Application to the U.S. Meat Packing Industry. *Journal of Agricultural Economics* 41: 362-370.

Baker, J. B. and T. F. Bresnahan. 1985. The Gains From Merger or Collusion in Product-Differentiated Industries. *Journal of Industrial Economics* 33(4): 427-443.

_____. 1988. Estimating the Residual Demand Curve Facing a Single Firm. *International Journal of Industrial Organization* 6: 283-300.

Bresnahan, T. F. 1989. Empirical Studies of Industries with Market Power. In *Handbook of Industrial Economics*, ed. R. Schmalensee and R. Willig. New York: North Holland. 1011-1057.

Buzzell, R. D. and B. T. Gale. 1987. *The PIMS Principles*. New York: Free Press.

Buzzell, R. D., B. T. Gale, and R. G. M. Sultan. 1975. Market Share—A Key to Profitability. *Harvard Business Review* January-February.

Clarke, R. and S. Davies. 1982. Market Structure and Price-Cost Margins. *Economica* 49(August): 277-287.

Clarke, R., S. Davies and M. Waterson. 1984. The Profitability-Concentration Relation: Market Power or Efficiency. *Journal of Industrial Economics* 32(June): 435-450.

Connor, J. 1990. Empirical Chalenges in Analyzing Market Performance in the U.S. Food System. *American Journal of Agricultural Economics* 72(5): 1219-1226.

Connor, J. and E. B. Peterson. 1992. Market-Structure Determinants of National Brand-Private Label Price Differences of Manufactured Food Products. *The Journal of Industrial Economics* 40(2):1-15.

Cotterill, R. W. 1991. *Food Retailing: Mergers, Leverged Buyouts, and Performance.* Food Marketing Policy Center Research Report 14. University of Connecticut.

Cotterill, R. W. and L. E. Haller. 1993. Barrier and Queue Effects: A Study of Leading U.S. Supermarket Chains Entry Patterns. *Journal of Industrial Economics*, Forthcoming.

Cowling, K. and M. Waterson. 1976. Price-Cost Margins and Market Structure. *Economica* 43(August): 267-274.

Demsetz, H. 1973. Industry Structure, Market Rivalry, and Public Policy. *The Journal of Law and Economics* 16(April): 1-10.

Domowitz, I. R. Hubbard and B. Petersen. 1987. Oligopoly Super Games: Some Empirical Evidence on Prices and Margins. *The Journal of Industrial Economics* 35(4): 379-398.

Fisher, F. M. and J. J. McGowan. 1983. On the Misuse of Accounting Rates of Return to Infer Monopoly Profits. *American Economic Review* 73(1): 82-87.

Fisher, F. M. 1984. The Misuse of Accounting Rates of Return: Reply. *American Economic Review* 74(3): 509-517.

French, B. C. 1977. The Analysis of Productive Efficiency in Agricultural Marketing: Models, Methods, and Progress in L.R. Martin, Ed. *A Survey of Agricultural Economics Literature.* Vol. 1, Minneapolis: Univ. of Minnesota Press.

Geroski, P. and R. Masson. 1987. Dynamic Market Models in Industrial Organization. *International Journal of Industrial Organization* 5(March): 1-14.

Gilbert, R. G. 1989. The Role of Potential Competition in Industrial Organization.

Journal of Economic Perspectives 3: 107-128.

Goldman Sachs. 1991. *Industrial Research: Packaged Food Quarterly*. New York: Goldman Sachs.

Gollop, F. M. and M. J. Roberts. 1979. Firm Interdependence in Ologopolistic Markets. *Journal of Econometrics* 10: 313-331.

Harris, F. 1988. Testable Competing Hypotheses from Structure-Performance Theory: Efficient Structure versus Market Power. *The Journal of Industrial Economics* 36(3): 267-280.

Kaufman, P. R. and C. R. Handy. 1989. *Supermarket Prices and Price Differences: City, Firm, and Store-Level Determinants*, Economic Research Service, U.S. Dept. of Agriculture. Tech. Bul. No. 1776. December.

Kmenta, J. 1971. *Elements of Econometrics*, New York: Macmillan Co.

Long, W. F. and D. J. Ravenscraft. 1984. The Misuse of Accounting Rates of Return: Comment. *American Economic Review* 74(3): 494-500.

Marion, B. W., W. F. Mueller, R. W. Cotterill, F. E. Geithman, and J. R. Schmelzer. 1979. *The Food Retailing Industry*, New York: Praeger.

____. 1977. The Price and Profit Performance of Leading Food Chains. *American Journal of Agricultural Economics* 61(August): 420-433.

Martin, S. 1988. The Measurement of Profitability and the Diagnosis of Market Power. *International Journal of Industrial Organization* 6: 301-321.

____. 1984. The Misuse of Accounting Rates of Return: Comment. *American Economic Review* 74(3): 501-506.

Mueller, D. C. Ed. 1990. *The Dynamics of Company Profits: and International Comparison*. New York: Cambridge University Press.

Perloff, J. 1991. Econometric Analysis of Imperfect Competition and Implications for trade Research. Occassional Paper 23, NC-194 Project, Dept. of Agricultural Economics, Ohio State University, Columbus, Ohio.

Porter, M. E. 1980. *Competitive Strategy: Techniques for Analyzing Industries and Competitors*. New York: Free Press.

Salinger, M. 1990. The Concentration-Margins Relationship Reconsidered. *Bookings Papers: Microeconomics*. Washington: Brookings Institution. p. 287-335.

Scherer, F. M. and D. Ross. 1990. *Industrial Market Structure and Economic Performance*. Boston: Houghton Mifflin Co.

Schmalensee, R. L. 1989. Inter-Industry Studies of Structure and Performance. In *Handbook of Industrial Economics*, ed. R. Schmalensee and R. Willig, 951-1009. New York: North Holland.

Schoeffler, S., R. D. Buzzell, and D. F. Heany. 1974. Impact of Strategic Planning on Profit Performance. *Harvard Business Review* (March-April).

Schroeter, J. R. and A. Azzam. 1990. Measuring Market Power in Multi-Product Oligopolies: The U.S. Meat Industry. *Applied Economics* 22: 1365-1376.

Sexton, R. Letter to author, April 3, 1992.

U.S. Congress. Joint Economic Committee. *The Profit and Price Performance of Leading Food Chains, 1970-1974*. Report prepared by Marion, B. W., W. F.

Mueller, R. W. Cotterill, F. E. Geithman, and J. R. Schmelzer. April 12, 1977a.
_____. Joint Economic Committee. *Prices and Profits of Leading Retail Food Chains, 1970-1974*. Hearings. 95th Cong., 1st sess. 1977b.
U.S. President. 1990. *Economic Report of the President*. Washington: U.S. Government Printing Office.
Weiss, L. W. ed. 1989. *Concentration and Price*. Cambridge: MIT Press.
_____. 1974. The Concentration-Profits Relationships and Antitrust. In *Industrial Concentration: The New Learning*, eds. H.J. Goldschmidt *et al.* Boston: Little Brown.

PART TWO

Market Structure Performance Studies

2

A PIMS Analysis of the Structure Profit Relationship in Food Manufacturing

Ronald W. Cotterill and Clement W. Iton

Introduction

When *Industrial Organization: The New Learning* was published in 1974, the consensus opinion was that a positive relationship exists between industry profits and seller concentration and that the relationship is due to the exercise of market power (Weiss 1974). In that compendium Harold Demsetz dissented, arguing that the superior efficiency of large share firms is the source of higher profits. Since then, the consensus has unraveled. In his most recent text, Scherer dismisses thirty years of empirical analysis of industry profits with the following statement:

> most, if not all, of the correlation between profitability and concentration found by Bain and his descendants (at least for the United States) was almost surely spurious-the result of aggregating a positive relationship between sellers' market shares and profitability to the industry level. . . . The positive profit-market share relationships observed in line of business studies represents a still-unknown mixture of temporary efficiency differences and more or less durable monopoly power. Disentangling the relative importance of the two effects . . . is the great challenge facing empirical industrial organization researchers (Scherer and Ross 1990, p. 434).

The empirical basis for this position is primarily research conducted in the 1980s using the Federal Trade Commission line of business data, and the Strategic Planning Institutes' Profit Impact of Market Strategy (PIMS) data base.

Using the F.T.C. line of business data, Ravenscraft (1983) found that the relationship between business unit profits and market share was positive and highly significant, but the relationship between business unit profits and industry concentration was negative and weakly significant. In an aggregate industry level model he found, as had numerous prior industry level studies, a significant positive relationship between concentration and profits. In another line of business study, Kwoka and Ravenscraft (1986) also report that market share has a significant positive relationship and concentration a significant negative relationship with profitability. Using the PIMS data base, Gale and Branch (1982) report that business unit profitability is positively related to market share and not related to industry concentration.

These business unit level studies cover the entire manufacturing sector. When Ravenscraft examined business unit profitability within twenty major industries, he found that both market share and concentration had a significant positive relationship with profitability in six industries, including food manufacturing. Earlier studies of the food manufacturing sector have also reported that both seller concentration and a firm's position have significant positive impacts on its profitability. The FTC (1969) analyzed ninety-seven publicly owned firms from the universe of 125 largest food manufacturers of 1950 and used an average annual rate of return on equity and a rate of return on investment (assets) for the five year period 1949-1953.[1] Imel and Helmberger (1971) studied ninety-nine large food manufacturers' after tax profit-sales ratios averaged over 1959-1967. Rogers (1978) analyzed sixty publicly owned large food manufacturers using five different measures of profitability averaged for the 1964-1970 period. In a recent study at the industry rather than firm level, Connor and Petersen (1992) report that the 1979 and 1980 markups of national brand products over comparable private label products are positively related to the Herfindahl index. Connor and Petersen did not analyze firm or business unit level performance to determine whether market share or concentration or both underpin their industry results.

This brief review of the literature suggests that the structure-performance relationships in food manufacturing industries are different than those in the majority of other industries. In this paper, we will explore this hypothesis using the PIMS data base. It contains four year average measures of return on investment (ROI) as well as numerous strategic and structural variables. These observations cover the 1972 to 1987 period. A particular business unit may be represented more than once, but individual observations in this data base are not identified in a way that enables one to do any time series analysis at the firm level. To determine if the food sector continues to be different, we will report some results for the entire PIMS data base (2,552 observations) as well as the food manufacturing subsample (88 observations).

One of the major criticisms of the structure performance studies cited, except Kwoka and Ravenscraft and Connor and Peterson, is the failure to deduce hypotheses explicitly from an economic model of the firm. This study does start with a formal model and produces some new insights. As a result, we are able to demonstrate that the prior empirical results cited above are consistent with a more general model and new empirical results that reinstate concentration as a significant source of profitability at the business unit level.

This work moreover suggests that the positive concentration profits relationship so extensively documented by Weiss and others at the industry level is not due to aggregation bias. Also, Demsetz's relative efficiency hypothesis is tested jointly in this model; and, it does not explain the concentration-profit result. Thus, the business unit level research reported here suggests that market power, not superior cost efficiency, is the source of the market share profit relationship in food manufacturing.

Our model also demonstrates that the efficient structure-collusion dichotomy first suggested as a model for empirical research by Cowling and Waterson (1976) has a major shortcoming for the analysis of market power. In essence, collusion, as measured by a positive conjectural variation parameter, is a sufficient but not necessary condition for the exercise of market power.[2] Focusing the analysis of market power solely upon collusion in this fashion is the "weak form" version of the "strong form" Chicago position that only explicit collusion, i.e., price fixing and government imposed barriers or cartels, create competitive inefficiencies. As we will demonstrate, it is possible to have a positive relationship between concentration and profits that indicates the exercise of market power when the conjectural variation parameter is zero or negative. The following section develops a more general specification of the oligopoly model. The last two sections present empirical results and conclusions.

Theory and Model Specification

The generalized Cournot oligopoly model for a homogeneous good in industry j with n_j firms that have possibly different but constant marginal costs gives the following equilibrium condition:[3]

$$\frac{P_j - c_{ij}}{P_j} = S_{ij}(1 + \lambda_{ij})/e_j \tag{1}$$

for $i = 1 \ldots n_j$ firms in industry j
$\quad j = 1 \ldots m$ industries

where:

$\quad p_j$ = market price in industry j

$\quad c_{ij}$ = constant marginal cost for firm i in industry j

$\quad S_{ij}$ = market share of firm i in industry j

λ_{ij} = conjectural variation of firm i in industry j

e_j = absolute value of the market elasticity of demand of industry j

A firm's price cost margin is a function of its own market share, its conjectural variation, i.e. its perceived estimate of how the output of all other firms in the market will change when it changes its output, and the market elasticity of demand. The distribution of market shares in this model is determined by the distribution of marginal cost levels with lower cost firms having higher shares (Clarke and Davies 1982).

Given this point, some researchers have inferred that the impact of market share in equation (1) on price cost margins measures only the Demsetz efficiency hypothesis. Most notably Clarke, Davies, and Waterson state:

> within each industry from [equation (1) above] profitability is proportionate to market share and, again as suggested by Demsetz, there are larger profit rates for the larger firms (Clarke, Davies and Waterson 1984 p. 439-440).

As the quote from Scherer and Ross at the advent of this chapter opined, this interpretation is not correct. Market share, independent of the conjectural variation parameter's value, measures market power as well as the differential efficiency of large firms. For example, when the generalized Cournot model is Cournot and industry j has n_j firms with identical and constant marginal costs, the result is a positive relationship between business unit profits and market share which increases as the number of firms declines.[4] Therefore when the identical cost constraint is relaxed, the business unit price cost margin is a function of market power related to market share as well as cost positions related to market share.[5] The models developed by Cowling and Waterson, and Clarke, Davies and Waterson ignore this fact. They equate the exercise of market power to the existence of a positive conjectural variation parameter, i.e. collusion. Kwoka and Ravenscraft emphatically reject this point stating:

> The idea that the Cournot solution represents the lower bound has gained credibility more from repetition than from economic theory or empirical evidence. The supposed analogy to "competitive independence" is fallacious (Kwoka and Ravenscraft 1986 p. 362).[6]

A less restrictive approach to the analysis of market power is clearly needed. As Assistant Attorney General James F. Rill recently explained when discussing the Justice Department's review of mergers:

> it is important to consider both coordinated and noncoordinated views of competitive effects when analyzing a merger of firms in a highly concentrated market where entry is not likely. The term "noncoordinated" refers to firms'

independent decisions about price and output-decisions that do not rely on the concurrence of rivals or on coordinated responses by rivals. In contrast, the term "coordinated" refers to such conduct as either tacit or overt collusion, price leadership, and concerted strategic retaliation-conduct that requires the concurrence of rivals to work out profitably. The Department considers both noncoordinated and coordinated effects, but often the parties to a merger or their counsel are prepared only to discuss collusion or other coordinated effects (Rill, 1990, p. 51).

Business strategists have added another perspective to the hypothesis that efficiency is not the sole source of the market share profit relationship. Market share is not the most appropriate measure for analysis of Demsetz efficiencies or other types of competitive advantages related to firm size. As Buzzel and Gale explain:

since businesses compete in many different kinds of served markets, an absolute market share of 15 may represent the market leader in a fragmented market or the number 4 competitor in a concentrated market. . . . Share relative to three largest competitors has been shown by our research to be the most useful measure of relative share for calibrating competitive advantage (Buzzel and Gale, 1987 p.71-72).[7]

Although it is not well known in the general economics profession, business strategists were not the first or only researchers to employ relative market share as a measure of firm position. In fact, the earliest practitioners of this approach are a group of agricultural economists associated with the University of Wisconsin who have documented the impact of relative market share and concentration upon profits and prices in the food sector of the economy. Studies in this empirical vein include F.T.C. (1969) and Imel and Helmberger (1971) and Rogers (1978) in food manufacturing, and Marion et al. (1979) and Cotterill (1986) in food retailing.

Thus, Cournot oligopoly theory and empirical research by business strategists and agricultural economists suggest that the influence of market share upon profits can be decomposed into a relative size component that measures Demsetz efficiencies and a market concentration component that measures the ability of leading firms to exercise market power as a group. For example, if four firm concentration is 40 percent and all firms in the industry have 25 percent relative market shares (10 firms at 10 percent market share) and if market concentration increases to 80 percent with all firms retaining the same 25 percent relative share position (5 firms at 20 percent market share) then an observed increase in profits, if any, is due to market power, not increased efficiency due to higher market shares relative to fringe firms. There are no fringe firms.

This decomposition is also desirable for statistical reasons. By definition, market share and concentration measures are correlated. In this case, multicollinearity is more than a sample problem. Standard techniques for mitigating multicollinearity, most notably gathering more data, will not be successful. Dividing market share by a measure of concentration produces a relative market share variable that is, in theory if not the sample, independent of concentration.[8]

In this research we will employ the four firm concentration ratio as the measure of seller concentration. To decompose market share, consider the following linear relationship:

$$S_{ij} = b_1 R_{ij} + b_2 C_j \qquad (2)$$

where:

S_{ij} = market share of firm i in market j

R_{ij} = relative market share of firm i in market j

C_j = four firm concentration ratio in industry j

and:

$$R_{ij} = \frac{S_{ij}}{C_j}. \qquad (3)$$

This decomposition is not mathematically exact; however, in the PIMS sample of 2552 observations relative share and four firm concentration explain 95% of the variation in market share. If we assume for the moment that all firms have zero (Cournot) conjectural variations and subsume the elasticity of demand in the estimated parameters, substituting equation 2 into equation 1 yields equation 4.

$$\frac{P_j - c_{ij}}{P_j} = \beta b_1 R_{ij} + \beta b_2 C_j = \beta_1 R_{ij} + \beta_2 C_j \qquad (4)$$

The firm level specification of relative market share and concentration rather than market share and concentration, as done in previous studies, seems even more plausible when one compares their industry or aggregate analogues. Aggregating the market share concentration model to the industry level gives the following relationship:

$$\frac{P_j - \overline{c}_j}{P_1} = \gamma_1 H_j + \gamma_2 C_j. \qquad (5)$$

The term $\overline{c}_j$ is the market share weighted marginal cost for the industry. Industry profits are a function of the Herfindahl index and the concentration ratio. This unattractive duplicative specification at the industry level reaffirms our point that share and concentration level are by definition correlated at the firm level, and thus should not be specified jointly. Aggregating the relative share concentration model to the industry level gives:

$$\frac{P_j - \overline{c_j}}{P_j} = \beta_1 \frac{H_j}{C_j} + \beta_2 C_j. \tag{6}$$

The decomposition of share at the firm level produces a decomposition of the Herfindahl index at the industry level into the four firm concentration ratio (C) and the variation in the Herfindahl index that is not due to variation in four firm concentration (H/C). This latter variable measures the competitive advantage related to differences in size.

Moving beyond the simple Cournot equilibrium, one can model an individual firm's conjectural variation as follows:

$$\lambda_{ij} = \alpha S_{ij} = \alpha_1 R_{ij} + \alpha_2 C_j. \tag{7}$$

A firm's conjectural variation is a function of its own market share, and that can, in turn, be decomposed into a function of relative market share and concentration. This specification allows a firm's conjectural variation to be a function of concentration as in the Cowling and Waterson model. If $\alpha_1 = 0$ and α_2 is nonzero then concentration by itself determines a firm's conjectural variation and conjectural variations are constant across firms in a given market. It allows the conjectural variation to be an unconstrained function of market share as compared to the Clarke, Davies and Waterson model wherein the conjectural variation was constrained to be negatively related to share (Schmalensee 1987). Finally, it allows for differences in leader and followers conduct as defined by relative market position.

Substituting equations 2 and 7 into equation 1 and subsuming the elasticity of demand in the estimated parameters produces the following specification of the general oligopoly model at the firm level.

$$\frac{P_j - c_{ij}}{P_j} = \beta_1 R_{ij} + \beta_2 C_j + \beta_1 \alpha_1 R_{ij}^2 + \beta_2 \alpha_2 C_j^2 \tag{8}$$
$$+ (\beta_1 \alpha_2 + \beta_2 \alpha_1) R_{ij} C_j$$

The oligopoly problem is now a general quadratic form. Equation 8 is overidentified with five estimated coefficients and only four parameters. One of the coefficients, therefore, is a nonlinear function of the other four. Nonlinear regression techniques can be used to impose this restriction and to test its significance, thereby testing how well the general oligopoly model fits observed conduct data.[9]

Denoting the left hand side of 8 as Π_{ij} and taking the derivatives with respect to R_{ij} and C_j gives equations 9 and 10 respectively.[10]

$$\frac{\partial \Pi_{ij}}{\partial R_{ij}} = \beta_1 + 2\beta_1 \alpha_2 R_{ij} + (\beta_1 \alpha_2 + \beta_2 \alpha_1)C_j \qquad (9)$$

$$\frac{\partial \Pi_{ij}}{\partial C_j} = \beta_2 + 2\beta_2 \alpha_2 C_j + (\beta_1 \alpha_2 + \beta_2 \alpha_1)R_{ij} \qquad (10)$$

Equation 9 establishes that the impact of an increase in a firm's relative share on its profits depends upon the level of concentration as well as its relative position. Equation 10 establishes that the impact of an increase in concentration on a firm's profits depends on its relative share as well as concentration. If the four parameters are positive then the slopes of the relative share-profit and concentration-profit relationships are positive and become more positive as share and concentration increase. Note, however, if α_1 and α_2 are negative, it is possible to obtain a positive convex slope for each relationship. In other words, the signs of conjectural variation parameters determine whether the structure-profit relationships are concave or convex, not whether they are positive or negative. Conjectural variation parameters, therefore, do not provide a complete answer to the market power question. For the impact of concentration one needs an estimate for β_2 as well as for λ_{ij}, which is determined by α_1 and α_2.

Research on the PIMS data have identified several other factors in addition to relative share and concentration that influence a firm's profitability. These factors can be introduced into the model as controls. Of particular interest is the PIMS measure of a firm's cost advantage relative to its three leading competitors (RELDC). Introducing it is a direct test for Demsetz's differential efficiency hypothesis. All the variables used in this analysis are listed below.[11]

ROI	=	return on investment (%)
S	=	market share (%)
R	=	relative firm market share: market share divided by four firm concentration (%)
C	=	the four firm concentration ratio for a business unit's served market (%)
RELDC	=	relative direct costs, i.e., the firm's cost position relative to its three leading competitors (%)

GBV	=	the ratio of gross book value of plant and equipment to net sales (%)
TOTMKT	=	the ratio of total marketing expenses to net sales (%)
TOTRD	=	the ratio of total research and development expenditure to net sales (%)
VI	=	the ratio of value added to net sales (%)
INDGR	=	the growth rate of the SIC industry (%)
CAPU	=	the capacity utilization (%)
EMPUN	=	the proportion of employees unionized
TOTIN	=	the ratio of finished goods inventory and raw materials and work-in-process inventory to net sales (%).

Empirical Results

The sample for this study was drawn from the PIMS/Strategic Planning Institute line of business research data base SPI4.[12] Buzzel and Gale describe the construction of the data base, its general composition and the results of numerous studies of business unit profitability. Marshall (1987) compares and contrasts the PIMS data to the FTC line of business data base. Following Gale and Branch, and Schmalensee (1987), we opt for return on investment (ROI) rather than return on sales (ROS) as the measure of firm profits at the strategic business unit level.[13] As Gale and Branch document, they are closely correlated (Gale and Branch, 1982), and Buzzel and Gale (1987) report similar results for PIMS models that alternatively employ ROS as the dependent variable. In this study market share, seller concentration and other market structure variables are measured in the PIMS-defined served market. Served markets are usually narrower than Census four digit SICs and seek to identify economic markets where firms compete, for example regional rather than national markets for some strategic business units.

Relative market share is less correlated with four firm concentration than is market share in the full PIMS sample ($r_{C,R}=0.22, r_{C,S}=0.57$) and in the food subsample ($r_{C,R}=0.22, r_{C,S}=0.46$). Table 2.1 analyzes the relationship of concentration, relative share, and market share to return on investment for the full PIMS sample. Equation 1 contains only market share. It has a highly significant positive impact on ROI and explains 12 percent of the variation in ROI. Equation 2 contains only relative market share. It performs slightly better than market share. In Equation 3, market concentration by itself has a highly significant positive impact upon ROI, but it only explains 2.37 percent of the variation in ROI.

Equation 4 introduces the linear decomposition of share into relative share and concentration. Each has a positive impact on ROI and is significant at the

TABLE 2.1 The Impact of Concentration (C), Relative Firm Market Share (R) and Market Share (S) on Return on Investment: Full PIMS Sample (2,552 Observations)

Eq	INTERCEPT	R	C	S	R^2	F-VALUE
1	10.7947			0.4684[a] (19.026)	0.1243	361.970[a]
2	7.7088	0.4578[a] (19.279)			0.1272	371.683
3	9.9035		0.1714[a] (7.874)		0.0237	62.005
4	3.4055	0.4364[a] (17.702)	0.0675[a] (3.158)		0.1306	191.481[a]
5	15.4825		-0.0827[a] (-3.280)	0.5254[a] (17.456)	0.1280	187.057[a]

[a] Significant at the 99% level
Numbers in parentheses are t-ratios

one percent level. R^2 increases slightly to 13 percent. Equation 5 illustrates that when share and concentration are specified jointly, concentration has a significant negative impact upon ROI and share maintains its significant positive impact. This is the result reported by Kwoka and Ravenscraft and others. Thus the choice of the theoretical model and specification does make a dramatic difference.

The market share concentration model underlying equation 5 suggests a radical revision in antitrust policy. According to this efficient structure-collusion model, large share firms have higher profits because they have lower costs and increases in concentration are beneficial because they induce negative collusion, i.e. rivalry among large firms. Hence, an antitrust policy that questions mergers when market shares or the Herfindahl index is high should be altogether scrapped.

The relative share-concentration model underlying equation 4 provides support for current merger policy. Firms with large relative shares enjoy higher profits, possibly due to their relative efficiency advantage, and markets with higher concentration have higher profits due to market power. If higher concentration, holding relative shares constant, is due to economies of scale relative to the size of the market, there would be a trade-off between efficiency and market power. This is the standard efficiency defense.

Table 2.2 reports similar results for the food subsample model. Equation 1 contains only market share. It has a highly significant positive impact on

TABLE 2.2 The Impact of Concentration (C), Relative Firm Market Share (R) and Market Share (S) on Return on Investment for the Food Subsample (88 Observations)

EQ.	INTERCEPT	R	C	RR	CC	S	R^2	F-VALUE
1	-0.8017					0.7182[a] (5.36)	0.2507	28.768[a]
2	-1.5167	0.6127[a] (4.742)					0.2073	22.487[a]
3	-22.1627		0.5111[a] (3.459)				0.1221	11.962[a]
4	-15.6167		0.2185 (1.426)			0.6188[a] (4.119)	0.2682	15.574[a]
5	-29.6506	0.5348[a] (4.189)	0.3825[b] (2.757)				0.2723	15.905[a]
6	-0.8512	-0.4174 (0.520)	0.0073 (-0.451)	0.0073 (1.181)	0.0063 (0.940)	-0.4643 (-0.573)	0.2951	6.867[a]

[a] Significant at the 99% level.
[b] Significant at the 95% level.
Numbers in parentheses are T-ratios.

ROI. The estimated coefficient is much higher than it is in the full PIMS sample (0.71 versus 0.47) and R^2 is twice as high (0.25 versus 0.12). This type of change is generally true for the other equations as well, suggesting that stronger structure-profit relationships do occur in the food sector, as previous studies have documented. In equation 2, relative share is regressed on ROI. Again, it is positive and statistically significant and explains 20 percent of the variation in ROI. Equation 3 has served market concentration as the only independent variable. It is positive and significantly related to ROI, and, by itself explains 12 percent of the variation in ROI.

Equation 4 in Table 2.2 shows that when market share and concentration are specified jointly in the food subsample, concentration becomes positively related to ROI but it is not significant and share maintains its significant positive impact. Thus in food, as Ravenscraft, and Kwoka and Ravenscraft documented with the FTC line of business data base, both market share and concentration appear to be positively related to profits. In Equation 5, when relative share and concentration are specified, both coefficients are positive and significantly related to ROI. The coefficient of determination and F-value are somewhat higher than those for equation 4.

Equation 6 is an unconstrained estimation of the full quadratic specification of the model. Although the constrained nonlinear regression model would not converge, the results for unconstrained are so weak compared to the linear model that little explanation is lost. F-ratio tests of the five specifications prior to equation 6 indicate that any one including the linear relative share and concentration specification, is preferable to 6. Consequently, the conjectural variation parameters α_1 and α_2 are not significantly different from zero.

Table 2.3 explores the structure profit relationship in food manufacturing further by introducing a set of control variables for other determinants of firm ROI, including relative direct cost (RELDC). In equation 1 market share retains its significant and positive impact on ROI. Adding the control variables increases R^2 from 0.25 (equation 1, Table 2.1) to 0.48. This level of overall explanation of ROI is in line with prior studies in food manufacturing except for Ravenscraft's analysis of the food businesses with the FTC LOB-data base.[14] That study explained only 12 percent of the variation in profits as measured by a gross profit sales ratio. Equation 2 introduces market share and concentration jointly with the set of control variables. Market share continues to perform strongly as in prior specifications and concentration has a positive but insignificant coefficient.

Specifying relative share and concentration in lieu of market share in equation 3 enables concentration to regain a positive and significant impact on ROI. A 10 percentage point increase in served market concentration results in a 2.7 percentage point increase, *ceteris paribus*, in ROI. Rogers' analysis of ROI at the company level for the years 1964-1970 compared to this 1972-1987 study produces a lower estimated coefficient for concentration, 0.20 (Connor, *et al.*, p. 336). Rogers' estimated coefficient for relative share, 0.23, is much lower than ours, 0.41. The difference may be due to sample selection or time period differences; however, it may also be true that working at the line of business level rather than using weighted aggregate structural variables to predict company profits, enables a more accurate look at structure performance relationships.

Equation 4 introduces a PIMS measure of the business units' cost level relative to the cost level of its three leading competitors. The coefficient for this cost advantage measure is negative as hypothesized but not significantly different from zero. Moreover, introducing this direct measure of cost advantage to the model does not reduce the significance or impact of relative market share on ROI. The Demsetz differential efficiency hypothesis is rejected as an explanation for the relative share profit relationship in food manufacturing. In other words, profits of firms with large relative market shares are not higher because c_i in equation 4 of the text is lower. This result directly suggests that share-related profits are due to the ability of larger firms to charge higher prices than smaller rivals.

Equation 5 displays the estimation results for the full quadratic specification. The quadratic form does no better here than it did in Table 2.2. As a result, it seems that zero conjectural variation (linear) models provide a sufficient theoretical base for structure-profit analysis. In fact, F-tests indicate that equations 3 and 4 are the most appropriate specifications. The zero parameter restrictions in them result in no significant reduction of explanatory power relative to equation 5.

These empirical results deserve perspective. Several different theoretical models, including differentiated product oligopoly models, may underpin these results. These models and the homogeneous product oligopoly model presented in this paper are static equilibrium models. The development of dynamic models could analyze how changes in cost and demand conditions, mergers, technology, and firm conduct influence market structure and performance over time. Such generalization would directly address the criticism that market structure is endogenous. If the observed level of concentration is a function of more fundamental determinants, a deconcentration or merger policy that does not change underlying determinants would not succeed. Over time the prior equilibrium level would reestablish itself. Yet, if merger policy is an underlying (exogenous) determinant of concentration and if it does change (i.e. becomes tougher), then the ex-post equilibrium will be at a lower level of concentration. This would be the case if firms know that the government will challenge mergers that increase concentration beyond a threshhold level.

Regarding dynamic efficiency, Salinger, (1990) and Peltzman (1977) have suggested that static inefficiency (market power) is affected by dynamic cost

TABLE 2.3 The Impact of Concentration (C), Relative Firm Market Share (R), Market Share (S) and Control Variables on Return on Ivestment (88 Observations)

VAR	Eq1	Eq2	Eq3	Eq4	Eq5
INTERCEPT	-16.6049	-26.9900	-36.6181	-12.4553	25.2601
R			0.4066[b] (3.255)	0.3891[a] (3.031)	0.7336 (1.027)
RR					0.0049 (0.831)
C		0.1513 (0.997)	0.2744[b] (2.021)	0.2788[b] (2.043)	-1.1322 (-1.302)
CC					0.0115[c] (1.765)
S	0.5426[a] (4.194)	0.4689[a] (3.146)			-0.8605 (-1.118)
RELDC				-0.2377 (-0.635)	-0.2635 (-0.706)
TOTMKT	-0.7863[a] (-3.250)	-0.7570 (-3.107)	-0.7707[a] (-3.185)	-0.7520 (-3.073)	-0.6536[b] (-2.605)
VI	0.8417[a] (4.030)	0.8111[a] (3.842)	0.8264[a] (3.954)	0.8092[a] (3.824)	0.8092[a] (3.816)
INDGR	0.8440	0.8552	0.8728	0.7382	0.7884

	(1.586)	(1.607)	(1.646)	(1.288)	(1.380)
TOTRD	-4.2093 (-0.860)	-3.0994 (-0.618)	-2.6408 (-0.530)	-1.9877 (-0.389)	-0.5944 (-0.114)
CAPU	0.0897 (0.674)	0.0952 (0.716)	0.0698 (0.524)	0.0716 (0.536)	0.1315 (0.918)
EMPUN	-0.0292 (-0.364)	-0.0141 (-0.173)	-0.0169 (-0.207)	-0.0020 (-0.022)	-0.0030 (-0.035)
TOTIN	-0.1288 (-0.705)	-0.2070 (-1.041)	-0.1986 (-1.002)	-0.2264 (-1.111)	-0.3602[c] (-1.667)
GBV	-0.4400[a] (3.759)	-0.4208[a] (3.546)	-0.4196[a] (-3.554)	-0.4042[b] (-3.343)	-0.3684[b] (-2.994)
R^2	0.4818	0.4884	0.4924	0.4951	0.5223
F-VALUE	8.057[a]	7.350[a]	7.471[a]	6.775[a]	5.702[a]

[a] Significant at the 99% level.
[b] Significant at the 95% level.
[c] Significant at the 90% level.
Numbers in parentheses are T-ratios.

reductions related to increases in market concentration. As concentration increases, consumer prices may decrease because cost decreases are greater than profit increases. In another approach Geroski and Masson (1987) and Mueller (1990) have analyzed how long it takes for market forces to erode observed supercompetitive profits in concentrated industries. They find that the adjustment process takes decades rather than months or years. The existence of the static structure profit relationship in food manufacturing during the 1950s, 1960s, 1970s, and 1980s, and, the documented increases in concentration (Connor *et al.*) through much of this period suggest that sector profits have increased over time. The performance of the stock prices of food manufacturers in the Standard and Poor's 500 Index provides a dramatic illustration of this point. The price of these stocks increased 900% during the 1980s.[15] A formal dynamic analysis of the persistence of profits and the underlying determinants of concentration in the food manufacturing sector would be a valuable extension of the analysis of this chapter.[16]

Conclusions

This research produces several conclusions that provide counsel for future research on structure profit relationships. First and perhaps most important, the theory developed here and the consequent empirical analysis suggest that the positive relationship between concentration and profits at the industry level reported by Weiss and others was not due to aggregation bias. As we have demonstrated, one can specify a model at the business unit level that produces this relationship. Moreover, previous results that find a negative relationship between concentration and profits, when market share is specified in the model, are consistent with the exercise of market power in more concentrated markets.

When collusion is defined as output restricting behavior greater than that resulting from Cournot behavior, then equating the exercise of market power to collusion is misguided. Market power can exist when conjectural variations parameters are zero (Cournot behavior) or when they are negative. If the conjectural variation parameter is related to elements of market structure, then in a model that decomposes market share as we have, it determines whether the structure profit relationship is concave or convex. Specifically, it is possible to have a positive (convex) structure profit relationship that is due to the exercise of market power when the conjectural variation parameter is negative. In this research, however, conjectural variations are not significantly different from zero, and the structure-profit relationships are linear.

Finally, the research that specified market share and concentration jointly at the line of business level has contributed to a de-emphasis on market share and concentration by antitrust enforcement agencies when reviewing mergers and acquisitions. Our analysis supports prior studies that find share *or* relative share and concentration are positively related to profitability [Ravenscraft (1983), Rogers (1978), Imel and Helmberger (1971) and FTC (1969). For food manufacturing the relationship is strongest for the relative share concentration specification. Including a direct measure for relative cost efficiency suggests that the profits of large relative share firms and firms that operate in highly concentrated markets are not due to lower costs. Our analysis also rejects the hypothesis that increased rivalry among large firms results in lower industry profit. Market share and concentration seem to persist as indicators of market power in the food manufacturing industries.

Notes

1. A pioneering study for food manufacturing that preceded these is Schrader and Collins (1960). See Connor, *et al.* (1985) p. 282-286 for a discussion of these firm level structure performance studies.

2. Shepherd, 1986a,b, and others have also made this point.

3. See Cowling and Waterson (1976), or Kwoka and Ravenscraft (1986) for a formal derivation of this basic model. A heterogenous goods model like that used by Connor and Petersen (1992) may be more appropriate since the food manufacturing sector has significant product differentiation. A homogeneous goods model, however, is sufficient to illustrate the market share related measurement problems that this analysis focuses upon. Moreover including marketing expenses and other contral variables produces a model that may capture most firm level effects of differentiation.

4. The business unit or firm level equilibrium is given by:

$$\frac{P_j - c_j}{P_j} = \frac{1}{n_j e_j}$$

where e_j is the absolute value of the price elasticity of demand for industry j.

5. For any given n_j the business unit equilibrium is:

$$\frac{P_j - c_{ij}}{P_j} = \frac{S_{ij}}{e_j}.$$

Given n_j, a firm's market share is related to its cost advantages; however, as n_j becomes smaller, a firm's share and price cost margin may also increase.

6. Others have criticized the focus upon collusion as being too narrow noting, that dominant firms and other leader-follower conduct patterns will generate market power (Shepherd, 1986, a,b, Scherer and Ross p. 221). In the Stackelberg model, for example, the conjectural variation parameter for followers is Cournot (zero) and it is negative for the leader.

7. Economists have traditionally preferred to measure cost advantage relative to the costs of the marginal firm, however, it is impossible to identify marginal firms in each industry. Defining a cost advantage relative to the leading firms is more feasible and does not negate the fact that fringe firms may exist with very small relative shares and

large cost disadvantages. See Cotterill "A Response to the FTC/ Anderson Critique" in this volume for a more extended analytical discussion of this point.

8. A proof of these propositions is available from the authors.

9. If one specifies relative share and concentration in logarithmic form and regards the profit rates as the antilog of an underlying performance variable, the generalized Cournot model is a translog production function, and the traditional linear/Cournot model nested in it is equivalent to a Cobb-Douglas production function.

10. These partial derivatives ignore the indirect effects that arise from the definition of relative share as a function of share and concentration. Expanding the derivatives to include those effects would not negate the conclusion concerning the second order effect of the conjectural variation parameters on the structure profit relationships. We prefer this form of the derivatives because, in our view, market share is an intermediate variable that we have decomposed into two components: relative share and concentration. Moreover it is consistent with the statistical model that we estimate.

11. See Buzzell and Gale (1987) for an explanation of the hypothesized effects of these variables on ROI.

12. 192 observations were not included due to lack of data on one or more variables, giving a sample with 2,552 observations.

13. Schmalensee (1987) derives the ROI model from the ROS model and demonstrates that it requires the addition of a capital intensity variable such as our control variable GBV. We acknowledge and reject the Fisher critique that accounting profits do not necessarily measure economic profits (see the discussion in the overview section of Chapter 1, this book). The PIMS data base is maintained by firms to analyze the profit impact of marketing strategy (PIMS). Since firms use the PIMS ROI variable as a performance measure, it is an appropriate indicator for the analysis of market power and efficiency questions.

14. See Connor, *et al.* (1985) p. 335-336 for a summary of prior studies results.

15. This computation is based upon data from Goldman Sachs (1991, p. 1) and the 1990 Economic Report of the President, p. 401.

16. The Strategic Planning Institute has a panel data base (business unit data on a yearly basis) that may be useful.

References

Breshnahan, T. 1989. Empirical Studies of Industries with Market Power. *Handbook of Industrial Organization.* Vol. 2. Amsterdam:North Holland.

Buzzell, R. D., and B. T. Gale. 1987. *The PIMS Principles.* New York: The Free Press.

Clarke, Roger, and Stephen Davies. 1982. Market Structure and Price-Cost Margins. *Economica* 49(Aug): 277-287.

Clarke, R., S. Davies, and M. Waterson. 1984. The Profitability-Concentration Relation: Market Power or Efficiency? *Journal of Industrial Economics* 32(June): 435-450.

Connor, J., and E. Petersen. 1992. Market-Structure Determinants of National Brand-Private Label Price Differences of Manufactured Food Products. *Journal of Industrial Economics* 40(2): 1-15.

Connor, J., R. T. Rogers, B. W. Marion, and W. F. Mueller. 1985. The Food Manufacturing Industries. Lexington: Lexington Books.

Cotterill, R. W. 1986. Market Power in the Retail Food Industry: Evidence from Vermont. *Review of Economics and Statistics* 68(Aug): 379-386.

Cotterill, R. W., and C. Iton. *Market Share, Concentration, and Market Power: A Rapproachment*. University of Connecticut, Food Marketing Policy Center, Research Report No. 18 (forthcoming 1992).

Cowling, K., and M. Waterson. 1976. Price-Cost Margins and Market Structure. *Economica* 43(Aug): 267-274.

Demsetz, H. 1974. Two Systems of Belief about Monopoly. In *Industrial Concentration: The New Learning*, ed. H. J. Goldschmid, H.M. Mann, and J.F. Weston, 164-184. Boston:Little, Brown.

Federal Trade Commission. 1969. *On the Influence of Market Structure In the Profit Performance of Food Manufacturing Companies,* (Washington, D.C.:U.S. Government Printing Office) Sept.

Fisher, F. M., and J. J. McGowan. 1983. On the Misuse of Accounting Rates of Return to Infer Monopoly Profits. *American Economic Review* 73(March): 82-97.

Gale, B. T. 1972. Market Share and the Rate of Return. *Review of Economics and Statistics* 54(Nov): 412-423.

Gale, B., and B. Branch. 1982. Concentration Versus Market Share: Which Determines Performance and Why Does It Matter?. *Antitrust Bulletin* 27(Spring): 83-105.

Geroski, P., and R. Masson. 1987. Dynamic Market Models in Industrial Organization. *International Journal of Industrial Organization* 5(March): 1-14.

Goldman Sachs. 1991. *Industrial Research:Packaged Food Quarterly*. Goldman Sachs:New York.

Imel, B., and P. Helmberger. 1971. Estimation of Structure-Profits Relationships with Application to the Food Processing Sector. *American Economic Review* 61(Sept): 614-627.

Kwoka, J. E. Jr., and D. J. Ravenscraft. 1986. Cooperation v. Rivalry: Price-Cost Margins by Line of Business. *Economica* 53(August): 351-363.

Marion, B. W., W. F. Mueller, R. W. Cotterill, F. E. Geithman, J.R. Schmelzer. 1979. *The Food Retailing Industry*. New York: Praeger.

Marshall, C. T. 1987. *PIMS and the FTC Line-of-Business Data: A Comparison*. Doctoral dissertation, Harvard University.

Martin, S. 1988. Market Power or Efficiency. *Review of Economics and Statistics* 70(2): 331-335.

Mueller, D. C. ed. 1990. *The Dynamics of Company Profits: An Internaitonal Comparison*. Cambridge: Cambridge University Press.

Peltzman, S. 1977. The Gains and Losses from Industrial Concentration. *Journal of Law and Economics* 20(October): 229-63.

Ravenscraft, D. 1983. Structure-Profit Relationships at the Line of Business and Industry Levels. *Review of Economics and Statistics* 65(February): 25-37.

Rill, J. J. 1990. Sixty Minutes with the Honorable James F. Rill, Assistant General, Antitrust Division, U.S. Department of Justice. *Antitrust Law Journal* 59: 45.

Rogers, R. T. 1978, as reported in Connor, J, R.T. Rogers, B.W, Marion, and W.F. Mueller. 1985. *The Food Manufacturing Industry*. Lexington: Lexington Books.

Salinger, M. 1990. The Concentration-Margins Relationship Reconsidered. *Brookings Papers:Microeconomics*. Washington:Brookings Institution.

Scherer, F. M., and D. Ross. 1990. *Industrial Market Structure and Economic Performance*. Boston:Houghton Mifflin Co.

Schmalensee, R. L. 1988. Inter-Industry Studies of Structure and Performance. In *Handbook of Industrial Economics*. Amsterdam: North Holland.

____. 1987. Collusion Versus Differential Efficiency: Testing Alternative Hypotheses. *Journal of Industrial Economics* 35(June): 399-425.

____. 1985. Do Markets Differ Much? *American Economic Review* 75(June): 341-351.

Schrader, L. and N. Colline. 1960. Relation of Profit Rates to Market Structure in the Food Industries. *Journal of Farm Economies* 42: 1526-1527.

Shepherd, W. G. 1972. The Elements of Market Structure. *Review of Economics and Statistics* 54(February): 25-35.

____. 1986. Tobin's q and the Structure-Performance Relationship: Comment. *American Economic Review* 76(December): 1205-1210.

____. 1986. On the Core Concepts of Industrial Economics. In *Mainstreams in Industrial Organization - Book I*, ed. H. W. deJong and W. G. Shepherd, Boston: Kluwer.

Stigler, G. 1964. A Theory of Oligopoly. *Journal of Political Economy* 72(1): 44-61

U.S. President. 1990. *Economic Report of the President*. Washington D.C.: U.S. Government Printing.

Weiss, L. W. 1974. The Concentration-Profits Relationship and Antitrust. In *Industrial Concentration: The New Learning*, ed. H. J. Goldschmid, H. M. Mann, and J. F. Weston. Boston: Little, Brown.

3

Vertical Coordination and the Competitive Performance of the U.S. Food Industries

Stuart D. Frank and Dennis R. Henderson

The structure and competitiveness of the U.S. food manufacturing industries is influenced in part by the exchange mechanisms used for vertical coordination. Traditionally, industry vertical structure has been studied using the theory of vertical integration. However, vertical integration is only one of many modes of vertical structure. Vertical coordination is a more comprehensive concept, capturing not only vertical integration, but is the "process by which the various functions of a vertical value adding system are brought into harmony" (Marion 1976, p. 180).

Many economic factors affect vertical structure and performance of the food industries. Theory suggests that in a market-hierarchy framework, the criterion for the organization of production is the minimization of production and transaction costs. Williamson (1979, p. 233) suggests that in this framework the use of various administered vertical exchange arrangements is motivated by transaction costs, stating that "if transaction costs are negligible, the organization of economic activity is irrelevant." Alternatively, in an environment congested with transactional inefficiencies, organized vertical coordination will attenuate and possibly eliminate transactional inefficiencies, thus improving industry performance.

Previous studies that examined vertical coordination in the food sector, qualitatively addressed the antecedents and implications of vertical coordination. These studies casually linked transaction costs to vertical coordination. Several empirical studies have examined the effects of transactional inefficiencies either on vertical integration or long-term contracts (see Joskow and Levy), but not on the broader concept of vertical coordination. These analyses do report transaction costs linkages to vertical coordination.

In the literature, there are many studies that have examined the U.S. food industries in the structure—conduct—performance framework. These studies have found linkages between the structure of industry and its competitive performance. However, the transaction costs effects on vertical coordination between farms and food manufacturing industries have not been empirically analyzed. Further, little empirical analysis has been reported on the linkages between vertical coordination, as a structural component, and competitive performance of the food industries.

To quantitatively analyze transactional inefficiencies and the effects of vertical coordination on food industry performance, a measure of vertical coordination must be devised. The purpose of this paper is to introduce an index measure of vertical coordination and use it to examine; (1) the linkages between transaction costs and vertical coordination in the food industries and (2) the effects of vertical coordination on food industry competitive performance. The research hypotheses are; (1) transaction costs motivate the increased use of various non-market vertical coordination methods and (2) vertical coordination, as a structural characteristic, has a positive relationship with food industry competitive performance.

The first three sections of the paper will discuss the theory and measurement of vertical coordination. The fourth and fifth sections discuss transaction costs and their effects on food industry vertical coordination. The sixth, seventh, and eighth sections will discuss the vertical coordination and food industry competitive performance linkages. The final section will present the paper's summary and conclusions.

Administered Coordination

Administered coordination involves the transfer of certain rights from one firm to another by using various coordinating mechanisms other than spot markets. Transfer of rights involves an increased consolidation of control by the contractor or integrator.

Administered coordination arrangements range from virtually no control to those that transfer almost complete control from one firm to another. These include both contracts and tacit arrangements in addition to vertical integration. Tacit arrangements (e.g. providing technical expertise and advice, increased credit, etc.) allow firms some control over vertically interdependent enterprises that are owned by others (Blois).

Williamson (1979) provides theoretical insight into the structure of contracts within the vertical coordination process. Using Macneil's contract law classifications, he advanced three classifications for contractual coordination; classical, neoclassical, and relational. Classical contracts are based on a set of legal rules with formal documents and self-liquidating transactions. Neoclassical contracts involve longer-term arrangements that do not cover all contingencies, but includes additional governance structures (i.e. arbitration). Relational contracts are agreements in principle on the entire spectrum of the contracting parties' relationship over time, including tacit as well as explicit arrangements.

Williamson argues that increases in transaction complexity, frequency, and uncertainty, accompanied by idiosyncratic investments, result in a shift in the coordination structure from classical to neoclassical to bilateral and finally to unilateral relational contracts. Typically in this progression, one party has increased consolidated control.

Specific to the food sector, Mighell and Jones discuss several administered arrangements utilized to vertically coordinate. They specify three general types of contracts; market specification, production management, and resource providing.

Market specification contracts are standardized contracts in which the supplier transfers part of the risk and management functions to the contractor. Transferred management only regards the decision of what to produce and when and where the product is to be delivered. Production management contracts are similar except the contractor has increased control over the production process. This may occur when the contractor is concerned with the quality of production. With this type of contract, the transfer of managerial decisions usually takes the form of resource specification. Finally, resource providing contracts are the closest to vertical integration. The contractor not only provides a market for the production, but also is a major provider of inputs into the production process.

Empirical Measurement of Vertical Coordination

A specification that includes spot markets, ownership, and contractual relationships for vertically interdependent firms or industries provides a more complete measure of vertical coordination than the traditional measures of vertical integration. Such a specification should include both the consolidation of control, as well as the degree of interdependency among firms and industries.

Empirical research has examined vertical organization primarily in the context of vertical integration. Studies by Adelman, Laffer, Levy, and Tucker and Wilder used variations of the value-added to sales ratio to calculate vertical integration. However, this ratio is influenced by such factors as firm profitability and the position of the firm in the production process. Moreover, it does not capture the partial consolidation of control between vertically related firms or industries through contracts and other agreements.

A second measure of vertical integration examines the linkages between industries through production functions. Maddigan advanced this measure, which considers the input-output interdependencies between firms. These interdependencies are captured by aggregate production functions and are expressed by physical input-output coefficients.

A viable starting point for measuring vertical coordination is Maddigan's Vertical Industry Connection (VIC) index. This index exploits the interactions of the Leontief input-output model. In the Leontief framework, each x_{ij} in the input-output transactions matrix X is the optimal value of industry i's output used as an input by industry j.

The input-output transactions matrix is manipulated to form the initial component of the up- and down-stream interdependent linkages matrix of the vertical coordination index. Two matrices, A and B, (equations 1 and 2) capture all net production interrelationships for an industry's input-output linkages.

$$A = I - [x_{ij} / (z_j - x_{jj})] + [y_{ij}] \qquad (1)$$

and

$$B = [x_{ij} / (z_i - x_{ii})] - [y_{ij}] - I \qquad (2)$$

where;

I = identity matrix, r x r,

x_{ij} = the value of the ith industry's output used as an input to the jth industry; $i, j = 1,...,r$,

z_j = total value of the output of industry j; $j = 1,...,r$,

y_{ij} = $[x_{ii} / (z_i - x_{ii})]$ if $i = j$; 0 if $i \neq j$; $i, j = 1,...,r$.

Each element of matrix A, a_{ij}, represents the percentage of the value of industry j's net output supplied by industry i. Each element of B, b_{ij}, represents the percentage of the value of industry i's output supplied to industry j. In short, matrix A represents an industry's up-stream connections and matrix B its down-stream connections. Notationally, inputs are negative as values used in production and outputs are positive.

In order to calculate vertical interdependence for each specific industry, two matrices, C_K and D_K are defined. Matrices C_K and D_K capture industry k's primary and secondary interindustry connections. The association of industry k with its interdependent industries is determined by the flow of net production. These matrices are constructed using the rows and columns of matrices A and B, specifically, the columns of A and the rows of B. Matrices C_K and D_K are represented by equations 3 and 4:

$$c_{ij} = a_{s(i)s(j)}i, \qquad j = 1...n \ (n \leq r) \qquad (3)$$

and

$$d_{ij} = b_{s(i)s(j)} \qquad i, j = 1...n \ (n \leq r) \qquad (4)$$

where;

$s(i)$ = industries with which industry k is associated, indexed by i,

c_{ij} = the percentage of the value of industry s(j)'s net output supplied by industry s(i),

d_{ij} = the percentage of the value of industry s(i)'s net output supplied to industry s(j).

For matrix C_K, for column j where j = k, industry k has a primary input relationship with industry i and for column j, j ≠ k, industry k has a secondary input relationship with industry i. It is the obverse for matrix D_K, for row i, i = k, industry k has a primary output relationship with industry j and for row i, i ≠ k, k has a secondary output relationship with industry j.

To complete the vertical coordination index, the degree of administrative control that is consolidated by the contractor/integrator must be specified. Administration of vertical interdependencies may be accomplished through ownership and/or a wide variety of contractual relationships. This implies the existence of a progressive relationship of consolidated control between the end points of none (spot markets) and complete integration. Along this continuum, the contractor/integrator consolidates increasing degrees of vertical control.

To capture the coordination mechanisms for an industry's primary and secondary interactions, matrices E_K and F_K are created. Each e_{ij} represents the measure of consolidated control for industry k with the up-stream industry i. Similarly, each f_{ij} represents industry k's consolidated control with down-stream industry j. To measure consolidation, each coordinating structure is assigned a value representing the percent consolidated control. Equation 5 represents the calculation for matrices E and F:

$$e_{ij} \text{ and } f_{ij} = \sum_{g=1}^{s} \sum_{h=1}^{t} L_{gh}O_{gh}N_{gh} \quad i,j = 1...n \ (n \leq r) \tag{5}$$

where;

g = number of products produced in each industry, g = 1...s,
h = type of coordinating mechanism, h = 1...t,
L = for e_{ij}, product g's percentage of industry j's input mixture and for f_{ij}, product g's percentage of industry i's output mixture,
O = assigned value of consolidated control,
N = percent of production coordinated by each transaction type.

With matrices C, D, E, and F, the Vertical Coordination index can be calculated. Equation 6 is the generalized formulation of the Vertical Coordination index for industry k:

$$VC_k = 1-[1/\prod_{i=1}^{n} (C^i)^P(D^i)^P(E^i)^P(F^i)^P] \tag{6}$$

where;

C^i = column i of industry k's up-stream connections matrix,
D_i = row i of industry k's down-stream connections matrix,
E^i = column i of industry k's up-stream control matrix,
F_i = row i of industry k's down-stream control matrix,

P = vector dot product,

n = number of industries which industry k is interdependent.

This specification of Vertical Coordination (VC) has several desirable properties.

1. VC increases (decreases) when an input industry becomes relatively more (less) important by accounting for a larger (smaller) percentage of the total value of output of another industry.
2. VC increases (decreases) when relatively more (less) of the output of an industry is used as an input to another industry.
3. VC increases (decreases) as an industry increases (decreases) its number of vertical interactions with other industries.
4. VC increases (decreases) as an industry exercises increased (decreased) up and/or down stream consolidated control.
5. The range of VC is between 0 and 1.

The components of VC are influenced, but not skewed, by several factors. Price and quantity changes affect the values of the input-output transactions matrix and also argument L in equation 5. If X_{ij} or argument L increases between industries i or j, the importance of the industries to each other increases. However, at the same time all other linkages of either industry are relatively less important. Thus, VC will increase or decrease depending on the relative magnitude of the changes.

Vertical coordination is also affected by the value added by an industry. As value added increases, industry input-output vector elements, X_{ij}, decrease. Therefore, the industry's primary and secondary connections become less important.

Calculation of Vertical Coordination for Food Manufacturers

To examine the vertical linkages between production agriculture and food manufacturers, the 1982 input-output transactions matrix was constructed using the four digit Standard Industrial Classification (S.I.C.) scheme to classify or group firms into industries.[1] The transactions matrix incorporated the interdependencies between the production agriculture (S.I.C. 0111 to 0291) and the food manufacturing (S.I.C. 2011 to 2099) industries. Vertical coordination for each specific food manufacturing industry was also calculated at the four-digit S.I.C. level.

To calculate VC for the food manufacturing industries, it is necessary to have data on the use of various coordinating methods in agriculture. While there is no systematic reporting of agricultural contract data, a number of researchers have provided various estimates on contract use consistent with the Mighell and Jones' classification. Five coordinating methods are used; (1) spot markets, (2) market specification contracts, (3) production management contracts, (4) resource providing contracts, and (5) integration.

Data were unavailable on the usage of each method to procure agricultural commodities by food manufacturers. To approximate the up-stream coordinating methods of food manufacturers, data on down-stream contracting by the farm sector were used.

The share of total farm-to-processor product flow for each of the five coordinating methods was estimated on the basis of previously reported information (Table 3.1). These shares were used to calculate the elements of matrix E, equation 5. Also, a decreasing marginality functional relationship was utilized for the consolidation of control associated with each coordinating method (captured in argument O of equation 5).[2] That is, the percentage of consolidated control increases at a decreasing rate for each successive coordination method, moving from spot markets (0%) to integration (100%).

The coordinating method data utilized for this study pose potential biases in the estimation of vertical coordination. The Mighell and Jones' contract classifications do not explicitly incorporate the dynamics of many vertical arrangements as captured by Williamson's governance scheme, thus understating the extent of common or shared control among interdependent firms.

Where there are intermediaries between farmers and food processors, the processor's amount of up-stream control (represented by the farm's down-stream linkage) is biased downward by the amount of the processor-intermediary linkage. Also, if the first handler is also a food processor, integration will be understated to the extent that internal transfer of procured farm commodities exceeds farmer-processor integration. In addition, many food manufacturing industries procure a portion of their inputs from other food processing industries. However, food manufacturing industry up-stream coordinating method data are not available. Furthermore, the F-matrix in the Vertical Coordination index (equation 6) cannot be calculated due to the unavailability of food manufacturer down-stream coordinating data.

In the absence of conceptually desirable data, the use of available data that understates the true magnitudes of the phenomena of interest, should not diminish the underpinnings of the transaction costs and vertical coordination linkages. Arguably, these linkages would be more pronounced absent such bias.

Measuring Transaction Costs

To examine the incidence of vertical coordination between farms and food manufacturing industries due to transactional inefficiencies, several industrial characteristics affecting transaction costs factors are examined.

TABLE 3.1 Share of U.S. Food Manufacturing Industries' Farm-Originated Inputs Coordinated by Various Methods.

Industry	Spot Market[a]	Market Specification	Contracts		Intergration
			Production Management	Resource Providing	
			(percent)		
Meat Packing	89.5	7.0	0.0	0.0	3.5
Sausages and other prepared meats	89.3	7.2	0.0	0.0	3.5
Poultry dressing	13.0	0.0	0.0	73.0	14.0
Poultry and egg processing	6.5	0.0	18.5	48.2	26.8
Creamery butter	17.7	81.0	0.0	0.0	1.3
Cheese, natural and processed	17.7	81.0	0.0	0.0	1.3
Condensed and evaporated milk	17.7	81.0	0.0	0.0	1.3
Ice cream and frozen desserts	19.0	70.8	3.6	0.0	6.6
Fluid milk	17.7	81.0	0.0	0.0	1.3
Canned specialties	30.1	6.2	38.8	0.0	24.9
Canned fruits and vegetables	35.4	10.4	30.6	0.0	23.6
Dehydrated fruits, vegetables and soups	24.9	14.0	33.4	0.0	27.7
Pickles, sauces, and salad dressings	36.9	1.3	40.3	0.0	21.5
Frozen fruits and vegetables	24.9	14.0	33.4	0.0	27.7
Frozen specialties	19.3	14.0	15.7	24.1	26.9
Flour and other mill products	91.7	7.8	0.0	0.0	0.5
Cereal breakfast foods	81.6	13.0	0.0	0.0	5.4
Rice milling	91.5	8.0	0.0	0.0	0.5
Wet corn milling	92.5	7.0	0.0	0.0	0.5
Dog, cat, and other pet food	93.3	6.0	0.0	0.0	0.7

Prepared feeds, n.e.c.	92.4	7.1	0.0	0.0	0.5
Bread, cake, and related products	40.0	35.0	0.0	0.0	25.0
Cookies and crackers	100.0	0.0	0.0	0.0	0.0
Raw & refined cane and beet sugar	0.0	0.0	69.0	0.0	31.0
Confectionery products	85.0	12.3	0.0	0.0	2.7
Chocolate and cocoa products[b]	100.0	0.0	0.0	0.0	0.0
Cottonseed oil mills	82.3	16.7	0.0	0.0	1.0
Soybean oil mills	89.5	10.0	0.0	0.0	0.5
Vegetable oil mills, n.e.c.	89.5	10.0	0.0	0.0	0.5
Animal and marine fats and oils	100.0	0.0	0.0	0.0	0.0
Shortening and cooking oils	100.0	0.0	0.0	0.0	0.0
Malt beverages	93.2	6.0	0.0	0.0	0.8
Malt	92.5	7.0	0.0	0.0	0.5
Wines, brandy, and brandy spirits	32.0	41.0	0.0	0.0	27.0
Distilled liquor, except brandy	92.5	7.0	0.0	0.0	0.5
Bottled and canned soft drinks[b]	100.0	0.0	0.0	0.0	0.0
Flavoring extracts and syrups, n.e.c.[b]	100.0	0.0	0.0	0.0	0.0
Canned and cured seafoods	100.0	0.0	0.0	0.0	0.0
Fresh or frozen packaged fish	96.0	3.0	0.0	0.0	1.0
Roasted coffee[b]	100.0	0.0	0.0	0.0	0.0
Macaroni and spaghetti	11.0	0.0	45.0	0.0	44.0
Food preparations, n.e.c.	79.0	16.0	0.0	0.0	5.0

[a]Residual values.

[b]Industry had no up-stream linkages to agricultural producers (SIC 0111-0291).

Source: Marion 1986; Krause, Crom, Lasley, Van Ardsall *et al.*, Flinchbaugh, Reimund *et al.*, and Buckley *et al.*

Uncertainty

Convention holds that uncertain demand/supply causes firms to rely more on non-market coordination methods. When transactions are conducted under uncertainty, it can become very costly and may be impossible to anticipate all contingencies. In which event, the use of alternative means of coordination that attenuate uncertainty may be more desirable. To measure the uncertainty of food manufacturers input supply, the percent change in farm output supply between 1981-1982 (PCFS) is used. If current farm supplies are anticipated to fall short, food manufacturers are motivated to use non-market coordination methods to attenuate supply uncertainty. A negative PCFS coefficient corresponds to an increase in vertical coordination activity. To measure unanticipated demand uncertainty (UNANT), the log of food industry sales were regressed on a time trend, equation 7.[3] The value for UNANT is the variance of the error term from equation 7.

$$LFIS_k = \alpha + \beta \, TT + \mu \qquad k = 1 \text{ to } n \qquad (7)$$

where:

$LFIS_k$ = log of food industry k's sales,
TT   = time trend.

Concentration

As the number of buyers and sellers in a market diminishes, small numbers bargaining problems become more prevalent. In such circumstances, firms utilize non-market coordinating methods to reduce potential opportunistic behavior. Several variables are used to proxy the number of current and potential future market participants. Anticipated demand growth (ADG) is used to represent the number of potential firms in the industry. Williamson (1979, pg. 260) states, "as generic demand grows and the number of supply sources increases, exchange that was once transaction specific loses this characteristic and greater reliance on market mediated governance is feasible." As future demand increases, the motivation to vertically coordinate by means of non-market methods diminishes. The value for ADG is the time trend coefficient from equation 7.

To measure the buyer concentration and market power for the food industries, the food manufacturing industries' four firm concentration ratio (CR4) is used. Two variables capturing seller (input supplier) concentration, one each for the farm output industries and food manufacturing industries, are calculated. The variable FSGC is a farm industry's GINI coefficient weighted by the net contribution of each farm industry as a supplier to each food industry.[4] Similarly, the variable FUSC is the weighted four firm concentration ratio of the food manufacturing industries that supply inputs to a given food manufacturing industry (e.g. meat packing industry supplying inputs to the sausage and prepared meats industry).

Idiosyncratic Investments

Firms that produce specialized or differentiated products and those with highly intensive technical production processes may have increased asset specificity, i.e. idiosyncratic investments. This may result in an increased need to vertically coordinate. To measure these differential characteristics, the industry advertising to sales ratio (AS) and industry ratio of research and development expenditures to sales (RD) are utilized.[5]

Costs of Administered Vertical Coordination

The costs of internalizing transactions is also a consideration. Firms will internalize transactions up to the point where the market costs of an activity just outweigh the cost of internalization. Several firm/industry characteristics may determine internal costs of administrative control. These include firm specialization, capital intensity, and flow economies. Variables to proxy these characteristics are the industry specialization ratio (SPCR), capital to sales ratio (KS), and a food production dispersion index (FPDI).

Stigler has demonstrated that as a firm specializes in a particular product, it vertically disintegrates to more fully capture increased scale economies. The greater the capital intensity, concomitant with uncertainty, firms will vertically coordinate to maintain production capacity. Firms have an incentive to vertically coordinate to capture flow economies in the production process. Physically closer stages of production can more readily capture such economies, thus they have greater incentives to vertically coordinate. Utilized as a proxy for flow economies, FPDI is calculated to quantify the proximity of enterprises with output-input linkages. A negative FPDI coefficient implies an increase in vertical coordination. The FPDI for industry k is:

$$FPDI_k = \sum_{c=1}^{m} W_c \left[\sum_{i=1}^{r} |F_i^c - P_i^k| \right] \qquad k = 1 \text{ to } n. \qquad (8)$$

where;

F_i^c = the percent of farm commodity c produced in region i,
P_i^k = the percent of processed food k manufactured in region i,
W_c = percent of commodity c's net contribution to food industry k.

Transaction Costs and Vertical Coordination Relationships

The analysis was based on 1982 data for 42 four-digit S.I.C. food manufacturing industries. Using variations of equation 6, two progressively comprehensive vertical coordination measures for the food manufacturing industries were specified; VC1, the up-stream consolidation of control associated with various coordinating methods and VC2, the up- and down-stream interdependencies plus up-stream consolidated control. As the amount of information increases, the value of the index increases, revealing the importance of each component of the most comprehensive measure.

Each OLS regression estimation was examined for heteroscedasticity and multicollinearity. The Breusch-Pagan test was used to confirm the presence of heteroscedasticity. To correct for heteroscedasticity, the coefficients and t-statistics were estimated using White's heteroscedastic-consistent covariance matrix. The Belsley-Kuh-Welsch procedure was conducted to test for multicollinearity. This test failed to suggest the presence of multicollinearity in the explanatory matrix.

For the two vertical coordination specifications, VC1 and VC2, the coefficients for PCFS, UNANT, FSGC, FUSC, RD, SPCR, and FPDI were generally of expected sign (Table 3.2, columns 1 and 2). These coefficients were statistically different from zero to at least the 10 percent level of significance. The coefficients for ADG and KS were not statistically significant in either equation, while only CR4 was of opposite sign and statistically different from zero. The variable AS was statistically positive in the VC1 estimation, but significantly negative in the VC2 estimation. Each of these three estimated relationships explained at least 71 percent of the total variation in vertical coordination as illustrated by the coefficient of determination (R^2). Moreover, the two equations do provide a strong statistical relationship between transaction costs and vertical coordination (F-values significant at the 1 percent level).

A comparison of the most complete vertical coordination measure, VC2, with previously used measures of vertical structure is also presented in Table 3.2. First, Maddigan's VIC index, capturing both up- and down-stream interdependencies, was examined with the same set of transaction cost explanatory variables. The coefficients for only FSGC and FPDI were of expected sign and statistically significant. The estimates for AS and KS were also statistically significant but of opposite sign. The estimated coefficients for the remaining independent variables PCFS, UNANT, CR4, ADG, FUSC, RD, and SPCR, were not statistically different from zero. Overall, VIC was not revealing. Its estimated equation achieved a relatively low coefficient of determination ($R^2 = 0.34$) and the overall test of the relationship (F-value = 1.38) was not significant. The vertical coordination measure VC2, which adds coordinating methods to VIC, performed considerably better. Thus, recognition

TABLE 3.2 Transaction Costs and Vertical Coordination Relationships

Explanatory Variables	Dependent Variables			
	VC1	VC2	VIC	VI
			(t-statistics in parenthesis)	
constant	1.82^c	1.18^c	0.06	0.12
	(3.83)	(3.22)	(0.15)	(0.53)
PCFS	-2.44^c	-1.00^b	0.49	-0.47^a
	(4.62)	(2.44)	(0.84)	(1.47)
UNANT	-0.50	0.83^a	0.46	-1.22^{bb}
	(0.57)	(1.31)	(0.50)	(2.43)
CR4	-0.005^{bb}	-0.0008	0.002	0.001^a
	(2.26)	(0.37)	(1.30)	(1.31)
ADG	0.33	-0.55	-0.54	0.42
	(0.25)	(0.46)	(0.41)	(0.74)
FSGC	0.84^c	1.01^c	0.35^c	-0.06
	(6.56)	(12.06)	(3.63)	(0.80)
FUSC	0.006^c	0.004^c	-0.001	0.003^c
	(4.39)	(3.89)	(1.07)	(4.99)
RD	24.09^c	16.96^c	2.31	6.46^b
	(3.55)	(3.80)	(0.41)	(1.84)
AS	3.31^c	-2.20^{bb}	-4.09^{cc}	2.47^b
	(2.51)	(2.08)	(3.64)	(2.41)
KS	0.17	-0.06	-0.30^{aa}	0.09
	(0.76)	(0.33)	(1.76)	(0.80)
SPCR	-0.02^c	-0.01^c	0.002	0.0001
	(4.91)	(3.80)	(0.42)	(0.03)
FPDI	0.02	-0.03^a	-0.05^b	0.008
	(1.06)	(1.31)	(1.72)	(0.63)
R^2	0.71	0.77	0.34	0.65
F-value	6.72^c	9.14^c	1.38	5.05^c
DF	30	30	30	30

Note: a, b, and c are significant at the 10, 5, and 1 percent level of significance for a one-tailed test, respectively.

aa, bb, and cc are significant at the 10, 5, and 1 percent level of significance for a two-tailed test, respectively.

of coordinating methods appears to be empirically important when examining vertical coordination.

The other dependent variable examined was a traditional measure of vertical integration, VI, defined as the value-added to sales ratio. The coefficients for PCFS, CR4, FUSC, RD, and AS were of expected sign and statistically significant. Only the estimate for UNANT was of opposite sign and significant. The coefficients for the remaining independent variables, ADG, FSGC, KS, SPCR, and FPDI were not statistically different from zero. The estimated equation for VI performed well relative to VIC with an R^2 of 0.65 and an F-value significant at the 1 percent level.

A comparison between the estimated equations for VI and VC2 is revealing. The estimation for VI and VC2 yielded five and seven, respectively, explanatory variables that were of the expected sign and statistically significant. Three independent variables, PCFS, FUSC, and RD have the same sign and were significant in both equations. Two variables, UNANT and AS, were significant in both equations, but with different signs. The variables, ADG and KS, were not statistically significant in either estimated equation. The estimated regression coefficients for all other explanatory variables, CR4, FSGC, SPCR, and FPDI were significant in one but not both equations.

This result supports the relative merits of a vertical organization measure focusing on inter-industry linkages rather than an intra-industry value-added measure in industry cross-section analysis. For instance, VC2 was influenced by up-stream concentration (FSGC and FUSC), uncertainty (PCFS and UNANT), and factors affecting scale economies (SPCR and FPDI). However, two of the statistically significant variables in the VI equation (CR4 and AS) are more strongly related to increased profitability and value-added, which are inherent in the traditional value-added to sales ratio specification of vertical integration. This has been a weakness of VI comparisons in cross-section industry studies.

Measuring Food Industry Competitive Performance and Structure

The results from the previous section demonstrate that a variable incorporating coordinating methods with technical input-output interdependencies bridges the gap in the dichotomy of market versus ownership coordination. With empirical evidence of the transaction costs and vertical coordination relationships supporting vertical coordination as a structural characteristic, the second research hypothesis may now be tested.

A predominant theory of industrial organization, the structure—conduct—performance paradigm, describes the linkages between the organization of markets and the performance of those markets. In the literature, an abundant number of studies have examined these linkages. Many have examined the relationship between market power and several performance measures such as profits, growth, or innovation. For example, using concentration as a proxy for monopoly market power, a broad base of literature suggests a positive relationship between concentration levels and profits.

As discussed in the previous section, vertical coordination is related to several economic and strategic factors. When these motivating factors are examined in the context of the structure—conduct—performance paradigm, theory suggests there are linkages between the degree of vertical coordination and industry performance.

To examine vertical coordination and competitive performance linkages, the food processing industries' return on capital and export propensity were examined. In addition to vertical coordination, the performance-vertical coordination models included several industry structural explanatory variables such as industry concentration, research and development, advertising, capital intensity, and labor productivity in order to account for performance variability associated with factors other than vertical coordination.

Competitive Performance Variables

Two variables are used to proxy food industry domestic and international competitiveness. The domestic variable is the food industries' return on capital and the international variable is the industries' export propensity. Food industry return on capital, RK, is defined as the difference between value-added and total payroll divided by capital (Census of Manufacturers). Export propensity, XMS, is equal to industry exports divided by domestic value of shipments (International Trade Administration).

Industry Concentration

The four firm concentration ratio, CR4, for each industry is a proxy to measure market power. It is hypothesized that CR4 is positively related to returns on capital and exports. Increased domestic concentration suggests increased domestic market power, providing firms greater resources, allowing them to export more products.

Technology and Skill Factors

Research and development (RD), the "technology" factor, represents leadership in creating new and improved products and production processes. The expected sign for research and development in the export equations is ambiguous. Marvel and Pagoulatos and Sorensen suggest research and development will improve export competitiveness. However, Handy and MacDonald suggest that research and development may in fact reduce industry export propensity. For return on capital, the expected sign is positive.

The variable MGMT, non-production employment payroll as a percent of total wages, is a measure of the industries' managerial labor force. Marvel and Baldwin suggest this variable approximates the managerial skill level of an industry's labor force. However, it may represent the managerial labor force's contribution to total labor cost. Therefore, its expected signs are ambiguous in relation to return on capital and export propensity.

The production workers impact upon trade flows is measured by the level of production worker average hourly earnings, PWHE. Again, Marvel and Baldwin suggest that the average wage is an indication of the skill level of the labor force. As an industry's skill level increases, its degree of competitive performance increases. However, PWHE, may be a proxy for the cost of labor. Therefore, its expected signs are ambiguous.

Barriers to Entry

To account for possible barriers to entry, the industry's capital to sales ratio (KS) and advertising to sales ratio (AS) are examined. The capital to sales ratio is equal to the end of year book value of depreciable assets divided by the value of industry shipments. It is a proxy for possible scale economy barriers to entry. This variable is expected to be negatively related to export propensity and return on capital.

The advertising to sales ratio was calculated as the value of advertising expenditures in the six measured media plus the value of industry advertising expenditures divided by industry shipments. Industry advertising expenditures is an approximation of the industry's product differentiation acting as a possible barrier to trade. Its sign is ambiguous for the export equation. Pagoulatos and Sorensen suggest a positive relationship while Handy and MacDonald suggest a negative relationship. However, it is expected to be positive for return on capital.

Product Transportability

The variable, GDI, is the geographical dispersion index of U.S. manufacturing suggested by Collins and Preston. The index is represented by equation 9:

$$SDI = \sum_{i=1}^{n} |s_i - p_i| \qquad (9)$$

where s_i is the share of domestic shipments originating in region i and p_i is the share of the population in region i. Increased values of this index are associated with increased concentration of regional manufacturing. Assuming the demand for an industry's output is constant across regions and production is located in one region, this suggests that the industry's production is easily transported. It is hypothesized that increases in GDI are associated with increased exports. Regional production data were available from the Census of Manufactures and population data from U.S. Department of Commerce statistics.

Foreign Trade Barriers

The final variable, FTB, is a proxy for possible trade barriers against U.S. exports representing the potential percentage increase in U.S. trade without trade barriers. The variable FTB, is the calculated percent increase between the actual exports of domestic food products and the estimated value of exports without trade barriers. Actual and estimated trade values were available from USDA. As FTB increases, the larger the trade barrier, therefore it is expected the sign of FTB is negative in the export equation.

Food Industry Return on Capital Results

Two equations, one using VC1 and the second using VC2, were estimated for return on capital (Table 3.3, columns 1 and 2). In general, the specified equations captured a large percentage of the total variation while the explanatory variables have expected signs.

Focussing on the vertical coordination variables, the coefficient for VC1, which captures the up-stream coordinating methods is positive and statistically significant. The sign for VC2 (up- and down-stream connections plus up-stream coordinating methods) is also positive, although not significant. This suggests that the component, up-stream coordinating methods as measured by VC1, is an important structural component influencing industrial performance.

As hypothesized, the coefficients for industry concentration (CR4), research and development (RD), and advertising (AS) were positive with CR4 and AS being statistically significant. Increased market power as captured by CR4 leads to higher returns on capital. It also appears that by creating a differential advantage through increased advertising generates improved rates of return.

The variable KS (capital to sales ratio) is negative and statistically significant with return on capital. Thus, it appears that the KS variable is capturing the average cost function, the lower the average cost structure, the greater the returns on capital.

The two labor variables, non-production payroll (MGMT) and production worker average wages (PWHE), are both positive, but only PWHE is significant. However, the direction of causation is unclear for the variables PWHE and returns on capital. It may be that profitable industries pay higher wages because they can afford to.

Food Industry Export Propensity Results

In general, the regression results do not support the *a priori* effects of vertical coordination on export propensity (Table 3.3, columns 3 and 4). Specifically, the estimated coefficients for the vertical coordination variables, VC1 (up-stream coordinating methods) and VC2 (up- and down-stream connections plus up-stream coordinating methods), are negative and statistically significant. Therefore, as the food industries increase the administrative control over their domestic inter-industry linkages, they tend to decrease their export activities.

However, calculation of the vertical coordination index incorporated only the domestic importance of inter-industry linkages and domestic institutions governing industry transactions, not direct foreign investment or foreign license agreements.   The negative relationship may be an indication that vertical coordination comes about because of increased transactional inefficiencies and that vertical coordination is not perfectly overcoming such inefficiencies. This suggests that the food industries without transactional inefficiencies do not need to vertically coordinate.   Therefore, industries with less transactional inefficiencies may be more competitive than industries with transactional inefficiencies and higher levels of vertical coordination.

Industries with increased research and development (RD) and advertising (AS) expenditures are less competitive in export markets. These results are similar to Handy and MacDonald suggesting that increased research and development and advertising generates a differential advantage (degree of market power) for the firm/industry.   Exporting innovative or highly technical production technologies may require some form of idiosyncratic investment. In such an economic environment, this evolves into a bilateral monopoly case, where opportunistic behavior increases, thereby possibly increasing direct foreign investment. Likewise, for intensive advertising industries, it is difficult to duplicate such an organization in foreign or culturally different markets. To avoid opportunism and increased transaction costs, firms may then enter foreign markets through direct investment or license agreements.

The two labor variables, production worker average hourly earnings (PWHE) and non-production worker payroll percentage (MGMT), are negatively and positively related to export propensity, respectively. However, only PWHE is statistically significant. The negative coefficient may suggest PWHE is a proxy for industry labor costs, suggesting that higher labor costs may make the industry less competitive.

The four firm concentration ratio (CR4) is positive, suggesting that market power does influence an industry's export propensity. Also, industries with higher capital-sales ratios (KS) tend to export more. The variable KS may be a proxy for an industry's average cost function, the lower the cost function, the more competitive. The variable capturing the "transportability" of output, GDI,

TABLE 3.3 Vertical Coordination and Competitive Performance Relationships

Explanatory Variables	Dependent Variables			
	RK	RK	XMS	XMS
		(t-statistics in parenthesis)		
constant	-0.73 (0.79)	-0.48 (0.47)	0.17^{bb} (2.46)	0.16^{bb} (2.27)
VC1	0.75^{b} (1.78)	n/a	-0.09^{bb} (2.69)	n/a
VC2	n/a	0.52 (0.95)	n/a	-0.10^{bb} (2.32)
CR4	0.029^{c} (2.50)	0.027^{b} (2.24)	0.001 (1.29)	0.001 (1.20)
RD	9.55 (0.42)	13.79 (0.54)	-3.67^{aa} (1.73)	-3.63^{aa} (1.77)
AS	9.83 (1.27)	14.24^{a} (1.51)	-1.37^{bb} (2.61)	-1.87^{cc} (3.19)
KS	-6.34^{c} (3.55)	-6.18^{c} (3.48)	0.11 (0.99)	0.09 (0.87)
MGMT	1.19 (0.76)	1.15 (0.73)	0.08 (0.88)	0.07 (0.74)
PWHE	0.18^{a} (1.31)	0.14 (1.14)	-0.02^{cc} (2.85)	-0.02^{bb} (2.60)
GDI	n/a	n/a	0.08^{c} (3.08)	0.09^{c} (3.58)
FTB	n/a	n/a	-0.00004^{b} (2.24)	-0.00002^{a} (1.62)
R^2	0.59	0.56	0.39	0.39
F-value	6.92^{c}	6.29^{c}	2.26^{b}	2.24^{b}
Degrees of Freedom	34	34	32	32

Note: a, b, and c are significant at the 10, 5, and 1 percent level of significance for a one-tailed test, respectively.

aa, bb, and cc are significant at the 10, 5, and 1 percent level of significance for a two-tailed test, respectively.

is positive and statistically significant to export propensity. This suggests that as product transportability increases, it becomes increasingly traded in foreign markets. Also, the variable capturing foreign trade barriers, FTB, is negative and significantly related to export propensity. This suggests that as foreign trade barriers become more restrictive, U.S. food manufacturers export less.

Summary and Conclusions

The structure and competitive performance of the food manufacturing industries is influenced in part by the exchange mechanisms used for vertical coordination. Previous sturdies that examined vertical coordination in the food sector, qualitatively addressed the mechanisms and implications of vertical coordination. In the literature, many studies have examined the structure—conduct—performance of the U.S. food industries. However, the transaction costs effects on vertical coordination and the linkages between vertical coordination and competitive performance in the food industries have not been examined.

Based upon a specification of vertical coordination that incorporates product flow linkages plus the use of coordinating methods between vertically interdependent industries, empirical analysis supports the hypothesis that transaction costs are a primary motivation for vertical coordination. The transaction cost factors found to be most influential are those specifically related to uncertainty (PCFS and UNANT), input supplier concentration (FSGC and FUSC), asset specificity (RD and AS), and scale economies (SPCR and FPDI).

Comparison of the results between the vertical coordination measures VIC, which captures only product flow interdependencies and VC2, which captures interdependencies plus coordinating methods, reveals the importance of non-market exchange mechanisms in attenuating transactional inefficiencies. In addition, comparison of the traditional VI measure and VC2 reveals differences in the factors affecting each. In previous studies of vertical industrial organization, the role of non-market exchange mechanisms along with vertical integration was not empirically examined. The results herein demonstrate that a variable incorporating coordinating methods with technical input-output interdependencies bridges the gap in the dichotomy of market versus ownership coordination.

While this is a promising start in specifying a robust measurement of vertical coordination, much work remains. Simply to improve the accuracy of the measure, much greater detail on the types of coordinating methods used and their relative importance is needed. Not only is information on coordinating methods between farms and food manufacturers needed, but also among manufacturers, down-stream distributors, and ultimately, consumers.

The results for the export propensity equations generally did not support the hypothesized effects. Specifically, both measures of the vertical coordination index were negative and statistically significant with export propensity. Also, the research and development, advertising, labor wage rate, and foreign trade barriers variables were negative and significantly related with export activity. The variables representing industry concentration, capital to sales ratio, non-production employment payroll, and product transportability were positively related to export propensity. Only product transportability was statistically significant.

The domestic industry return on capital equations explained the greatest amount of variability. Three of the four vertical coordination variables had positive coefficients but, only the up-stream governance contract variable was statistically significant. Industry concentration, research and development, advertising, non-production employment payroll, and labor wage rates were positive and generally significant. The capital-sales ratio was negative and statistically significant with food industry returns on capital.

The domestic food processing industries with higher levels of vertical coordination generally export less and attained greater returns on capital. Theory suggests that vertical coordination activities come about where transactional inefficiencies exist. Through vertical coordination, industries should be able to eliminate transactional inefficiencies, becoming more competitive. The results, however, demonstrate a negative relationship between vertical coordination and export propensity. Therefore, it can now be hypothesized that vertical coordination results in only an attenuation of, not the elimination of transactional inefficiencies. This implies that the theoretical understanding of vertical coordination and its relationship to performance in both domestic and foreign markets is incomplete.

Theory suggests that industries with low or no transactional inefficiencies do not need to utilize vertical governance structures to be competitive. Industries operating in an environment of many transactional inefficiencies have economic incentives to utilize non-market methods of vertical coordination. However, based on these findings, non-market vertical coordination appears not to reduce market inefficiencies enough to be as "competitive" as industries without market such inefficiencies. To the extent that the existence of high vertical coordination activities in domestic markets is indicative of transactional inefficiencies, and if the same type of inefficiencies are present in international markets, domestic industries that need to utilize non-market methods of vertical coordination may be at a competitive disadvantage in foreign markets relative to those industries with fewer transactional inefficiencies.

Notes

1. An input-output transactions matrix was provided by Alward.
2. Several specifications for the degree of consolidated control associated with various coordinating structures were examined. These included decreasing marginality, constant marginality, and increasing marginality. In an analysis of the three relationships, the decreasing marginality specification proved superior. Refer to Frank 1990, pp. 38-44 and 61-71 for a detailed discussion.
3. Equation specified by Levy.
4. The farm industry GINI coefficient is calculated from Lorenz curves based upon the ratio of cumulative percent of output to cumulative percent of farms in each size classification, using Census of Agriculture data.
5. Advertising expenditures were provided by Rogers and research and development expenditures as reported in Scherer.

References

Adelman, M. A. 1955. Concept and Statistical Measurement of Vertical Integration. in *Business Concentration and Price Theory*. Ed. G.J. Stigler, Princeton. Princeton University Press.

Alward, G. U.S. Forest Service, Fort Collins, Co.

Baldwin, R. E. 1971. Determinants of the Commodity Structure of U.S. Trade. *American Economic Review* 61:126-146.

Belsley, D. A., E. Kuh, and R. E. Welsch. 1980. *Regression Diagnostics*. New York: Wiley.

Bureau of Census. 1982. Census of Agriculture. U.S. Government Printing Office. 1(51).

Blois, K. J. 1972. Vertical Quasi-Integration. *Journal of Industrial Economics* 20:253-272.

Breusch, T. S. and A. R. Pagan. 1979. A Simple Test for Heteroscedasticity and Random Coefficient Variation. *Econometrica* 47:1287-1294.

Buckley, R., S. Hamm, B. Huang, and G. Zepp. 1988. *U.S. Fruit and Vegetable Processing Industries*. USDA, ERS, Commodity Economics Division.

Collins, N. and L. R. Preston. 1968. *Concentration and Price-Cost Margins in Manufacturing Industries*. Berkeley, California.

Crom, R. 1988. *Economics of the U.S. Meat Industry*. USDA, ERS, Number 545.

Flinchbaugh, B. L. 1989. *Who Will Control U.S. Agriculture—Revisited*. Kansas State University.

Frank, S. D. 1990. *The Structure and Performance of the U.S. Food Manufacturing Industries: Measuring and Analyzing Vertical Coordination*. unpublished Ph.D. Dissertation. The Ohio State University.

Handy, C. and J. MacDonald. 1989. Multinational Structures and Strategies of U.S. Food Firms. *American Journal of Agricultural Economics* 71:1246-1254.

Joskow, P. L. 1988. Asset Specificity and the Structure of Vertical Relationships. *Journal of Law, Economics, and Organization* 4:95-117.

Krause, K. 1987. *Corporate Farming, 1969-1982*. USDA, ERS, Ag. Report Number 578.

Laffer, A. 1969. Vertical Integration by Corporations, 1929-1965. *Review of Economics and Statistics* 51:91-93.

Lasley, F. 1983. *The U.S. Poultry Industry*. USDA, ERS, Ag. Economic Report Number 502.

Leontief, W. W. 1951. *The Structure of American Economy, 1919-1939*. New York. Oxford University Press.

Levy, D. 1985. The Transactions Cost Approach to Vertical Integration: An Empirical Examination. *Review of Economics and Statistics* 67:438-445.

Macneil, I. R. 1978. Contracts: Adjustment of Long-Term Economic Relations Under Classical, Neoclassical, and Relational Contract Law. *Northwestern University Law Review* 72:854-905.

Maddigan, R. 1981. The Measurement of Vertical Integration. *Review of Economics and Statistics* 63:328-335.

Marion, B. W. 1976. Vertical Coordination and Exchange Arrangements: Concepts and Hypotheses. In *Coordination and Exchange in Agricultural Subsectors*. N.C. Project 117, Monograph 2.

_____. 1986. *The Organization and Performance of the U.S. Food System*. Lexington, MA:Lexington Books.

Marvel, H. P. 1980. Foreign Trade and Domestic Competition. *Economic Inquiry* 18:103-122.

Mighell, R. and L. Jones. 1963. *Vertical Coordination in Agriculture*. USDA, ERS, Agricultural Economic Report No. 19.

Pagoulatos, E. and R. Sorensen. 1975. Domestic Market Structure and International Trade: An Empirical Analysis. *Quarterly Review of Economics and Business* 16:45-59.

Reimund, D., R. Martin, and C. Moore. 1981. *Structural Change in Agriculture: The Experience for Broilers, Fed Cattle, and Processing Vegetables*. USDA, Economics and Statistics Service. Number 1648.

Rogers, R. Department of Agricultural and Resource Economics, University of Massachusetts, Amherst, MA.

Scherer, F. M. 1982. Using Linked Patent and R & D Data to Measure Inter-Industry Technology Flows. Federal Trade Commission, Bureau of Economics Working Paper No. 57.

Stigler, G. J. 1951. The Division of Labor is Limited by the Extent of the Market. *Journal of Political Economy* 59:185-193.

Tucker, I. and R. Wilder. 1977. Trends in Vertical Integration in the U.S. Manufacturing Sector. *Journal of Industrial Economics* 26:81-94.

U.S. Department of Agriculture, FAS. 1988. *Trade Policies and Market Opportunities for U.S. Farm Exports*.

U.S. Department of Commerce, Bureau of Census. 1982. Census of Manufacturers.

U.S. Department of Commerce, International Trade Administration. Various annual statistics.

U.S. Department of Commerce. 1988. *State Population and Household Estimates, With Age, Sex, and Components of Change 1981-1987.* Series P-25, No. 1024.

Van Arsdall, R. and K. Nelson. 1984. *U.S. Hog Industry.* USDA, ERS, Number 511.

Williamson, O. 1979. Transactio Cost Economics: The Governance of Contractual Relations. *Journal of Law and Economics* 22:233-262.

White, H. 1980. A Heteroscedasticity-Consistent Covariance Matrix Estimator and a Direct Test For Heteroscedasticity. *Econometrica* 48:817-838.

PART THREE

New Empirical
Industrial Organization Studies

4

Imperfect Competition in Multiproduct Food Industries with Application to Pear Processing

Joyce Wann and Richard J. Sexton

Evidence on food industry market structure suggests that many food product markets are not perfectly competitive (Connor *et al.*, 1985). Food processing industries often comprise relatively few processors who purchase a raw farm product from many local producers and transform it into usually multiple product forms and sell to a number of consumers. Such an industry structure may result in imperfect competition on both the buying and selling sides of the market. Attendant impacts include distorting the farm-retail margins for food, thus affecting the welfare of both farmers and consumers.

This study attempts to improve understanding of market behavior in food processing by developing and estimating a generalized model of farm-retail price spread determination that reflects these key structural characteristics of agricultural markets. The model assumes the existence of an identifiable raw product input market and allows for multiple processed product output markets and for imperfect competition in both output and raw product markets.

A key feature of the model is its ability to distinguish input market power from output market power, based on the assumption that there is a perfectly competitive "benchmark" processed product form. The marketing margin for the benchmark product is used to estimate oligopsony power. Oligopoly power in the other processed product markets is estimated by comparing the margin for these products with the margin for the benchmark product.

The California pear processing industry was chosen for application of the

---

Significant portions of this paper have been published in *American Journal of Agricultural Economics*, November, 1992. Permission has been given by the *American Journal of Agricultural Economics* editors.

conceptual model. Estimation results indicate this industry has exercised market power in both its farm input market and the markets for canned pears and fruit cocktail.

Prior Work

Prior to 1980, analyses of market power in food industries were generally based upon the structure-conduct-performance paradigm. These studies usually involved interindustry analyses of profitability or price-cost margins as functions of concentration ratios and other structural measures. This work is aptly summarized in Connor *et al*. Modern variations on this theme specify models to explain price (not profits) as a function of market structure variables and usually focus on the behavior of single industries. Recent studies by Cotterill and by MacDonald illustrate the approach, and a good summary is provided by Weiss.

The present study is in the evolving tradition of what has become known as NEIO, the new empirical industrial organization (Bresnahan). NEIO studies usually focus on a single industry and involve econometric estimation of firm or industry marginal cost and conduct parameters. A cornerstone methodology of the NEIO is to analyze firm or industry conduct through the estimation of conjectural elasticities.[1] In a model with quantity-setting firms, such elasticities are computed as $\theta^i = (\partial Q / \partial q^i)(q^i / Q)$, where Q denotes industry output and q^i is output of the ith firm in the industry. Interpreted literally, θ^i is said to measure the firm's expectation of the percentage industry output change in response to its own output change. Alternatively, θ^i may be interpreted simply as an index of market structure.

The initial applications of the conjectural variations framework were concerned exclusively with oligopoly power and included studies by Gollop and Roberts, Appelbaum (1979, 1982), and Roberts. The fundamental approach to these studies involved estimating θ^i (or an industry-wide counterpart) as part of a system model. The model consisted of consumer demand, firm or industry supply, and factor demand behavior derived from profit maximization conditions.

As noted, the structural characteristics of many raw agricultural product markets also suggest that handlers may exercise oligopsony power. Agricultural products are often costly to transport, dictating that raw product markets are local or regional in scope and, hence, more highly concentrated than comparable markets for the finished products. The conjectural variations framework was first extended to consider the joint exercise of oligopoly and oligopsony power by Schroeter and by Schroeter and Azzam (1990). An important limitation of these analyses is the assumption that conduct as measured by the conjectural elasticity is identical in the raw product (input) and processed product (output) markets.

The technical reasons for Schroeter's and Schroeter and Azzam's (1990) identical conjecture assumption are worth noting. Within the dual cost function framework of Appelbaum, behavior in individual input markets cannot be distinguished unless the cost function is made separable in some of the inputs through invoking a fixed proportions assumption on the production technology. This assumption makes it impossible in the Schroeter and Schroeter and Azzam (1990) formulations to distinguish behavior in the input market from that in the output market because the conjectural elasticities are necessarily the same.

Input- and output-market conjectures do differ, however, if the scope of the markets differ, e.g., regional vs. national (Schroeter and Azzam 1991). The empirical relevance of this observation hinges upon whether the specific markets can be isolated and the appropriate data generated. Azzam and Pagoulatos suggest resolving the problem by adopting a primal production function framework. Here, individual inputs are distinguished in the first-order profit maximization conditions and no fixed proportions assumption is needed. Generalizing this approach to multiple outputs is not straightforward because the production technology is represented by a set rather than a differentiable function. Econometric considerations also usually favor the dual over the primal formulation of production behavior.

The primary goal of the present paper is to extend the methodologies described above to analyze market power simultaneously in multiple processed-product and raw-product markets and to apply the new methodology to the California pear industry. A secondary goal is to test Sexton's hypotheses about a "competitive yardstick" effect cooperatives may exercise in input markets.

The Model

Assume that a hypothetical food manufacturing industry processes a homogeneous farm product R. The farm input may be processed into multiple forms $q = \{q_1,...,q_k\}$, which may be sold in imperfectly competitive markets. Each of the product forms requires the material input in a fixed proportion: $q_j = \gamma_j R_j$, $j = 1,...,k$, where γ_j is the coefficient converting R_j amount of material input into q_j level of the jth output.

The model is developed for the general case of k processed forms, the market for product 1 being designated as the competitive benchmark. A representative firm's profit function is

$$\Pi^i = P_1 \gamma_1 R_1 + \sum_{j=2}^{k} P_j(Q_1,Q_2,...,Q_k) \gamma_j R_j^i$$

$$- C^i(q_1^i,...,q_k^i,w_m,F) \tag{1}$$

$$- w^i(R^i,L^i(R^i))R^i, \qquad R^i = \sum_j^k R_j^i,$$

where P_1 is the parametric output price for product 1; Q_j, $j = 2,...k$, represents the industry output of product form j; $P_j(\cdot)$, $j = 2,...,k$, is the industry inverse demand function for product form j; and R^i is the total amount of farm input used by the ith firm in processing.

Processing inputs are assumed nonsubstitutable for the raw product. Hence, processing costs, represented by $C(q_1^i,...,q_k^i,w_m,F)$ where w_m is a vector of variable input prices and F is a vector of fixed input quantities, are separable from raw product costs. These latter costs are modelled in a spatial market framework (Greenhut, Norman, and Hung; Sexton), in which the raw product input "mill" price w_i paid by the ith firm, depends upon its purchase level R^i and its market radius L^i. The latter, in turn, depends upon R^i and rivals' reactions to the ith firm's behavior.

The decision problem for the ith processor is to determine the optimal uses of the raw farm product, that is to choose R_j^i, $j = 1,...,k$ in order to maximize (1). The first order condition for the benchmark product, $q_1 = \gamma_1 R_1$, can be rearranged to yield the following relative price-spread formulation:

$$\frac{\gamma_1(P_1 - C_1) - w^i}{w^i} = \left(\frac{R^i}{w^i}\right)\left(\frac{dw^i}{dR^i}\right) = \frac{R^i}{w^i}\left(\frac{\partial w^i}{\partial R^i} + \frac{\partial w^i}{\partial L^i}\frac{dL^i}{dR^i}\right) = \eta_{wi,Ri}, \tag{2}$$

where C_j, $j = 1,...,k$, denotes marginal cost of processing the jth product form. Equation (2) states that the relative markup, after adjusting for processing costs for the competitive benchmark product, equals the firm's perceived total flexibility of raw product price with respect to its volume of raw product purchases, $\eta_{wi,Ri}$. From (2), $\eta_{wi,Ri}$ can be expressed as

$$\eta_{wi,Ri} = \epsilon_{wi,Ri} + \epsilon_{wi,Li} \cdot \eta_{Li,Ri} \tag{3}$$

where $\epsilon_{wi,Ri} = (\partial w^i/\partial R^i)(R^i/w^i)$ is growers' price flexibility of supply for the raw product, $\epsilon_{wi,Li} = (\partial w^i/\partial L^i)(L^i/w^i)$ is the elasticity of the firm's mill price w^i to its market radius, and $\eta_{Li,Ri} = (dL^i/dR^i)(R^i/L^i)$ is the firm's conjecture of the elasticity of its market area to its change in output.[2] The value of $\eta_{Li,Ri}$ is the key determinant of competitiveness in the raw product market. For example, if $\eta_{Li,Ri} = 0$, then $\eta_{wi,Ri}$ equals $\epsilon_{wi,Ri}$, the growers' supply flexibility, and processing firms can act as monopsonists within their market areas. $\eta_{wi,Ri} = 0$ corresponds to the competitive case. That is, the firm perceives that its purchase quantity R^i has no effect on the raw product price w^i it pays. Thus, in general, $\eta_{wi,Ri} \in [0, \epsilon_{wi,Ri}]$.[3] $\eta_{Li,Ri}$ and $\eta_{wi,Ri}$ can be estimated empirically to provide a test of the degree of spatial market power.

First-order conditions to (1) for the k - 1 nonbenchmark product forms can be arranged in elasticity form to yield the following relative price-spread formulations:

$$\frac{\gamma_j(P_j - C_j) - w^i}{w^i} = \eta_{w^i,R^i} - \frac{P_j\gamma_j}{w^i}\theta_j^i\left(\eta_{P_j,Q_j} + \sum_{\substack{m=2 \\ m \neq j}}^{k} \frac{S_m^i}{S_j^i}\eta_{P_m,Q_j}\right)$$

$$+ \frac{P_j\gamma_j}{w^i}\left(\sum_{m=2}^{k} \frac{S_m^i}{S_j^i} \sum_{\substack{s=2 \\ s \neq j}}^{k} \eta_{P_m,Q_s}\theta_{s,j}^i\right), \quad j=2,...,k. \tag{4}$$

In (4), $\eta_{Pj,Qj}$ is the inverse market demand elasticity (price flexibility) of product form j, j = 2,...,k; $\theta_j^i = (dQ_j/dq_j^i)(q_j^i/Q_j)$ is the ith firm's quantity conjectural elasticity for product j, j = 2,...,k; and $S_j^i = P_jq_j^i$, j = 1,...,k, so that S_m^i/S_j^i measures for firm i the revenue share of product form m relative to product form j.

If an output market is competitive, a change in the firm's output will not induce any net change in market quantity; hence, $\theta_j^i = 0$, which is the case for the benchmark product. In contrast to perfect competition, θ_j^i is unity under a pure monopoly $(q_j^i = Q_j)$. Therefore, $\theta_j^i \epsilon [0, 1]$, and its value can be estimated empirically for testing market structure. Note in the multiproduct case that the effect of θ_j^i on the price spread depends not only on own-price flexibility $\eta_{Pj,Qj}$ but also on the share-weighted cross-price flexibilities $\eta_{Pm,Qj}$, m $\neq$ j.

The final set of terms in (4) measure cross-price effects, namely the manner in which production changes in product j influence production of other product forms s $\neq$ j. In particular, η_{Pm,Q_s} is a cross-price flexibility measuring the percentage price response of product form m to a one percent change in production of product form s. The $\theta_{s,j}^i = (\partial Q_s/\partial q_j^i)(q_j^i/Q_s)$ are cross-product conjectural elasticities which, interpreted literally, measure the ith firm's expectation of the percentage response of output in product form s due to a one percent change in its output of product form j.

Evaluating behavior for the nonbenchmark products relative to the benchmark merely requires substituting (2) into (4) to obtain:[4]

$$\frac{\gamma_jP_j - \gamma_1P_1}{\gamma_jP_j} = \frac{\gamma_jC_j - \gamma_1C_1}{\gamma_jP_j} - \theta_j^i\left(\eta_{P_j,Q_j} + \sum_{\substack{m=2 \\ m \neq j}}^{k} \frac{S_m^i}{S_j^i}\eta_{P_m,Q_j}\right)$$

$$+ \sum_{m=2}^{k} \frac{S_m^i}{S_j^i} \sum_{\substack{s=2 \\ s \neq j}}^{k} \eta_{P_m,Q_s}\theta_{s,j}^i, \quad j = 2,...,k. \tag{5}$$

Equation (5) indicates that after adjusting for processing cost differentials, markups of nonbenchmark products over the benchmark are due to oligopoly power in the nonbenchmark markets as measured by the last two sets of terms on the right-hand side of (5).[5]

In the presence of imperfect competition, consideration of impacts among multiple processed products is likely to raise the price spread. In particular, most processed products from a given raw product are likely to be substitutes, so $\eta_{Pm,Qj} < 0$ in (5). Moreover, if higher output and, hence, lower prices for a given product j result in greater output of competing product forms, then $\theta^i_{s,j} > 0$, causing the second term in curly brackets in (5) to be negative and also to contribute to a greater margin.

Equations (2) and (5) together provide a complete base for testing the presence of competitive behavior/market power of processing firms in both input and output markets. The processor has market power in the raw product market if the hypothesis H_0: $\eta_{wi,Ri} = 0$ is rejected. The presence of imperfect competition in the nonbenchmark output markets can be examined by testing the hypotheses H_j: $\theta_j = 0$, $j = 2,...,k$.

Aggregation Issues

Some applications of the conjectural elasticity methodology have utilized cross-sectional, firm-level data (Gollop and Roberts, Roberts). Such data are most often unavailable. The application to pear processing in the present study is facilitated by time series cost data calculated each year as a weighted average across firms in the industry. Therefore, our empirical results apply directly to the behavior of an average firm in the industry—the interpretation preferred by Bresnahan (p. 1030).

Because most applications utilize aggregate data, it is useful to consider the problems presented by aggregation in the multiproduct case. For cost functions to be well-defined at the industry level, the firm-level cost functions must be of the Gorman polar form (Appelbaum 1982), with constant and identical marginal costs but fixed costs possibly varying among firms. A further assumption necessary in the multiple-product case is nonjointness in production (Hall, Schroeter and Azzam 1990), which implies that the marginal costs of a given product j are unaffected by production level of other products $i \neq j$.

The remaining aggregation problem in the multiple-product model concerns interpretation of the conjectural elasticities. Schroeter and Azzam (1990) suggest a procedure whereby the aggregate industry conjectural elasticity is derived as a quantity-share weighted average of individual firm conjectures. To illustrate a problem with this approach, let $j = 2$ and $k = 3$, with product 1 again acting as the benchmark. Schroeter and Azzam's aggregation procedure then implies:

$$\theta_2 = \sum_{i=1}^{N} \frac{q_2^i \theta_2^i}{Q_2} = \sum_{i=1}^{N} \frac{q_3^i \theta_2^i}{Q_2}, \tag{6}$$

$$\theta_{3,2} = \sum_{i=1}^{N} \frac{q_2^i \theta_{3,2}^i}{Q_2} = \sum_{i=1}^{N} \frac{q_3^i \theta_{3,2}^i}{Q_2}, \tag{7}$$

Inspection of (6) and (7) reveals that consistent aggregation is assured to hold only if each firm's share is identical across product forms or if all firms entertain the same conjectures. The first condition is unlikely to hold exactly but may be an acceptable approximation in some cases. Appelbaum argued the second condition holds *ex post*, but Bresnahan has criticized this position, claiming that firms normally would not exhibit the same conduct. Although Bresnahan is correct in general on this point, Appelbaum is correct logically for the case where firms produce homogeneous products and, hence, face identical prices and identical marginal costs. That is, when the latter conditions hold, optimizing behavior compels that *ex post* firms' conjectures are identical. Because aggregation of costs to the industry level entails assuming constant and identical marginal costs across firms, the assumption of identical conjectures in equilibrium is achieved at no additional cost in generality.

The California Pear Processing Industry

The California pear processing industry reflects prototypical structural characteristics of modern food processing. A large number of pear growers (over 1,100 in 1987) sell to relatively few pear processors, who transform raw pears into various processed forms including grade pack pears (25-40 percent of total production), fruit cocktail (40-60 percent of total production), mixed fruits, fruit salad, baby food, and juiced pear products. California produced about 60 percent of the U.S. pear supply in the 1980s, followed by Washington (26 percent), and Oregon (13 percent). Previous work on the pear industry (O'Rourke and Masud and the references they quote) has not focused on competitive conditions in the industry.

In California pear processing is handled either by cooperatives or independent firms. Most of the processors are multiple-product canners, often packing both pear halves and fruit cocktail. The number of pear processors in California has decreased from 26 firms in the early 1950s to 11 in the late 1980s. Presently, about two-thirds of California's canning pear tonnage is processed by two cooperatives. Most of the remaining one-third is purchased by three private firms (one of which is dominant among the three). The other six firms constitute a fringe, processing substantially smaller amounts for either baby food, frozen pears, or nectar.

The domestic market has been the primary outlet for U.S. canned pear products. Foreign sales now account for less than one percent of total grade pack pear movement. Annual exports of canned fruit cocktail have accounted for 10-20 percent of total movements since World War II. Limited quantities of fresh pears have been imported into the U.S. annually, mainly to supplement domestic supplies during March-June, the off season for domestic pears.

For purposes of empirical analysis the California pear industry was assumed to produce three product forms: fresh pears, canned pears, and canned fruit cocktail, since other product forms are of minor importance. The market for fresh pears was considered to be the competitive benchmark market. California fresh Bartletts have been marketed by a relatively large number of handlers (over 20 packing houses in 1989), and none has had a dominant position in shipping fresh Bartletts.[6]

Empirical Specification

Our empirical model includes specifications for growers' raw product supply, market demand for processed products, and the relative margins separating raw pear prices from prices for the uses. Annual time series data were collected for the years 1950-1986.

Growers' Supply

Variables used to explain intertemporal variations in California Bartlett pear bearing acreage (BA) were similar to those French and Willet specified.

$$
\begin{aligned}
lnBA_t = \alpha_0 + \epsilon_{R,w} lnw_t + \alpha_1 lnBA_{t-1} + \alpha_2 lnRU_{t-7} \\
+ \alpha_3 ln(RU_{t-7})^2 + \alpha_4 DR_{t-1} + \alpha_5 T,
\end{aligned}
\tag{8}
$$

where w_t is the average mill price received by California Bartlett pear growers, RU is the cash return to pear bearing acreage, DR is a dummy variable indicating years with serious pear decline (a disease), and T is a time trend. Pear trees usually begin bearing six years from planting, so current bearing acreage is assumed to depend on the return of the seventh previous year. The reciprocal of $\epsilon_{R,w}$ is substituted into (3) for the growers' supply price flexibility $\epsilon_{w,R}$. Given (8), total California production of Bartletts was expressed in terms of the identity: $R_t \equiv BA_t \cdot YD_t$, where YD (yield) was treated as exogenous.

Output Demand

As the analysis was conducted at the processor level, the relevant industry demand functions are the wholesale demands for canned pear products at the shipping point. Because export demand and government purchases have been very low relative to domestic consumption, they were treated as exogenous. Double-log inverse demand functions in (9) and (10) express wholesale prices of canned pears (P_P) and fruit cocktail (P_C) as functions of (i) quantities demanded for own product (Q_P or Q_C, where P and C subscripts denote canned pears and fruit cocktail, respectively) (ii) quantities of substitute products (Q_C or Q_P), (iii) U.S. per capita income (Y), (iv) beginning stocks (PINV or CINV), and (v) a time trend (T70) beginning in 1970 to capture the change in canned fruit consumption since the early 1970s (French and King, 1988):[7]

$$
\begin{aligned}
lnP_{Pt} = d_{10} &+ \eta_{PP,QP}lnQ_{Pt} + \eta_{PP,QC}lnQ_{Ct} \\
&+ d_{11}lnPINV_t + d_{12}lnY_t + d_{13}T70_t,
\end{aligned}
\tag{9}
$$

$$
\begin{aligned}
lnP_{Ct} = d_{20} &+ \eta_{PC,QC}lnQ_{Ct} + \eta_{PC,QP}lnQ_{Pt} \\
&+ d_{21}lnCINV_t + d_{22}lnY_t + d_{23}T70_t.
\end{aligned}
\tag{10}
$$

Marketing Margins

In addition to raw pears, labor (L), sugar (S), and canning material (M) were assumed to be the variable inputs in pear processing, and capital stock (K) was assumed to be a fixed factor in the short run. Given that substitution elasticities in pear processing are believed *a priori* to be low, the generalized Leontief multiproduct cost function, known to perform well when substitution elasticities are low (Guilkey, Lovell, and Sickles, 1983), was chosen for the processing cost function:[8]

$$
\begin{aligned}
C_1(q_P, q_C, w_L, w_M, w_S, K) = &\sum_i^2 \sum_j^2 \sum_m^3 \sum_s^3 \beta_{ijms}(q_i q_j w_m w_s)^{1/2} \\
&+ \sum_i^2 \sum_j^2 \sum_m^3 \beta_{ijmK}(q_i q_j w_m K)^{1/2} \\
&+ \sum_i^2 \sum_j^2 \beta_{ijK}(q_i q_j)^{1/2}K + \sum_m^3 \beta_m w_m + \beta_K K \\
&+ TREND \sum_i^2 \sum_m^3 \beta_{iTm}q_i w_m + TREND \sum_i^2 \beta_{iTK}q_i K,
\end{aligned}
\tag{11}
$$

where subscripts i and j denote outputs P (canned pears) and C (fruit cocktail), and subscripts m and s stand for variable processing inputs L,M, and S. A time trend was added to C_1 to serve as an indicator of technical change and to allow for variations in marginal cost over time. Symmetry restrictions $\beta_{ijms} = \beta_{jims}$ for all i,j, and $\beta_{ijms} = \beta_{ijsm}$ for all m,s are imposed *a priori*. Restrictions associated with output nonjointness, $\beta_{PCms} = 0$ for all m,s, $\beta_{PCmK} = 0$ for all m, and β_{PCK}

= 0 are also imposed. Marginal cost functions for grade pack pears and canned fruit cocktail are obtained by differentiating C_1 with respect to q_P and q_C, respectively. Input demand functions for the variable inputs also can be obtained from (11) via Shephard's lemma. For example, the labor (L) input demand function is as follows:

$$
\begin{aligned}
L_t = & \, q_{P_t}\beta_{PPLL} + 2q_{P_t}[\beta_{PPLM}(w_{M_t}/w_{L_t})^{1/2} + \beta_{PPLS}(w_{S_t}/w_{L_t})^{1/2} \\
& + \beta_{PPLK}(K_t/w_{L_t})] \\
& + q_{C_t}\beta_{CCLL} + 2q_{C_t}[\beta_{CCLM}(w_{M_t}/w_{L_t})^{1/2} + \beta_{CCLS}(w_{S_t}/w_{L_t})^{1/2} \\
& + \beta_{CCLK}(K_t/w_{L_t})] \\
& + \beta_L + \beta_{PTL}q_{P_t}TREND_t + \beta_{CTL}q_{C_t}TREND_t.
\end{aligned}
\tag{12}
$$

Costs of fresh pear packing were estimated separately since the packing process for fresh pears differs greatly from that for canning pears. Labor and packing materials are the two principal inputs used in packing fresh pears. The cost function for fresh packing was also defined as a generalized Leontief because packing labor and material are not likely to be good substitutes:

$$
\begin{aligned}
C_2(q_F, wF_L, wF_M) = & \, q_F[\beta_{FLL}wF_L + \beta_{FMM}wF_M + 2\beta_{FLM}(wF_L wF_M)^{1/2}] \\
& + \beta_{FL}wF_L + \beta_{FM}wF_M,
\end{aligned}
\tag{13}
$$

where wF_L and wF_M are the wage rate for packing house workers and the price of pear packing material, respectively. Capital stock is not included as an input because fresh packing requires very little capital equipment. From (13), the marketing margin for the benchmark fresh pears at time t can be stated explicitly in price-dependent form as follows:

$$
\begin{aligned}
PF_t = & \, [\beta_{FLL}wF_{Lt} + \beta_{FMM}wF_{Mt} + 2\beta_{FLM}(wF_{Lt}wF_{Mt})^{1/2}] \\
& + w_t[1 + e_{w,R} + e_{w,L}\eta_{L,Rt}]/\gamma_F,
\end{aligned}
\tag{14}
$$

where PF is the f.o.b. wholesale price of fresh Bartletts.[9] Equation (14) is the empirical analog of (2) for the benchmark product. The price flexibility $\epsilon_{w,R}$ of supply in (14) is restricted to be the inverse of $\epsilon_{R,w}$ in (8), while $\epsilon_{w,L}$ was treated as a parameter to be estimated. Among hypotheses to be tested are that cooperative processors and a cooperative bargaining association have had a procompetitive effect on the raw pear input market. To formulate these tests, the spatial market conjectural elasticity $\eta_{L,R}$ was specified as a linear function of variables to reflect the cooperatives' and bargaining association's involvement:

$$\eta_{L,R} = g_0 + g_1 BGA_t + g_2 MO_t + g_3 CO59_t$$
$$+ g_4 CO72_t + g_5 CO78_t + g_6 CO81_t, \tag{15}$$

where BGA_t is the percentage of pears marketed annually through the bargaining association, MO_t is a dummy variable reflecting presence of a federal marketing order for processed Bartletts since 1967, and the four "CO" variables are indicators to reflect years in which cooperatives have gained significant market share in the pear market.[10]

From (11) the price-dependent margin equation for grade pack pears is:

$$P_{P_t} = \beta_{PPLL} w_{L_t} + \beta_{PPMM} w_{M_t} + \beta_{PPSS} w_{S_t} + \beta_{PPK} K_t$$
$$+ \beta_{PTL} TREND_t w_{L_t} + \beta_{PTM} TREND_t w_{M_t} + \beta_{PTS} TREND_t w_{S_t}$$
$$+ \beta_{PTK} TREND_t K_t + 2\beta_{PPLM} (w_{L_t} w_{M_t})^{1/2} + 2\beta_{PPLS} (w_{L_t} w_{S_t})^{1/2}$$
$$+ 2\beta_{PPLK} (w_{L_t} K_t)^{1/2} + 2\beta_{PPMS} (w_{M_t} w_{S_t})^{1/2} + 2\beta_{PPMK} (w_{M_t} K_t)^{1/2} + 2\beta_{PPSK} (w_{S_t} K_t)^{1/2}$$
$$+ \frac{(\gamma_F/\gamma_P)[PF_t - \beta_{FLL} wF_{L_t} + \beta_{FMM} wF_{M_t} + 2\beta_{FLM} (wF_{L_t} wF_{M_t})^{1/2}]}{1 + \theta_P[\eta_{PP,QP} + (S_C/S_{P_t})\eta_{PC,QP}]}, \tag{16}$$

where S_C/S_P denotes sales revenue share of fruit cocktail relative to canned pears. Apart from including an additional term to represent per-unit cost of other fruits used in fruit cocktail, the fruit cocktail margin equation is similar to (16) and is omitted for brevity.[11] Equation (16) is the empirical analog of (5).

Own- and cross-price flexibilities $\eta_{Pi,Qi}$ and $\eta_{Pj,Qi}$ are obtained from the inverse output demands in (9) and (10), and θ_P represents the pear output market conjectural elasticity. To simplify estimating equations, cross conjectural elasticities $\theta_{s,j}$ for processed pear products were assumed zero.[12]

Data and Estimation

Data for 1950-1986 were obtained from a number of sources and are discussed in detail in Wann. Farm production data on bearing acreage, average yield, and quantity of raw pears utilized for processing, fresh sales, and residual uses were obtained from the California Agricultural Statistics Service (CASS). The mill price w was the average annual price paid to growers (CASS).

Domestic f.o.b. prices of grade pack pears and fruit cocktail were available from Kuznets and from the private label f.o.b. prices published by the American Institute of Food Distribution. Data on total pack, movement, and inventory (PINV, CINV) are from the California League of Food Processors.

The primary data source for the processing cost function was the annual cost study of the pear processing industry conducted by Touche Ross & Co. Total costs per standard case of grade pack pears and fruit cocktail were decomposed by type of processing inputs and services, including production and warehouse labor, raw product, cans, labels, cases, sugar, energy, and water as variable inputs. Because energy and water cost was negligible, variable inputs for pear processing were considered to be labor, canning material, and sugar. Input prices used for w_L were the July, August, and September average wage rate in fruit processing, for w_S the wholesale price of sugar, and for w_M the combined costs of cases, cans, and labels per case.

The quantity of fixed capital stock was based on the perpetual inventory method by using average annual investment expenditures on depreciation, repairs, rent, and factory supplies in the Touche Ross annual studies. Wholesale prices of fresh Bartletts at San Francisco published by the Federal-State Market News Service were used as the f.o.b. price PF of fresh pears. Packing cost data for fresh pears were collected from the California Tree Fruit Agreement.

In summary the empirical model for the California pear industry consists of (a) the bearing acreage function (8) and the acreage-yield identity, (b) the wholesale demand functions (9) and (10), (c) two of the three processing input demand functions—see (12), (d) the fresh pear packing cost function (13), and (e) the margin equations for fresh pears (14), canned pears, and fruit cocktail—see (16). Each equation was assumed associated with an additive error to capture unexplained factors. The stochastic nature of the margin equations was assumed due to errors in optimization. The model is closed by identities defining equilibrium in the farm-processing and processing-wholesale sectors.[13] A convenient summary of the full model is provided by Wann (1990, Table 5.1, pp. 94-96).

To accommodate the large number of parameters and attendant multicollinearity problems, the acreage supply function,(8), and wholesale demand functions (9) and (10) were estimated separately from the rest of the equation system (Gollop and Roberts, Schroeter and Azzam (1990), Azzam and Pagoulatos). Estimates of (8) were obtained using OLS, while (9) and (10) were estimated jointly using maximum likelihood estimation. The margin equations, cost functions, and input demand functions were estimated as a system using full information maximum likelihood. Errors in this equation system were assumed jointly normally distributed with mean zero and nonsingular variance-covariance matrix. A first-order autoregressive parameter ρ was introduced into this system and the demand system to adjust for serial correlation.

Equation (17) provides OLS estimation results for the acreage response function (standard errors are in parentheses):

$$lnBA_t = \begin{array}{l} -0.680 + 0.029lnw_t + 0.966lnBA_{t-1} - 0.015lnRU_{t-7} \\ (0.428) \quad (0.031) \quad\quad (0.071) \quad\quad\quad (0.008) \end{array}$$

$$+ 0.307ln(RU_{t-7})^2 - 0.030lnDR_{t-1} - 0.0004T \tag{17}$$
$$(0.158) \quad\quad\quad (0.015) \quad\quad (0.0005) ,$$

$$R^2 = 0.93.$$

As expected, the short-run elasticity of supply $\epsilon_{R,w} = 0.029$ is small since a high price in the current period is likely to have only a small effect on orchard removal decisions. This inelastic supply implies a price flexibility of supply of $\epsilon_{w,R} = 1/\epsilon_{R,w} = 34.0$. Other estimated coefficients in the acreage response equation had the anticipated effects.

Estimated parameters and asymptotic standard errors of the wholesale demand functions are:

$$lnP_{P_t} = \begin{array}{l} 1.791 - 0.500lnQ_{P_t} - 0.214lnQ_{C_t} - 0.005lnPINV_t \\ (0.264) \quad (0.100) \quad\quad (1.133) \quad\quad (0.010) \end{array}$$

$$+ 0.571lnY_t + 0.002T70 \tag{18}$$
$$(0.109) \quad\quad (0.006) ,$$

$$lnP_{C_t} = \begin{array}{l} 1.234 - 0.342lnQ_{C_t} - 0.127lnQ_{P_t} -0.004lnCINV_t \\ (0.216) \quad (0.115) \quad\quad (0.086) \quad\quad (0.008) \end{array}$$

$$+ 0.430lnY_t + 0.011T70 , \quad \rho = \begin{array}{c} 0.373 \\ (0.132) \end{array} \tag{19}$$
$$(0.090) \quad\quad (0.005)$$

The value of the likelihood ratio statistic to test the overall significance of the system is $\mathcal{L} = 205.7$, which exceeds substantially the $\chi^2_{0.05}$ critical value 18.3 for 10 restrictions. Based on estimates of the own price flexibilities, $\epsilon_{PP,QP} = -0.500$ and $\epsilon_{PC,QC} = -0.342$, canned pears and fruit cocktail have elastic demands. Negative cross price flexibilities in the two inverse demand equations indicate that the two products are substitutes as anticipated. Converting the income flexibilities in (18) and (19) to elasticities yields estimated income elasticities of $\epsilon_{QP,Y} = 1.14$ for canned pears and $\epsilon_{QC,Y} = 1.26$ for fruit cocktail.

Maximum likelihood parameter estimates and standard errors of the conjectural elasticities and cost, input demand, and margin functions are reported in Table 4.1. The likelihood ratio statistic for the full system is $\mathcal{L} = 427.7$,

TABLE 4.1 Nonlinear Full Information Maximum Likelihood Estimation Results

Parameter	Estimated Coefficient	Asymptotic Standard Error
Fresh Packing Cost Function:		
β_{FLL}	-0.26871E-02[a]	0.13920E-02
β_{FMM}	-0.59912E-02[a]	0.28492E-02
β_{FLM}	0.47312E-02[a]	0.20076E-02
β_{FL}	0.42391	1.0223
β_{FM}	-1.4596	2.4357
Canning Cost Function:		
β_{PPLL}	-0.43727E-03	0.73773E-03
β_{PPLM}	0.10535E-04	0.70539E-05
β_{PPLS}	-0.26677E-04	0.19109E-04
β_{PPLK}	0.65326E-05	0.76296E-05
β_{PTL}	0.13922E-04	0.22018E-04
β_{CL}	-0.31547[a]	0.15063
β_{CCLM}	-0.81817E-03	0.20545E-02
β_{CCLS}	0.70683E-02	0.47978E-02
β_{CCLK}	0.68783E-02[a]	0.15837E-02
β_{CTL}	0.68948E-02[a]	0.38831E-02
β_{PPMM}	-0.36753E-04	0.58626E-04
β_{PPMS}	-0.13960E-05	0.87773E-05
β_{PPMK}	0.82949E-06	0.61740E-06
β_{PTM}	0.27384E-05	0.19058E-05
β_{CCMM}	0.15617[a]	0.19676E-01
β_{CCMS}	0.52299E-03	0.22984E-02
β_{CCMK}	-0.41109E-03[a]	0.18868E-01
β_{CTM}	-0.12802E-02[a]	0.63946E-02
β_{PPSS}	-0.29320E-03	0.38807E-03
β_{PPSK}	0.11137E-05	0.79077E-06
β_{PTS}	0.17193E-04	0.12464E-04
β_{CCSK}	-0.88988E-04	0.21534E-03
β_{CTS}	-0.39354E-02	0.38703E-02
β_{PK}	0.12557E-01[a]	0.77649E-03
β_{PTK}	0.77703E-04[a]	0.26089E-04
β_{CK}	0.31166E-01[a]	0.13676E-02
β_{CTK}	0.33852E-02[a]	0.46697E-03

(*continues*)

TABLE 4.1 (*continued*)

Parameter	Estimated Coefficient	Asymptotic Standard Error
Spatial Conjectural Elasticity Function:		
$\epsilon_{w,L}$	-10.415[a]	0.99684
g_0	3.4548[a]	0.55041
g_1	3.8858[a]	0.72909
g_2	0.77917[a]	0.33295
g_3	-1.5107[a]	0.41120
g_4	-0.78089[a]	0.22489
g_5	-0.10515	0.13155
g_6	-0.23415[a]	0.08404
Output Conjectural Elasticities:		
θ_P	0.07603[a]	0.01333
θ_C	0.48221[a]	0.02094
p^b	0.95109[a]	0.01555

[a] Significantly different from zero at 0.05 level
[b] p is the first-order autocorrelation coefficient

substantially above the $\chi^2_{0.05}$ critical value of 50.9 for 36 restrictions. The majority of the 44 estimated parameters in table 1 are also significantly different from zero at the 5 percent level. Many of the insignificant parameters are cross-product coefficients in the canning cost function and might be attributed to limited substitution possibilities between the variable canning inputs. Monotonicity and concavity properties of the cost function are satisfied at sample means and for most observation points in both the fresh packing and processing cost functions.

Our estimates of $\epsilon_{w,L}$, θ_P, and θ_C are plausible and consistent with theory. The estimated elasticity of the mill price to market radius, $\epsilon_{w,L} = -10.42$, is negative as expected and significantly different from zero. The output conjectural elasticities θ_P and θ_C, which measure the degree of competition, both fell significantly into the (0, 1) range based on 95 percent confidence intervals.[14] Thus, the hypotheses that the wholesale markets for grade pack pears and fruit cocktail have been perfectly competitive over the study period ($\theta_P = 0$ and $\theta_C = 0$, respectively) are rejected. Correspondingly, a hypothesis that either product has been characterized by collusive, monopoly behavior ($\theta_P=1$, $\theta_C = 1$) also would be rejected. The estimated conjectural elasticity for grade pack pears, $\theta_P = 0.076$, is considerably smaller than that for fruit cocktail, $\theta_C = 0.482$, implying the market for fruit cocktail has been less competitive than the grade pack pear market. This finding is consistent with the fact that fruit cocktail has been packed exclusively in California, while grade pack pears also have been produced in Oregon and Washington, making the selling side of the grade pack pear market less concentrated relative to that of fruit cocktail.

Given our estimates of price flexibilities of demand and output conjectural elasticities for grade pack pears and fruit cocktail, the markup (after adjusting for processing cost differentials) of the nonbenchmark relative to the benchmark products can be estimated from (5) for each year of the sample. Assuming $\theta_{s,j}=0$, $s \neq j$, the markup for canned pears becomes:

$$\frac{\gamma_P P_P - \gamma_F P_F}{\gamma_P P_P} - \frac{\gamma_P C_P - \gamma_F C_F}{\gamma_P P_P} = -\theta_P \left(\eta_{P_P,Q_P} + \frac{S_C}{S_P} \eta_{P_C Q_P} \right).$$

The average markups over benchmark, as a percentage of the processed product prices, were 7.1 percent and 19.7 percent for canned pears and fruit cocktail, respectively. These relatively modest markups in an industry with a strong cooperative presence are consistent with Wills' result that cooperatives have been less successful in enhancing prices than have comparable proprietary firms.

The hypothesis that cooperatives have had a procompetitive effect on the farm product market can be analyzed by testing each coefficient g_3 through g_6 associated with major historical changes in processing cooperatives' pear market activity. All estimated cooperative coefficients were negative and all were statistically different from zero except g_5, which corresponds to the establishment of Glorietta Foods in 1978. Recall that lower values of the spatial conjectural elasticity $\eta_{L,R}$ are associated with a greater degree of oligopsony power (e.g., $\eta_{L,R} = 0$ corresponds to the monopsony case). Therefore, because indicator variables associated with g_3 through g_6 all correspond to growth in cooperatives' market position, the results imply that cooperative growth in the California pear processing industry did not enhance competition among processors in purchasing raw pears. Indeed, it may have reduced farm input market competition. This outcome is consistent with predictions from Sexton's model when cooperatives have closed membership policies, a usual practice in the California pear processing industry. The industry bargaining cooperative, however, does appear to have enhanced competition since $g_1 = 3.89$ ($t = 5.32$).

Estimates of spatial conjectural elasticities $\eta_{L,R}$ range from 2.7 to 5.2 over the sample period. The mean value is 3.9 (standard error 0.73), causing rejection of the hypothesis that the raw pear market has been characterized by monopsony or Loschian behavior ($\eta_{L,R} = 0$). However, the mean value of $\eta_{w,R}$ in (3) was 1.46 (standard error 0.72), rejecting also the hypothesis that the raw product market was on average competitive ($\eta_{w,R} = 0$) over the sample period.

Conclusions

Many agricultural markets exhibit structural characteristics suggestive of oligopoly power on the selling side and oligopsony power on the buying side. We have developed and estimated a model of farm-retail price spread determination for these types of markets. A key feature of the model is its ability to distinguish power in the raw product input market from power in multiple processed output markets. The key to accomplishing this decomposition of market power is the assumption that there is a competitive "benchmark" processed product form that can be used to estimate oligopsony power based upon the margin between the benchmark product price and the raw product price. Oligopoly power in each of the other processed forms is estimated by analyzing the margin between its price and the benchmark price.

Application of the model to the California pear industry revealed modest price enhancement above the competitive norm in both canned pear and fruit cocktail markets. The hypothesis of competition in the raw pear input market also was rejected. Increases in the share of product handled by marketing cooperatives did not appear to increase competitiveness in the raw product market.

Notes

1. For alternative NEIO methodologies, see Baker and Bresnahan, Atkinson and Kerkvliet, and Holloway and Hertel.

2. Our model does not depend upon a spatial markets formulation. Raw product supply may be specified as a general function $w(R)$, where R is aggregate raw product volume purchased in the market. Then first-order conditions for the benchmark product could be stated simply as

$$\partial \pi / \partial R^i_1 = \gamma_1 P_1 - C_1 - w^i - R^i (\partial w / \partial R)(\partial R / \partial R^i) = 0,$$

where $(\partial R / \partial R^i)(R^i / R)$ would denote the conjectural elasticity in the raw product market.

3. The values of $\eta_{Li,Ri}$ and $\eta_{wi,Ri}$ for "intermediate" modes of competition, including Cournot and Bertrand (Hotelling-Smithies) competition, are developed in Sexton.

4. Because the sums and double sums in (4) and (5) may be confusing, we present (5) for product form 2 for the case of $k=4$ with product 1 as the competitive benchmark:

$$\frac{\gamma_2 P_2 - \gamma_1 P_1}{\gamma_2 P_2} = \frac{\gamma_2 C_2 - \gamma_1 C_1}{\gamma_2 P_2} - \theta^i_2 \left(\eta_{P_2,Q_2} + \frac{S^i_3}{S^i_2} \eta_{P_3,Q_2} + \frac{S^i_4}{S^i_2} \eta_{P_4,Q_2} \right)$$

$$+ \frac{S^i_2}{S^i_2} \left(\eta_{P_2,Q_3} \theta^i_{3,2} + \eta_{P_2,Q_4} \theta^i_{4,2} \right)$$

$$+ \frac{S^i_3}{S^i_2} \left(\eta_{P_3,Q_3} \theta^i_{3,2} + \eta_{P_3,Q_4} \theta^i_{4,2} \right)$$

$$+ \frac{S^i_4}{S^i_2} \left(\eta_{P_4,Q_3} \theta^i_{3,2} + \eta_{P_4,Q_4} \theta^i_{4,2} \right).$$

5. A matrix formulation of the entire system is also possible. Define $M_j = (\gamma_j P_j - \gamma_1 P_1)/\gamma_j P_j - (\gamma_j C_j - \gamma_1 C_j)/\gamma_j P_j$, $j = 2,...,k$ as the percent mark up over benchmark after adjusting for processing cost differentials. Given $S_j^i = P_j q_j^i$, $M_j S_j^i$ denotes percent of sales revenue from oligopoly power for product form j. Denote $MS_1 = (M_2 S_2^i,...,M_k S_k^i)'$ as the column vector of mark up terms. This vector can be solved as:

$$MS^i = \theta^i \eta_{P,Q} S^i,$$

where $\theta^i_{(k-1)x(k-1)} = [\theta^i_{s,j}]$ is a matrix of firm i's own- and cross-conjectural elasticities, $\eta_{P,Q(k-1)x(k-1)} = [\eta_{P_s,Q_j}]$ is the matrix of own- and cross-price flexibilities for the nonbenchmarks, and $S^i = (S_2^i,...,S_k^i)'$ is a column vector of sales revenues from the nonbenchmarks for the ith firm.

6. These structural conditions do not necessarily guarantee perfect competition in fresh pear marketing. Two types of biases are introduced in applications where the product chosen as benchmark is itself subject to oligopoly markups. First, the oligopoly markup for the benchmark will be attributed incorrectly to oligopsony power. Second, markups in the nonbenchmark industries will be evaluated relative to a supracompetitive rather than competitive benchmark price. This effect biases results in the conservative direction of accepting the null hypothesis of perfect competition in the nonbenchmark industries.

7. A variable to measure the volume of fresh pear sales was also included in the demand functions but was omitted in the final specifications because of statistical insignificance in both equations. Its deletion had little impact on the estimated price flexibilities.

8. As noted, the empirical model applies to the weighted average firm in the industry. Superscripts to denote firms are omitted in this section to simplify notation.

9. Based on industry standards, the following equivalence factors were used in (14) and (16) to convert one ton of raw pears into processed product: $\gamma_F = 52.2$ no. 36 cartons of fresh pears, $\gamma_P = 45$ cases of grade pack pears, and $\gamma_C = 120$ cases of fruit cocktail.

10. Specifically, CO59, CO72, and CO78 correspond to the establishment of the processing cooperatives California Canners and Growers, Pacific Coast Producers, and Glorietta Foods, respectively. CO81 denotes an expansion by the leading processing cooperative, Tri Valley Growers. Glorietta foods joined Tri Valley in 1981, and when California Canners and Growers filed for bankruptcy in 1983, most of its operations were assumed by Tri Valley, thereby maintaining the cooperative share of the market (Thor and Moen).

11. Fruit cocktail also contains peaches, grapes, pineapples, and cherries. The USDA specifies necessary ranges for the proportion of each fruit. In practice, manufacturers have adopted a relatively fixed combination with pears (40%) and peaches (35%) as the major ingredients.

12. Schroeter and Azzam (1990) found the cross product conjectures for beef and pork to be very small and statistically insignificant.

13. Because the model applies to the California pear industry, canned pear production in Oregon and Washington is subtracted from the market demand in specifying the processing-wholesale equilibrium. Fruit cocktail is packed exclusively in California.

14. Assuming that the θ_j are distributed normally, their 95 percent confidence intervals are: $\theta_P \in [0.050, 0.102]$, $\theta_C \in [0.441, 0.523]$. Confidence intervals and hypothesis tests for the nonlinear system are conditional on the values of $\epsilon_{R,W}$, $\epsilon_{PP,QP}$, and $\epsilon_{PC,QC}$ estimated from (17), (18), and (19).

References

American Institute of Food Distribution. *The Food Institute Report.* weekly issues.

Appelbaum, E. 1979. Testing Price Taking Behavior. *Journal of Econometrics* 9:283-294.

_____. 1982. The Estimation of the Degree of Oligopoly Power. *Journal of Econometrics* 19:287-299.

Atkinson, S. E. and J. Kerkvliet. 1989. Dual Measures of Monopoly and Monopsony Power: An Application to Regulated Electrical Utilities. *Review of Economics and Statistics* 71:250-257.

Azzam, A. and E. Pagoulatos. 1990. Testing Oligopolistic and Oligopsonistic Behavior: An Application to the U.S. Meat Packing Industry. *Journal of Agricultural Economics* 4:362-70.

Baker, J. and T. Bresnahan. 1985. The Gains from Merger or Collusion in Product-Differentiated Industries. *Journal of Industrial Economics* 33:427-444.

Bresnahan, T. F. 1989. Empirical Studies of Industries with Market Power. R. Schmalensee and R. Willig, eds. *Handbook of Industrial Organization*, Amsterdam: North Holland. 1012-1057.

California Agricultural Statistics Service. *Fruit and Nut Statistics.* Annual issues.

California League of Food Processors. *California Canned Stocks and Movements.* Periodic reports.

California Tree Fruit Agreement. Sacramento California. Annual Reports.

Connor, J. M., R. T. Rogers, B. W. Marion, and W. F. Mueller. 1985. *The Food Manufacturing Industries: Structures, Strategies, Performance and Policies.* Lexington, MA: Lexington Books.

Cotterill, R. W. 1986. Market Power in the Retail Food Industry: Evidence from Vermont. *Review of Economics and Statistics* 68(Aug):379-386.

French, B. C. and G. A. King. 1988. Dynamic Economic Relationships in the California Cling Peach Industry. Giannini Foundation Research Report No. 338. University of California.

Gollop, F. M. and M. J. Roberts. 1979. Firm Interdependence in Oligopolistic Markets. *Journal of Econometrics* 10:313-331.

Greenhut, M. L., G. Norman, and C. Hung. 1987. *The Economics of Imperfect Competition: A Spatial Approach.* Cambridge: Cambridge University Press.

Guilkey, D. K., C. A. K. Lovell, and R. C. Sickles. 1983. A Comparison of the Performance of Three Flexible Functional Forms. *International Economic Review* 24:591-616.

Hall, R. E. 1973. The Specification of Technology with Several Kinds of Output. *Journal of Political Economy* 81:878-892.

Holloway, G. J. and T. W. Hertel. 1991. Comparing Hypotheses about Competition. Working Paper No. 91-15. Dept. of Agric. Econ., University of California, Davis.

Kuznets, G. M. 1981. *Pacific Coast Canned Fruits F.O.B. Price Relationships 1980-81*, Giannini Foundation. University of California. Oct.

MacDonald, J. M. 1987. Competition and Rail Rates for the Shipment of Corn, Soybeans, and Wheat. *Rand Journal of Economics* 18:151-163.

O'Rourke, A. D. and S. M. Masud. 1980. The U.S. Pear Market. *American Journal of Agricultural Economics* 62:228-233.

Roberts, M. J. 1984. Testing Oligopolistic Behavior. *International Journal of Industrial Organization* 2:367-383.

Schroeter, J. R. 1988. Estimating the Degree of Market Power in the Beef Packing Industry. *Review of Economics and Statistics* 70:158-162.

Schroeter, J. R. and A. Azzam. 1990. Measuring Market Power in Multi-Product Oligopolies: The U.S. Meat Industry. *Applied Economics* 22:1365-1376.

_____. 1991. Marketing Margins, Market Power, and Price Uncertainty. *American Journal of Agricultural Economics* 73:990-999.

Sexton, R. J. 1990. Imperfect Competition in Agricultural Markets and the Role of Cooperatives: A Spatial Analysis. *American Journal of Agricultural Economics* 72:709-720.

Thor, E. and D. Moen. 1990. *A History of California Grade Pack Pears*. Tri Valley Growers. San Francisco, CA.

Wann, J. J. 1990. *Imperfect Competition in Multiproduct Industries with Application to California Pear Processing*. Ph.D. dissertation. University of California, Davis.

Weiss, L. W. 1990. *Concentration and Price*. Cambridge, MA: MIT Press.

Wills, R. L. 1985. Evaluating Price Enhancement by Processing Cooperatives. *American Journal of Agricultural Economics* 67:183-192.

5

Oligopsony Potential in Agriculture: Residual Supply Estimation in California's Processing Tomato Market

Catherine A. Durham and Richard J. Sexton

The food processing sector generally has significantly greater economies of size than the farm sector and, hence, the number of processing firms is small relative to the number of farmers or consumers. Studies of the structure of agricultural markets have focused primarily on the market power of processors as sellers (Connor *et al.*, 1985). However, monopsony power may also be a concern in food product industries. Many raw agricultural products are bulky and/or perishable and, therefore, costly to transport. Given that processors are few due to size economies, high raw product transportation costs inhibit competition in the input market by limiting the geographic space within which farmers can reasonably sell. The result is farmers often have few selling alternatives within a reasonable hauling distance.

This paper develops an empirical model to analyze the potential for exercise of monopsony power in food markets. The model is applied to the processing tomato industry in California. Our approach is based on estimation of residual supply functions facing a processor or group of processors. Residual supply is the raw product supply facing a processor or group of processors after rivals' demands for the product have been accounted for.

Prior Research

The 1980s were a period of considerable activity in the analysis of market power. Studies based on the structure—conduct—performance (SCP) paradigm

Significant portions of this paper have been published in *American Journal of Agricultural Economics*, November, 1992. Permission has been given by the *American Journal of Agricultural Economics* editors.

were modified to account for the critiques made by Demsetz and others. Alternative approaches were also developed. These approaches have been termed the new empirical industrial organization (NEIO) and differ from SCP approaches primarily by explicitly modeling firm behavior. Discussions of the SCP paradigm literature are found in Schmalensee (1989) and Scherer and Ross (1990). Geroski and also Bresnahan provide surveys of the NEIO.

One of the most popular models of the NEIO is based on Appelbaum's work (1979, 1982), where firm behavior is modeled using a dual framework. The firm's problem is to choose y_i to maximize

$$\Pi_i = p(Y,Z)y_i - c(y_i,w),\tag{1}$$

where $p(Y,Z)$ is industry demand for the homogeneous product Y, Z represents exogenous demand shifters, y_i is firm i's production, $c(y_i,w)$ is the cost function, and w is a vector of variable input prices. First-order conditions for the maximization of (1) can be rearranged to obtain the behavioral equation $p = -(\partial p/\partial Y)(\partial Y/\partial y_i)y_i + \partial c(y_i,w)/\partial y_i$. Via Shephard's Lemma the factor demands are derived as $x_k = \partial c(y_i,w)/\partial w_k$. These equations are estimated in conjunction with the market demand equation, $p = p(Y,Z)$. The markup term of price over marginal cost, $(\partial p/\partial Y)(\partial Y/\partial y_i)y_i$, contains both the slope of the demand curve, $\partial p/\partial Y$, and the so-called "conjectural variation," $\partial Y/\partial y_i$. Several studies have extended this approach to agricultural input markets to examine monopsony power, including Schroeter (1988), Schroeter and Azzam (1990), Azzam and Pagoulatos (1990), Lopez and Dorsainvil (1990), and Wann (1990).

Another approach of the NEIO has been to examine the residual demand faced by a firm or group of firms. Baker and Bresnahan (1985) used this framework to examine oligopoly in differentiated product markets and Scheffman and Spiller 1987) used it in spatial markets. Residual demand models investigate whether demand facing a firm or group of firms, having incorporated rivals' behavior, is sufficiently inelastic to enable the firm or group of firms to exercise market power. If a firm is in a perfectly competitive market, the residual demand price elasticity is very large and the firm cannot affect market price. The present study extends the work of Scheffman and Spiller and Baker and Bresnahan to the supply side.

In analysis of residual demand, a summary statistic is estimated that incorporates in one parameter the relevant market demand elasticity and conjectural elasticity, which a conjectural variation model is able to separate. Because the residual demand elasticity does not separate its two components, its use in analyzing actual firm or market conduct is limited. The residual demand elasticity does indicate whether a firm can exercise market power; the methodology is also useful in considering mergers of firms within an industry and the impact of other structural changes such as entry or exit.

The residual demand model may also be applied to a group of firms. A relatively inelastic residual demand elasticity for a group of firms indicates the group has the collective ability to exercise market power. In other words, external competition does not mitigate the market power which a merger would provide or the market power which firms within the group could exercise collusively.

Modeling Residual Supply

This analysis focuses on the residual supply of a farm product to a processing firm (often referred to as a processor) or group of firms. To estimate residual supply, it is necessary to develop a model incorporating raw product demand by processors and supply by growers in the geographic area under consideration, as well as processor demand and farm supply in outside locations. The goal is to derive a supply function in which quantity demanded by outside processors at each price has been substituted out of the equation. The next subsections present the components of the residual supply model.

Farm Supply

Assume a multi-crop farmer who maximizes profit given the prices of the crops produced, prices of variable inputs, and level of fixed resources. The jth grower's multi-product profit function can be written as:

$$\pi_j = wR_j + \omega Q_j - C_j(\theta, R_j, Q_j, E_j), \tag{2}$$

where w is the price received for the raw product R_j, ω is a vector of other farm prices, Q_j is a vector of other outputs, and $C_j(\theta, R_j, Q_j, E_j)$ represents variable production costs, where θ is a vector of variable input prices, and E_j is the quantity vector of fixed inputs.[1] The farmer maximizes profit with respect to R_j and Q_j given w, ω, θ, and E_j. The farmer's raw product supply function derived from maximizing (2) with respect to R_j and Q_j therefore incorporates the exogenous factors in (2):

$$R_j = f_j(w, \omega, \theta, E_j). \tag{3}$$

The raw product supply function facing a processor depends on aggregated supplies of individual growers as well as behavior of rival processors. It will be useful to represent in its inverse form the supply facing a given processor i, where price w_i that must be offered depends on the firm's purchase volume, its rivals' purchase volume represented by the vector $R_{\sim i} = \{R_1, ..., R_{i-1}, R_{i+1}, ..., R_N\}$, and on factor shifting grower supply:

$$w_i = w_i(R_i, R_{\sim i}, \omega, \theta, E). \tag{4}$$

The key to estimating residual supply is to use the equilibrium conditions for rival processors to substitute R_{-i} out of (4).

Processor Demand

Let the processors technology be represented as quasi-fixed proportions. Output of the finished product is produced according to the production function

$$Q_i = \min\{\lambda R_i, h(X_i)\}. \tag{5}$$

Equation (5) allows no substitution between raw farm input R_i and the vector of nonfarm inputs $X_i = \{Xv_i, K_i\}$, where the Xv_i are variable processing inputs and K_i denotes fixed inputs. Coefficient λ is the finished-to-raw-product conversion ratio.[2] Given (5) and $w_i > 0$, cost minimization requires that $Q_i = \lambda R_i$.

The profit function for a single product processor can then be written as

$$\Pi_i = p\lambda R_i - w_i(R_i, R_{-i}, \omega, \theta, E)R_i - C_i(v, R_i, K_i) - t(R_i). \tag{6}$$

Equation (6) assumes the processor is a competitor in its output market and, hence, faces a parametric price p, although the model easily can be generalized to allow imperfect competition in this market.[3] Variable processing costs are written as $C_i(v, R_i, K_i)$, where v is the vector of prices of variable processing inputs, Xv_i.

Function $t(R_i)$ in (6) specifies the cost to the firm of transporting raw product to its plant, where $t' > 0$ and $t'' > 0$ because increasing R_i implies hauling the product from ever more distant producing areas. Specification of transport costs in this manner implies the existence of a uniform spatial pricing scheme (Greenhut, Norman, and Hung—GNH) whereby the processor offers all growers the same uniform price and bears shipment costs himself. We adopt such a pricing system for modeling purposes because it is the system used in the California tomato processing industry.[4] The common alternative to uniform pricing is FOB or "mill" pricing, wherein the processor offers a uniform mill price and growers bear transport costs.[5]

The first-order condition of (6) yields the condition that the processed product price, adjusted for the conversion factor, must equal the marginal costs of acquiring, transporting, and processing the raw product:

$$p\lambda = w_i + \sum_{n=1}^{N} \frac{\partial w_i}{\partial R_n} \frac{\partial R_n}{\partial R_i} R_i + \partial C_i / \partial R_i + t'(R_i). \tag{7}$$

The summation term in (7) indicates the firm's raw product purchases may affect w_i both through their direct impact in the supply function and also through feedback effects on the elements of $\mathbf{R}_{-i}$.

Because many raw farm products (including processing tomatoes) are processed into multiple product forms, it is important to extend the model to a multiple processed products setting. In developing this extension we continue to assume that processing is characterized by quasi-fixed proportions. Given $w_i > 0$, cost minimizing behavior then implies that $Q_i^j = \lambda^j R_i^j$ for any processed product form $j = 1,\ldots,k$. Define $R_i = \Sigma_{j=1}^k R_i^j$ as aggregate raw product volume purchased by firm i. (Superscripts denote multiple products and subscripts denote multiple firms.) Raw product price (inverse supply function) w_i facing the ith firm is a function of R_i and $\mathbf{R}_{-i}$, the vector of raw input use of all other firms. Processing costs of firm i are a function of the processed product mix $Q_i^1,\ldots,Q_i^k$. Equation (8) is a generalization of (6) to multiple products:

$$\Pi_i = \sum_{j=1}^k p^j \lambda^j R_i^j - w_i(R_i, \mathbf{R}_{-i}, \omega, \theta, E) \sum_{j=1}^k R_i^j \tag{8}$$
$$- C_i(v_i, R_i^1, R_i^2, \ldots, R_i^k, K_i) - t(R_i).$$

Equation (9) is the multiproduct analog of (7) for each product j of the ith firm:

$$p^j \lambda^j = w_i + \sum_{n=1}^N \frac{\partial w_i}{\partial R_n} \frac{\partial R_n}{\partial R_i} R_i + \partial C_i / \partial R_i^j + t'(R_i), \quad j = 1,\ldots,k. \tag{9}$$

To depict the firm's aggregate raw product demand relationship, sum across the k first order conditions and divide by k to obtain:

$$\sum_{j=1}^k [p^j \lambda^j - MC_i^j]/k - t'(R_i) = w_i + \sum_{n=1}^N \frac{\partial w_i}{\partial R_n} \frac{\partial R_n}{\partial R_i} R_i, \tag{10}$$

where $MC_i^j = \partial C_i / \partial R_i^j$ denotes marginal processing cost of the jth product. The left-hand side of (10) represents net marginal revenue product of the raw product averaged across the k product forms. In equilibrium this value must equal the marginal factor cost of R, represented by the right-hand side of (10). Input demands for the individual processed product forms can be obtained from (9), given the form of the profit function.

Residual Supply for a Single Firm

Residual supply can now be derived, given the basic input demand and supply relationships in (4) and (10). To simplify the derivation, assume initially there are two producing regions, each containing many raw product producers and one processing firm. Generalization to multiple competitors and regional groups of firms is considered subsequently.

Analogous to (4), (11), and (12) are the inverse supplies facing firms 1 and 2.

$$w_1 = f_1(R_1, R_2, \Theta_1) \quad \text{and} \tag{11}$$

$$w_2 = f_2(R_2, R_1, \Theta_2), \tag{12}$$

where $\Theta_1 = (\omega_1, \theta_1, E_1)$ and $\Theta_2 = (\omega_2, \theta_2, E_2)$ represent farm supply shifters. The goal is to express R_2 in (11) in terms of the exogenous factors determining equilibrium of firm 2 by simultaneously solving (12) and firm 2's raw product demand relationship. From (10) the latter can be written as:

$$\sum_{j=1}^{k} [p^j\lambda^j - MC_2^j]/k - t'(R_2) = w_2 + [\partial w_2/\partial R_2 + (\partial w_2/\partial R_1)(\partial R_1/\partial R_2)]R_2. \tag{13}$$

Solution of the system consisting of (12) and (13) obtains a value of R_2 in terms of R_1 and the exogenous variables determining equilibrium of firm 2:

$$R_2 = R_2^*(R_1, P_2, \Delta_2, \Theta_2), \tag{14}$$

where P_2 is the vector of prices of the processed products produced by firm 2. Δ_2 represents other demand-side factors affecting firm 2's equilibrium, including processing input prices v_2, fixed input levels K_2, and factors such as fuel costs which shift transportation costs $t(R_2)$.

Equation (14) can now be used to substitute for R_2 in (11) to obtain:

$$w_1 = f_1[R_1, \Theta_1, R_2^*(R_1, P_2, \Delta_2, \Theta_2)] = f_1^*(R_1, \Theta_1, P_2, \Delta_2, \Theta_2). \tag{15}$$

Equation (15) is the residual supply function facing firm 1. Outside demand has been accounted for through inclusion of factors determining equilibrium of the outside processor. If demand by the outside processor for production in firm 1's region is very elastic, the residual supply facing firm 1 will, in turn, be very elastic.

Because R_1 and w_1 are determined simultaneously, estimation of residual supply must take place in conjunction with (16), the demand relation of firm 1, which is adapted from (10):

$$\sum_{j=1}^{k} [p^j\lambda^j - MC_1^j]/k - t'(R_1) = w_1 + [\partial w_1/\partial R_1 + (\partial w_1/\partial R_2)(\partial R_2/\partial R_1)]R_1. \tag{16}$$

Equation (14) can be used to substitute for R_2 whenever it appears in (16).

Equations (15) and (16) comprise a system of simultaneous equations in which the only endogenous variables are the volume of raw product purchased by firm 1 and the price paid. To examine identification, rewrite (16) in general functional form as

$$w_1 = g_1(R_1, P_1, \Delta_1), \tag{16'}$$

where P_1 is the vector of prices of processed products produced by firm 1. Δ_1 includes processing cost factors impacting firm 1's demand, including v_1, variable processing input prices K_1, levels of fixed processing inputs, and factors which shift the transportation cost function $t(R_1)$.

Comparison of (15) and (16') indicates that identification is achieved if a processed product produced by firm 1 is not also produced by firm 2, or if firm 1 incurs shifts in processing (v_1, K_1) or transportation costs not also experienced by firm 2. The specific conditions enabling identification in our application are discussed in the results section.

The analysis is easily extended to allow multiple competing firms. Let $2, \ldots, N$ denote firms competing with firm 1 for raw product. Equilibria for these firms can be expressed in terms of R_1 and the exogenous variables by solving simultaneously the firms' demand and supply relationships to obtain:

$$R_i = R_i^*(R_1, P_1, \Delta_1, \theta_1), \qquad i = 2, \ldots N. \tag{14'}$$

Generalizing (15) to allow for multiple competitors obtains:

$$w_1 = f_1^{**}[R_1, \theta_1, R_2^*(R_1, P_2, \Delta_2, \theta_2), R_3^*(R_1, P_3, \Delta_3, \theta_3), \ldots, \tag{17}$$
$$R_N^*(R_1, P_N, \Delta_N, \theta_N)].$$

The residual supply flexibility η_1^R facing firm 1 measures the percentage response of w_1 to a one percent increase in R_1, taking all feedback effects into account. η_1^R can be decomposed by totally differentiating (17) and converting the resulting expression to elasticities:

$$\eta_1^R = (\partial w_1/\partial R_1)(R_1/w_1) + \sum_{i=2}^{N} (\partial w_1/\partial R_i)(R_i/w_1)(\partial R_i/\partial R_1)(R_1/R_i) \tag{18}$$

$$= \eta_{11} + \sum_{i=2}^{N} \eta_{1i}\, \epsilon_{i1}.$$

Equation (18) shows the residual supply price flexibility η_1^R consists of the direct supply flexibility $\eta_{11} = (\partial w_1/\partial R_1)(R_1/w_1)$ and also the product of the cross-price flexibilities $\eta_{1i} = (\partial w_1/\partial R_i)(R_i/w_1)$, and rivals' reaction function $\epsilon_{i1} = (\partial R_i/\partial R_1)(R_1/R_i)$. $\epsilon_{i1} < 0$ is expected because normally $\partial R_i/\partial R_1 < 0$. That is, as firm 1 increases demand for the raw product in region 1, w_1 increases and outside firms normally will demand less product in the region. The stronger this effect is, the less flexible or more elastic is the residual supply firm 1 faces. In the limiting case where $\partial R_i/\partial R_1 = 0$ for all $i \neq 1$, the supply flexibility facing firm 1 is simply the flexibility η_{11} of the growers' supply function, and firm 1 can act as a monopsonist in its market.

Residual Supply for a Group of Firms

The flexibility of the residual supply function jointly facing a group of processors within a region provides a test of the collective market power of the group. If the residual supply flexibility facing the group is significantly greater than zero, the firms could benefit from collusive behavior, which becomes an increasing possibility as firm numbers shrink. Residual supply analysis also is relevant in considering the competitive impact of proposed mergers and entry or exit within the group.

Extension of the methodology to consider the residual supply facing a group of firms is straightforward, and it is the basis of the subsequent analysis of the processing tomato market. In adapting the analysis to regional groups of firms, observe that relations like (11) and (12) can describe grower supply to a group of processors as well as to a single processor. To construct residual supply for a given group of processors i, conditions determining equilibrium for rival group(s) j must be established analogous to (14) for the single-firm case. The process of determining equilibrium for a rival group involves, of course, equating demand and supply relations facing the group. If it can be assumed that individual firms within the rival group behave competitively, then the multiple-firm analog of (13) is simply the group's aggregate raw product demand function, obtained by summing demand across group members.

Application to the California Processing Tomato Industry

Processing tomatoes are grown over a broad area in California and into Arizona along the Colorado River. Processing tomatoes are used only for processing and do not have the properties desirable for fresh tomatoes. California's 22 tomato processors together operate about 40 plants. Plant sizes vary widely, and growers often have only a few selling alternatives within a 50 mile radius. Transportation costs represent about 25% of raw product value.

Two cooperative processors handle about 15% of production, and the California Tomato Growers Association (CTGA), a bargaining association, currently claims about 70% grower membership. Cooperative processors do not have open membership or acreage. The growers' association bargains with each processor individually and does not attempt to regulate the volume its members produce.

Tomato paste is the most important processed tomato product produced in California. It is sold in bulk form to food manufacturers or used as an intermediate input in production of catsup and sauces. Various sauces including puree, ground tomatoes, chile, and pizza comprise the next largest category, followed by whole peeled tomatoes.

The majority of processing tomatoes are harvested from July through September, although some areas may harvest as early as the end of May or as late as November. Weather conditions determine the percentage of tomatoes that ripen for harvest simultaneously, so that per unit production costs rise in cooler weather. Rainfall, due either to the damage to tomatoes or because of interference with harvesting, has a similar impact.

Data and Econometric Model

A key to estimating residual supply of processing tomatoes is to distinguish between tomato tonnage processed by local firms and that which is "exported" to more distant firms. Tomatoes are transported in trucks; each shipment is inspected prior to processing to determine tonnage, tomato condition, and the soluble solids in the tomatoes.[6] The inspections invoice indicates county of origin of the shipment and processor destination.

Inspections data are proprietary, but permission was obtained to use the data over the period 1985-89 provided transactions of individual firms were not released. This stipulation required that the data be aggregated into six regional groups of firms/plants prior to release. The geographic composition of the regions is indicated in figure 1. For purposes of the study, shipments were summed into weekly observations because delivery arrangements between firms and growers are arranged on a loads-per-week basis. Between 67 and 87 weekly observations were available for the six different regional groups of processors.

Grower prices were taken from booklets published by the CTGA, which list prices on a firm-by-firm basis, including premiums and discounts for early and late season deliveries and for quality factors such as soluble solids content. A major firm's pricing scheme was chosen to represent prices in each firm group. The pricing schemes between firms within a group were generally similar.

A three-step procedure was utilized to estimate the residual supply functions facing each of the six processor groups. The first step involved a limited dependent variable procedure because not all processor groups contracted locally grown tomatoes in every week of their processing season. Unless corrective steps are taken, this truncation of the dependent variable introduces a bias in OLS estimation results. The method used here was based on Lee's adaptation to accommodate multiple equations of Amemiya's two-step estimation procedure.

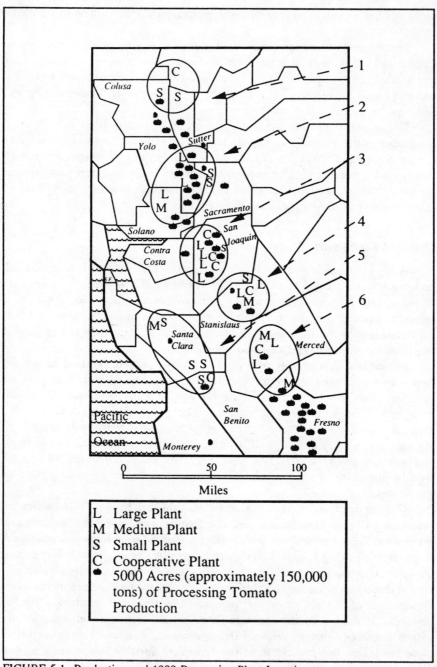

FIGURE 5.1 Production and 1989 Processing Plant Locations

Under this method the probability of an observation (volume of locally produced tomatoes contracted in a week) being zero or positive is calculated as the inverse *Mills' ratio* in a probit regression of the decision whether or not to contract for locally grow tomatoes. The Mills' ratio is then used as an instrument in the processor demand equation to eliminate truncation bias.[7]

In specifying the probit models, a processor's decision whether to contract tomatoes in a given week from the local production area was hypothesized to be influenced by factors affecting timing of production in the local region and alternative producing regions and by factors influencing total production desired by the processor. Temperature (TMP), computed as the average of the preceding four weeks' temperatures in the local and competing production region, was utilized to explain production timing. Processed product prices **P** and production capacity **K** were utilized as factors influencing total demand. The probit model can be expressed as

$$C_{it} = h(TMP_i, TMP_j, P_{lt}, K_{lt}),$$

where $C_{it} = 1$ if local tomatoes are contracted, and $C_{it} = 0$ otherwise.[8]

In the second step, raw product price and volume of raw tomatoes purchased within the processor group's production area were estimated as functions of all exogenous variables, including the Mills' ratios. In the final step, the residual supply and input demand relationships were estimated using as instruments the estimated prices and quantities from the previous stage. Thus, in essence, the estimation procedure was two-stage least squares with the addition of the probit estimation at the outset to account for the limited dependent variable problem.

An autocorrelated error structure is possible when observations are taken over the harvest season. This possibility was not rejected by Durbin-Watson tests. To correct for the problem, the final regressions were estimated using a

FIGURE 5.1 *(continued)*

Group 1 (three plants): 100,000-150,000 tons annually per average plant. Group 2 (six plants, 5 firms): 250,000-300,000 tons annually. Group 3 (eight plants, 6 firms): 215,000-280,000 tons annually. Group 4 (five plants): 125,000-160,000 tons annually. Group 5 (six plants): 100,000-160,000 tons annually. Group 6 (five plants): 170,000-265,000 tons annually.

covariance matrix constructed from autocorrelation parameters estimated separately as

$$\sum_{t=2}^{T} e_t e_{t-1} / \sum_{t=2}^{T} e_t^2$$

for each year. The covariance matrix constructed from these estimates is block diagonal with each block of size t by t, where t is number of weeks in that year's harvest period.

The residual supply and input demand relations were estimated in linear form. The estimating equation for the (inverse) residual supply function can be expressed as:

$$w_{it} = \alpha_i + \beta_i R_{it} + \gamma_i MR_{jt} + \delta_i \Theta_{it} \qquad (19)$$
$$+ \mu_i P_{jt} + \tau_i \Delta_{jt} + e_{it}, \quad i = 1,...,6,$$

where MR_j is the Mills ratio for competing region j, Θ_i are supply shifters for region i [see (11) and (12)], P_j are processed product prices in the competing region, and Δ_j denotes other factors shifting competing region j's demand for raw product in region i—see equation (14). Variables utilized in estimating (19) are discussed below.[9]

Results

Residual supply estimates for the six processor regions are reported in Table 5.1. The effect of tonnage purchased on raw product price was positive in all but one instance and was statistically significant in three groups. In the single case where the estimated tonnage coefficient was negative, the effect was very small and not statistically significant.[10]

The primary hypothesis to be tested is that processing firms in California have the potential to exercise market power in tomato procurement in producing regions near their plant(s). A price flexibility of zero in the residual supply equations indicates a group of processors lacks the collective ability to exercise market power in procuring raw tomatoes. Conversely, a positive price flexibility indicates the group acting collectively could influence the price it pays for tomatoes. A related hypothesis is that groups of firms located at relatively greater distances from rivals should face more flexible (less elastic) residual supplies.

In considering the impact of mergers, the U.S. Department of Justice Merger Guidelines of 1984 delineated an (output) antitrust market as the smallest group of firms which can affect a "small but significant" price increase above the competitive level. Earlier Guidelines (1982) were more specific, indicating that an antitrust market was defined by the collective ability to raise and

TABLE 5.1 Parameter Estimates for Residual Supply Equations

Group	(1) Tons	(2) Avg. Temp.	(3) Lagged Rain	(4) Harvest Wage	(5) Fuel Index	(6) Finished Product Price	(7) Plant Indicator Variable	(8) Soluble Solids	(9) Rival's Mill's Ratios	(10) Stock/ Move	(11) Cans & Labor Index	(12) Trans. Service	(13) Con- stant	(14) R² (adj)
1	.000247 (1.15)	-.39296 (1.58)	1.8477 (1.74)	11.183 (2.94)	-.0440 (0.67)	-.1695 (1.03)	35.66 (2.00)	-.9355 (1.03)		-162.8 (2.26)	2.522 (1.13)	6.260 (2.63)	-560.64 (1.77)	.67 (.61)
2	.000033 (1.47)	-.03897 (1.36)	1.5606 (1.09)	2.9491 (1.38)	-.0734 (1.03)	.28527 (3.67)		-3.5577 (1.59)		11.3 (1.00)	.656 (1.26)	.456 (0.84)	-157.02 (2.56)	.70 (.65)
3	.000065 (2.75)	-.00781 (0.19)	2.8788 (4.55)	6.5908 (7.47)	.0410 (1.02)	.17879 (9.88)	1.321 (0.73)		-.8505 (2.06)	5.3 (1.00)	.220 (0.50)	.410 (1.13)	-121.16 (3.43)	.92 (.90)
4	-.000047 (0.75)	-.00245 (0.61)	-1.0625 (0.77)	7.4733 (8.10)	-.1113 (3.79)	.24158 (6.71)	-20.12 (2.49)	.2767 (0.36)		-59.6 (2.87)	-2.686 (2.60)	3.015 (4.40)	107.17 (1.05)	.89 (.87)
5	.000059 (2.35)	-.19163 (1.43)	-1.0729 (1.43)	5.6452 (5.86)	.1581 (3.18)	.09534 (6.77)	-.981 (0.37)	-.0747 (1.11)			.237 (1.10)	.270 (1.71)	-65.61 (2.39)	.76 (.73)
6	.000139 (2.57)	-.28604 (3.51)	1.7828 (0.750)	1.8862 (0.91)	-.0347 (0.48)	.07120 (1.14)			-.7025 (1.40)	2.8 (0.22)	-.0283 (0.06)	1.340 (2.44)	-8.470 (0.14)	.77 (.73)

t-statistics in parentheses

maintain prices 5 percent above the competitive level. This guideline, in turn, implies that firms within the proposed market faced collectively a residual demand or supply elasticity of 10 or less for the case of linear functions (Scheffman and Spiller, 1987, p. 131). For a price flexibility, the comparable benchmark is a value or 0.10 or greater. Under that benchmark, mergers of firms which faced a joint residual supply price flexibility of 0.10 or more would be subject to scrutiny.

Table 5.2 presents price flexibilities of residual supply and associated 95 percent confidence intervals evaluated at the mean of volume and raw product price for observations with non-zero tonnage. Estimated price flexibilities of groups 1 and 6 are greater than 0.10. Confidence intervals were derived from first-order Taylor's series expansions to estimate variances of the point flexibility estimates (Dorfman, Kling, and Sexton, 1990). Examination of the confidence intervals indicates the supply flexibilities are significantly greater than zero in processor groups 3 and 6. These groups, therefore, would have some influence over their raw product prices if they acted jointly in setting price. However, Table 5.2 also indicates the lower bound of the 95 percent confidence interval is less than 0.10 in each region. Thus, there is no statistical assurance that any of the regions constitutes an antitrust market under the Justice Department's 5 percent rule.

As hypothesized, flexibility of residual supply increases with distance from competitors. The spatially most isolated groups in the study are Group 1 and Group 6 (Figure 5.1), which both have measured flexibilities greater than 0.10. Groups closer to the center of the State have lower flexibilities.

Group 1's flexibility, though it is the second highest, is not significantly different from zero, perhaps in part because it was estimated from the smallest number of observations. Group 2 has the third highest measured flexibility and the second smallest number of observations in the sample. With a confidence interval of 86 percent the lower bound on its flexibility also would remain above zero.

Group 6 includes Fresno County, which is the state's largest producing region and has the longest production period in the state. This group has the highest price flexibility, despite having one of the smaller percentages of local use (Table 5.2, column 6). It may be observed from the map that any user of Fresno County production must pay to ship the tomatoes a greater distance than must Group 6 firms. From spatial theory, this factor gives Fresno-area processors some power over price despite the apparent presence of significant competition in that region from outside processors.[11]

Estimation results for the other variables in the residual supply functions fall into two broad categories: (1) variables affecting grower supply, including weather conditions and farm input prices, and (2) variables influencing rival processors' demand for raw product, including their processed product prices, supply characteristics of the rivals' local production area, and transportation

TABLE 5.2 Residual Supply Price Flexibilities, Price Elasticities, and Confidence Intervals

Group	(1) Price Flexibilities η	(2) Elasticities $\epsilon = 1/\eta$	(3) Variance of η	(4) 95% Confidence Intervals Lower	(5) Upper	(6) 1987 % of Local Use
1	.1105	9.05	.0092	-.0771	.2982	35
2	.0417	23.98	.0008	-.0138	.0972	43
3	.0379	26.42	.0002	.0109	.0648	35
4	-.0069	-144.69	.0001	-.0250	.0111	22
5	.0180	55.49	.0001	.0030	.0331	55
6	.1163	8.6	.0021	.0277	.2048	29

costs. Estimated coefficients and t-statistics of variables affecting grower supply are in columns (2)—(5) of Table 5.1, and coefficients generally have the expected sign when significant. Average temperature (column 2) accounts for progression of the harvest season. Growers plant more tomatoes for harvest as temperatures rise. Thus, the supply function should shift outward: for a given quantity, a lower price is necessary. Though not always significant, the temperature variable does have the expected negative sign. Rainfall (column 3) should have the opposite affect, causing an inward shift in supply. This intuition is contradicted in regions 4 and 5, but in neither instance is the effect significant.

The variables in columns (4) and (5) in Table 5.1 represent grower production costs, and, thus, both are expected to be positively correlated with w. With only one exception, the lagged harvest wage coefficient has the expected positive sign and the nonpositive coefficient is insignificant. An index of the price of diesel fuel, was intended to measure growers' energy costs but is often of the wrong sign. This result appears to have its root in the inclusion of the transportation services index (Column 12) which enters residual supply as one of the parameters affecting outside demand. The latter variable was included to account for rival processing firms' transportation costs and has an unexpected positive coefficient. It may be that the transportation services index reflects farm-level energy costs more accurately than does the diesel fuel index.[12]

Columns (6) through (12) represent variables entering residual supply through outside processors' raw product demand relations. The first is an index of processed product prices for neighboring groups, with weights based on the importance of the various products in each producer region. As prices of outside firms' processed products rise, their raw product demand should increase. The finished product price index has the expected positive sign for all groups except Group 1, and in this case the effect was insignificant.

Two plant openings and one closure took place during the period of this analysis. A plant indicator variable was constructed to take a value of 1.0 when the relevant plant was open. These variables represent an increase in local capacity and are expected to increase both local and external demand for tomatoes--an effect is observed in regions 1 and 3. Residual supply in region 1 was influenced by a plant opening in region 2, while the indicator variable for Group 3 accounts for both the closure of a plant to the south (Group 4) and the reopening of one to its north (Group 2). The plant indicator variable in Group 4's and Group 5's residual supply pertains to the opening of a plant in Group 6. The reasons for a negative sign on this variable are not clear, but it is worth noting that the latter plant is owned by a multiplant processor who also operates a plant in Group 5.

A higher soluble solids level in a competing region should decrease outside demand for a local region's production, as it increases preference for the outside region's own production. These coefficients (column 8) are negative as expected but not significant. The Mill's ratios (column 9), calculated from the probit regressions, enter residual supply when a region provides the alternative location of production for an outside processor group. These coefficients are negative as expected. The Mill's ratio enters Group 3's (San Joaquin and Contra Costa Counties) residual supply from its southern neighbor (Group 4), which receives production from San Joaquin county when local tonnage is not being used. Most other groups use region 6 (Fresno and Merced Counties) as sources of outside supply. The Mills ratio for the largest of these users, Group 3, was used in estimating Group 6's residual supply.

Column 10 represents the ratio of January stocks of canned tomatoes to the preceding year's movement of canned tomatoes between January and July. January is when processing tonnage intentions are usually announced. The expected sign of the stock/movement variable is negative, because if stocks are high relative to expected use, firms are expected to decrease upcoming production levels. In both cases, when the stock effect is significant it has the expected sign.

Columns (11) and (12) represent operating costs of the processing firms and, thus, are expected to have a negative effect on raw input demand. Column 11 is an index combining can prices and processing labor wages. It is negative in the single instance in which it is significant. An index of transportation services (column 12) does not have the expected sign. These last two variables change only on an annual basis and this may contribute to their insignificance. As noted above, the transportation services index may be picking up the influence of fuel costs on grower supply.

For reasons of space, the input demand equations are not discussed in this paper (estimation results are discussed in Durham). Two demand-side variables were found to consistently and significantly affect input demand and therefore to identify residual supply. They were (i) the indicator variables for measuring regional changes in plant capacity due to plant openings or closings, and (ii) the Mills' Ratio. In each input demand equation, one or both of these variables were significant and enabled identification of the residual supply equation.

Conclusions

Objectives of this research were to extend the residual demand model of output market power to analyze of markets for agricultural inputs, and to examine the potential for imperfect competition in the processing tomato market. Many agricultural products exhibit processing concentration and transportation cost characteristics similar to those making the processing tomato market a candidate for input market power. The model constructed in this research may provide a useful alternative or supplement to the empirical models based on Appelbaum's work.

Estimation results showed that market power potential in the California processing tomato market is limited. This result contrasts somewhat with the Just and Chern (J&C) conclusions. However, two regions did exhibit less elastic residual supply price elasticities than those implied by the Department of Justice's 5 percent rule for identifying antitrust markets, indicating these markets would bear examination if plants within them considered merging.

In general, though, results imply that rivalry between neighboring markets is adequate to make them quite competitive.[13] Although processor numbers have remained in the twenties since 1975 (the last year of J&C's analysis), the industry no longer has a single dominant processor which J&C argued was acting as an oligopsony price leader. Moreover, the grower bargaining association has bargained continuously for price since 1973 and has maintained about 60% membership on average over this period. It, too, may be responsible for mitigating some of the oligopsony power J&C found prior to 1975. Finally, tomatoes are now hauled over much longer distances than during the period of the J&C analysis. Reasons for the increased haul relate to shifts in production areas, processors' desire to expand the processing season, and the greater durability and hence haulability of post-harvester tomato varieties. Thus, interregional competition in today's processing tomato industry is more extensive than it was 15 years ago.

Notes

1. Our analysis requires that individual output supply functions slope upward. Modest restrictions on the technology insure this result. Define T_j as the production possibilities set: $T_j = \{(Z_j, R_j, Q_j \mid E_j)$ such that R_j and Q_j can be produced from Z_j, given $E_j\}$, where Z_j is the vector of variable inputs. T_j is assumed to be nonempty, closed, and bounded from above. It is further assumed if some element of (R_j, Q_j) is strictly positive, then some element of Z_j is also strictly positive. These conditions on T_j insure that the profit function $\pi_j(w, \omega, \theta; E_j)$ exists and is convex and continuous in all prices. Finally, it is assumed π_j is differentiable, and, therefore, the raw product supply function derived using Hotelling's lemma is upward sloping as a consequence of convexity of π_j.

2. The advantage of the quasi-fixed proportions assumption is it enables raw product costs to be separable from other processing costs, in turn simplifying derivations. The assumption appears reasonable for most agricultural industries.

3. If firms are imperfect competitors in output markets, output price p is no longer parametric but, rather, must be represented as a function of the output level produced, e.g., $p = p(Q)$ where $Q = \sum_{n=1}^{N} Q_n$ and N denotes the number of processors. In this case the optimization process must take into account how expansion of raw product purchases and, hence, output affects not only raw product price but also processed product prices.

4. Uniform pricing is an extreme form of spatial price discrimination that benefits growers located at a greater distance from the processing plant at the expense of proximate growers. Existence of discriminatory (i.e., non-FOB) pricing is preliminary evidence of market power because price discrimination would be eliminated under perfect competition (GNH p. 136). However, tomato industry experts offer plausible alternative explanations for uniform pricing. Transactions costs are reduced by having truckers contract with a few processors instead of many growers. Also transactions costs from processor moral hazard (e.g., letting tomatoes rot at the factory gate) are eliminated by early transferral of title. Because uniform pricing maximizes price at the market boundaries, GNH (p.123) argue that it may actually emerge as a consequence of "extreme price competition."

5. The methodology is not sensitive to particular forms of spatial pricing. For example, transportation costs and distance would be incorporated into growers' supply relations under FOB pricing instead of processor demand relations as under uniform pricing.

6. Soluble solids are essentially the non-water percentage of the raw tomato. The solids level is important for condensed tomato products such as paste, puree, and sauce, because less energy is needed to reduce a high-solids tomato to desired density, thereby reducing processing costs and increasing value to processors.

7. The inverse Mills' ratio is the ratio of the probability density function to the cumulative density function of the standard normal as evaluated from the probit regression parameters.

8. Because no direct interest centers on the probit estimation results, they are not reported or discussed here. Estimation results may be obtained from the authors.

9. A third equation was estimated to account for the payment of continuous soluble solids premiums by processors in three of the regions depicted in Figure 5.1 (see note 6). In these cases the soluble solids level of tomatoes is also an endogenous variable because growers may influence the level by their cultural practices.

10. This case occurred in group 4, where a major processor declined to participate in the study. Omission of its data clouds interpretation of group 4 results, but note that the firm's omission effectively treats it as an outsider, making a flat residual supply for remaining firms a quite plausible outcome.

11. Zero-value observations for the quantity variable made use of logarithmic models undesirable in general, but to investigate the robustness of results for the linear models, double log models were estimated for groups 1 and 6, where zero-value observations were unimportant. In both cases adjusted R^2 was higher for the linear formulations. The estimates of the residual supply flexibilities were lower in both cases for the double log model: .05 vs. .11 for group 1 and .075 vs. 0.116 for group 6, but with either functional form the Department of Justice's 5% profitable price increase criterion is met.

12. A variable measuring value of competing crops was not significant and was eliminated from the residual supply function. The same decision was also made by Just and Chern.

13. Overlap of market areas is the rule in the processing tomato industry and occurs for two primary reasons: to obtain production before or after one's own region is in peak production, and to spread production to hedge against crop failure in the local production

region. The degree of market overlap may be a critical determinant of competitiveness of spatial markets.

References

Amemiya, T. 1974. Multivariate Regression and Simultaneous Equation Models When the Dependent Variables are Truncated Normal. *Econometrica* 42:999-1012.

Appelbaum, E. 1979. Testing Price Taking Behavior. *Journal of Econometrics* 9:283-294.

_____. 1982. The Estimation of the Degree of Oligopoly Power. *Journal of Econometrics* 19:287-299.

Azzam, A. and E. Pagoulatos. 1990. Oligopolistic and Oligopsonistic Behavior: An Application to the U.S. Meat Packing Industry. *Journal of Agricultural Economics* 41:362-370.

Baker, J. and T. Bresnahan. 1985. The Gains from Merger or Collusion in Product-Differentiated Industries. *The Journal of Industrial Economics* 33:427-44.

_____. 1987. Estimating the Elasticity of Demand Facing a Single Firm. Working Paper, Stanford University.

Bresnahan, T. 1989. Empirical Studies of Industries with Market Power. In *Handbook of Industrial Organization Volume II*, ed. R. Schmalensee and R.D. Willig Amsterdam: Elsevier Science Publishers.

Connor, J. M., R. Rogers, B. Marion, and W. Mueller. 1985. *The Food Manufacturing Industries*, Lexington, MA: D.C. Heath & Co.

Demsetz, H. 1972. Industry Structure, Market Rivalry and Public Policy. *Journal of Law and Economics* 16:1-9.

Dorfman, J. H., C. L. Kling, and R. J. Sexton. 1990. Confidence Intervals for Elasticities and Flexibilities: Re-evaluating the Ratios of Normals Case. *American Journal of Agricultural Economics* 72:1006-1017.

Durham, C. A. 1991. Analysis of Competition in the Processing Tomato Industry. Ph.D. diss., University of California, Davis.

Geroski, P. A. 1988. In Pursuit of Monopoly Power: Recent Quantitative Work in Industrial Economics. *Journal of Applied Econometrics* 3:107-23.

Greenhut, M. L., G. Norman, and C. Hung. 1987. *The Economics of Imperfect Competition: A Spatial Approach*. Cambridge: Cambridge University Press.

Just, R. E., and W. S. Chern. 1980. Tomatoes, Technology, and Oligopsony. *Bell Journal of Economics* 11:584-602.

Lee, L. 1978. Simultaneous Equations Models with Discrete and Censored Dependent Variables. In *Structural Analysis of Discrete Data with Econometric Applications*, ed. P. Manski and D. McFadden, Cambridge, Mass.:MIT Press.

Lopez, R. A. and D. Dorsainvil. 1990. An Analysis of Pricing in the Haitian Coffee Market. *Journal of Developing Areas* 25:93-106.

Scheffman, D. T. and P. T. Spiller. 1987. Geographic Market Definition Under the U.S. Department of Justice Merger Guidelines. *Journal of Law and Economics* 30:123-147.

Scherer, F. and D. Ross. 1990. *Industrial Market Structure and Economic Performance.* Boston: Houghton Mifflin Co.

Schmalensee, R. 1989. Inter-Industry Studies of Structure and Performance. In *Handbook of Industrial Organization Volume II*, ed. R. Schmalensee and R.D. Willig. Amsterdam: Elsevier Science Publishers-B.V.

Schroeter, J. 1988. Estimating the Degree of Market Power in the Beef Packing Industry. *The Review of Economics and Statistics* 70:158-162.

Schroeter, J. R. and A. Azzam. 1990. Measuring Market Power in Multi-Product Oligopolies: The U.S. Meat Industry. *Applied Economics* 22:1365-1376.

Wann, J. J. 1990. Imperfect Competition in Multiproduct Industries With Application to California Pear Processing. Ph.D. diss., University of California, Davis.

6

Dynamic Models of Oligopoly in Agricultural Export Markets

Larry S. Karp and Jeffrey M. Perloff [1]

A dynamic model with adjustment costs is used to identify, estimate, and test the market structure and the types of strategies for two agricultural commodities. The static oligopoly models used to estimate market structure used in most previous papers (e.g., Iwata, 1974; Gollop and Roberts, 1979; Sumner, 1981; Appelbaum, 1982) are inappropriate where there are substantial adjustment costs in training or in capital accumulation, or where there is learning over time. By incorporating costs of adjustment in a dynamic model, we can estimate and test the competitiveness of a market under different assumptions about the rationality of firms.

The game-theoretic literature abounds with dynamic models of oligopoly that are too general to be usable in estimation. Our model is restricted to a specific family of equilibria to facilitate estimation. This family includes the collusive, price-taking, and Nash-Cournot models among others. The justifications for collusive and price-taking behavior are obvious.

There are two reasons for including the Nash-Cournot model. First, including this model allows us to compare our results to earlier empirical studies. Second, the dynamic version of the Nash-Cournot model may be reasonably motivated. Suppose that there are discrete time periods (such as growing seasons) during which firms cannot vary their output levels. Thus, the firm is correct when it makes the Nash-Cournot assumption that its competitors cannot respond to changes in its output levels within a time period. Nonetheless, firms can respond over time.

We consider two types of equilibria. In the open-loop equilibrium, firms choose their trajectory of output levels at the initial time. They do not expect to revise their strategies after an unexpected shock (such as bad weather) affects the output levels of various firms. This failure to anticipate revision is irrational. In the subgame perfect feedback equilibrium, firms choose rules which determine their output as a function of the state.

We start by defining terms, examine the restrictions to our model that lead to the standard static model, and then describe our dynamic model. Next, we discuss the method of nesting the competitive, collusive, and Nash—Cournot models. A qualitative analysis of the model is provided in the third section, and the the details necessary for econometric implementation are given in the fourth section. In the following section, an econometric application to the international rice market is presented. Following the summary and conclusions, there is a technical appendix.

The Basic Model

We start by defining some terms and describing the linear demand curve and quadratic cost functions that are used in our model. Next we examine the restrictions that convert our dynamic model into a static model and discuss the interpretation of market power in the static model. We then show how an analogous interpretation of market power can be made in the dynamic model.

Basic Assumptions and Definitions

There are $n + 1$ ($n \geq 1$) firms in an industry.[2] Each Firm i ($i = 1, ..., n + 1$) exports at the price taker level, at the collusive level, or at an intermediate, oligopolistic level that lies between the two extremes.

For tractability, we assume that there is a linear demand curve and quadratic costs. Specifically, in period t, the firms face the inverse residual linear demand curve

$$p_t = a(t) - \sum_{i=1}^{n+1} q_{it} = a(t) - bQ_t, \qquad (1)$$

where p_t is the price in period t, q_{it} is the output of Firm i in period t, Q_t is the total exports of all firms, $a(t)$ includes the effects of various exogenous variables including exports of other countries, and b is a positive coefficient.

Each Firm i has a constant marginal cost, θ_i, with respect to contemporaneous exports[3], q_{it}, and a quadratic cost of adjustment, $(\gamma_i + .5\delta_i/u_{it})u_{it}\epsilon$, where $u_{i,t}\epsilon \equiv q_{it} - q_{i,t-\epsilon}$ is the change in a firm's export level from period t-ϵ to period t, where ϵ is a length of time and u_{it} is a rate. In contrast, in a static model, δ_i equals zero.

Static Model

In most empirical static models of oligopoly, aggregate or firm level data are used to estimate a parameter v that reflects the markup of price over marginal cost. In this approach, given demand equation (1) for a homogeneous product, Firm i's effective marginal revenue curve (the marginal revenue given the degree of market power actually exercised) is $MR_i(v_i) = p + (1 + v_i)p'q_i = p - (1 - v_i)bq_i$, where we suppress the subscript t (because there is only one period in a static model).

Suppose there is a common marginal cost for all firms, $MC = MC_i = \theta$, which could be a function of various exogenous variables such as weather, and that $v = v_i$ is the same for all firms. To determine v, the demand curve, equation (1), and equilibrium equations for each firm, $MR_i(v) = MC \equiv \theta$,

$$p = \theta_i + (1 + v)bq_i, \tag{2}$$

are estimated.[4] Dividing the coefficient on the q_i term in equation (2) by the estimate of b from the demand curve, equation (1), and subtracting one gives an estimate of v. The markup between price and marginal cost, $p-MC=(1+v)bq_i$, depends on v. For example, if $v = -1$, marginal revenue equals price ($MR = p$) and there is no markup; whereas, if $v = 1$, marginal revenue is less than price [$MR = p + p'(2q_i) = p + p'Q$] and the monopoly markup is observed. Intermediate solutions, such as the Nash-Cournot where $v = 0$, are also possible.

Some economists interpret v as a firm's constant conjectural variation about its rival: $v \equiv dq_j/dq_i$. We prefer the neutral interpretation that the gap between marginal cost and price—a measure of market power—is determined by v rather than a conjectural variation. We now turn to our dynamic models.

The Linear-Quadratic Dynamic Model

To make the estimation problem tractable, our model is restricted to a specific family of equilibria, which is indexed by a parameter v as in the static model. We estimate the family of equilibria under the assumption that firms use subgame perfect, Markov (feedback) strategies. That is, firms choose strategies (rules) that determine their exports as a function of the state variables.

For comparison, we also estimate the same family of equilibria under the assumption that firms use open-loop strategies. That is, firms choose, at the initial time, a path that they intend to follow thereafter. In the open-loop model, firms do not expect to revise their decisions after an unexpected shock (such as bad weather) affects production. This failure to anticipate revision is irrational. In contrast, the Markov equilibrium is subgame perfect.

In order to estimate the Markov model, we use a variation of the well-known solution to the linear-quadratic game (Starr and Ho, 1969).[5] A general open-loop model can be estimated, but using a linear-quadratic specification enables us to compare easily the open-loop and Markov equilibria.[6]

If we were certain that the firms behave noncooperatively, theoretical considerations would favor the assumption of Markov rather than open-loop strategies. Given sufficient information to test overidentifying restrictions, it would be possible to discriminate empirically between the two types of behavior (Karp and Perloff, 1989a). Because we do not have that information, we estimate both models in order to determine the sensitivity of the measure of market structure to the maintained hypotheses regarding behavior. Our two applications suggest that the open-loop and Markov perfect equilibria are similar. Such a comparison has implications for future empirical work because it provides a guide for determining whether the advantages of using more general functional forms offset the disadvantage of assuming that agents use open-loop strategies.

Our feedback and open-loop models explicitly include four common models: collusion, price taking, open-loop Nash-Cournot, and Markov Nash-Cournot. Other export paths that lie between those of collusion and price-taking could be produced by a number of other dynamic oligopolistic games. For example, it may be the case that the firms imperfectly collude (e. g., the folk theorem). Another possibility is that export levels are chosen by government exporting agencies subject to political pressures (e. g., one group wants to maximize export revenues whereas another wants to maximize labor demand), which causes a deviation from Nash-Cournot or collusive equilibria. Rather than try to model explicitly each of these games, we use an index v, which allows for intermediate paths and steady-state exports. This index is the dynamic analog of the price-marginal cost wedge used in static models of oligopoly.

Estimation relies on the discrete time model in which the length of a period is ϵ. The continuous model, obtained as $\epsilon \to 0$, is used to obtain most of our analytic results (see Karp and Perloff, 1988).

In each period of the dynamic model, Firm i's revenues, R_i, are $p_t q_{it} \epsilon$. Given a discount factor of β, the objective of Firm i is to maximize its discounted stream of profits,

$$\sum_{t=1}^{\infty} \beta^{(t-1)} \left[(p_t - \theta_i) q_{it} - \left[\gamma_i + \frac{\delta_i}{2} u_{it} \right] u_{it} \right] \epsilon . \tag{3}$$

In matrix form, the ith firm's objective (3) is

$$\sum_{t=1}^{\infty} e^{-r(t-1)\epsilon} \left[a e_i'(q_{t-\epsilon} + u_t \epsilon) - \frac{1}{2}(q_{t-\epsilon} + u_{t-\epsilon} \epsilon)' K_i(q_{t-\epsilon} + u_t \epsilon) - \frac{1}{2} u_t' S_i u_t \right] \epsilon, \tag{4}$$

where $u_t = (u_{1t}, u_{2t})$, the discount factor (β) is written in terms of the discount rate (r), e_i is the ith unit vector, e is a column vector of 1's, $S_i = e_i e_i' \delta$, and K_i

$= b(ee'_i + e_i e') + \theta e_i e'_i$. That is, K_i is a matrix with zeros everywhere except that all elements of the ith row and column are b other than the (i, i) element, which is $2b + \theta$. As $\epsilon \to 0$, this expression approaches

$$\int_0^\infty e^{-rt} \left[ae'_i q_t - \frac{1}{2} q'_t K_i q_t - \frac{1}{2} u'_t S_i u_t \right] dt. \tag{5}$$

Two Families of Equilibria. Both the open-loop and feedback families of equilibria are indexed by a parameter, v, that measures the degree of market power exercised. This parameter can be defined as $v = \partial\mu_{j,t}/\partial\mu_{i,t}$ for $i \neq j$ and is directly related to the gap between price and marginal cost.

The use of an indexing parameter is justified on pragmatic, empirical grounds. The leading cases where $v = -1/n$, 0, or 1 result in the price-taking, Nash-Cournot, and collusive (if all firms are identical) equilibria, respectively. The estimated v is a measure of the closeness of the observed market to a particular ideal market. If $v = -1/n$, each firm acts as if it believes its rivals will exactly offset its own deviation from equilibrium. Because the good is homogeneous, the firm acts as a price taker. If $v = 1$ and firms are identical, each firm acts as if its rivals will punish it for deviating from the equilibrium by making equal changes in their own output. This assumption is equivalent to a market-sharing agreement and leads to the collusive outcome.

With Markov strategies, Firm i chooses changes in exports, u_{it}, as a function of its current information: its own and its rival's lagged exports. Let $J_i(q_{t-1}; v)$ be the value of country i's program, given the state vector $q_{t-1} \equiv (q_{1,t-1}, q_{2,t-1})$ and an index of market power, v. Firm i's dynamic programming equation is

$$J_i(q_{t-1}; v) = \max_{u_t} (p_t - \theta_i) q_{it} - \left[\gamma_i + \frac{\delta}{2} u_{it} \right] u_{it} + \beta J_i(q_t; v). \tag{6}$$

The first-order condition for this problem is

$$p_t = \theta + (1 + v) b q_{it} + \gamma_i + \delta u_i - \beta \left[\frac{\partial J_i(q_t; v)}{\partial q_i} + v \frac{\partial J_i(q_t; v)}{\partial q_j} \right], \tag{7}$$

where $p - (1 + v) b p_{it}$ is marginal revenue and the term in brackets is the discounted shadow value of an extra unit of current exports. The terms are grouped in equation (7) to emphasize its similarity to equation (2), the equilibrium condition from the static model. The gap between marginal cost and price is the same function of v as in the static model.

For simplicity, we assume in the rest of this section $\theta_i = \theta$, $\delta_i = \delta$, and

$\gamma_i = 0$. The last equality implies that adjustment costs are minimized when there is no adjustment. As a result, the steady-state levels of output in the open-loop, collusive, noncooperative Nash-Cournot, and price-taking equilibria are equal to their static analogs. This equality holds for general cost and revenue functions and not simply the quadratic ones assumed here.

In the open-loop model, each firm chooses a sequence of changes in output, given a particular behavioral assumption, v. The equilibrium levels can be expressed in feedback form; in this case, strategies are open loop with revisions that are unanticipated. When players choose their current levels, they act as if they were also making unconditional choices regarding future levels.

Characteristics of the Model. In Karp and Perloff (1988), we derive a number of properties of these models analytically and through simulations. Six of these properties are briefly summarized here.

1. If $v = -1/n$ or $v = 1$, the open-loop and feedback trajectories, control rules, and equilibria are identical because, if firms either take price as given or share the market in each period, it does not matter whether they choose levels or control rules. Simulation results (Karp and Perloff, 1988) indicate this condition is also necessary.

2. For $v \in (-1/n, 1)$ and for given symmetric initial output level q, output at t is greater along the feedback trajectory than along the open-loop trajectory.[7] The feedback Nash-Cournot equilibrium is farther from the monopoly solution than is the open-loop Nash-Cournot equilibrium. The intuition is that, under the feedback assumption, rivals' investments are discouraged by greater capacity. Therefore, firms have a greater incentive to invest today as a means of preempting their rivals' future investment. Thus, they develop larger capacities and hence larger output levels.

3. Industry profits are higher (and social surplus lower) in the open-loop equilibria. That is, feedback strategies are relatively procompetitive.

4. For given $v \in (-1/n, 1)$, the steady-state equilibrium output is dependent on ϵ in the feedback game but is independent of ϵ in the open-loop game. Consider, for example, the Nash-Cournot case where $v = 0$. Under the feedback model, a firm expects its rivals to react to its current decision only after an interval of ϵ, so its current decision depends on ϵ. It can be shown using simulations that the steady-state feedback output is decreasing in ϵ. When ϵ is very small, a firm expects a rapid response (for $v = 0$) from its rivals, thereby increasing the firm's preemptive incentive and resulting in larger outputs. The estimated index of market structure, therefore is dependent on the assumed period of adjustment in the empirical analysis. For the open-loop model, a firm expects no response on the part of its rivals (for $v = 0$), and the equilibrium is, therefore, independent of ϵ.

5. Given $\theta = 0$ and the normalization $\delta = (n + 1) c$ (where $c > 0$ is a constant), the open-loop and feedback Nash-Cournot equilibria converge to the

competitive equilibrium as n → ∞; the price-taking and collusive equilibria are invariant to n. As n becomes large, the adjustment cost for each firm becomes infinite so each firm makes only infinitesimal adjustments and thus captures only an infinitesimal share of the market.

6. Under both open-loop or feedback policies, output decreases in v.

Estimation

Our objective is to obtain a consistent estimate of the index of market structure, v. In the process we also estimate the adjustment parameter, δ. For estimation, it is convenient to allow the parameters γ_i and θ_i to vary across firms and to be time dependent. We also allow the demand intercept, a_i, for each firm to vary as it will if there are certain types of quality differences or differences in transportation costs. A set of restrictions are eliminated by this flexibility, but the remaining restrictions implied by the model are sufficient to identify the characteristics of the market.

Throughout the estimation, we assume that the discount factor β is known and common to all firms, that all firms have the same δ (symmetry among firms, i. e., homogeneous output and identical adjustment cost functions), and that v is common to all firms. However, the discussion in the next section considers the more general case of differentiated products or values of δ or v that vary across firms.

We estimate the adjustment equation,

$$q_t = g(t) + G q_{t-1}, \tag{8}$$

where G is a $(n + 1) \times (n + 1)$ matrix and g(t), a $(n + 1) \times 1$ vector, is an unrestricted function of exogenous variables. That is, we make no assumptions regarding whether firms have rational expectations about the *exogenous* variables nor do we impose assumptions about whether the inverse demand intercepts and affine costs are constant over time and across firms.[8]

The most obvious reason for this estimation strategy is its simplicity. The elements of G are used to infer the parameter v. The type of market structure is logically distinct from the rational expectations hypothesis. If the rational expectations hypothesis is true, there is a loss in efficiency from ignoring it, but our estimates are consistent.

Open Loop

Define v_i as an $n + 1$ dimensional column vector with 1 in the *i*th position and $v_{ij} = \partial\mu_{jt}/\partial\mu_{it}$ for $(i \neq j)$ elsewhere. (This approach generalizes the previous section where $v_{ij} = v$ for $i \neq j$ was assumed.) Given an assumed value of β, an estimated matrix G and demand slope b (and hence K_i), v_i and δ for the open-loop equilibrium satisfy (see the Appendix)

$$K_i v_i = \left[G^{-1}(I - G)(I - \beta G)\right]' e_i \delta. \qquad (9)$$

The derivation of (9) does not depend on the assumption that firms are symmetric, but the solution of (9) requires either symmetry or a similar assumption. The matrix K_i is of rank 2 so, in general, there are either infinitely many solutions to (9) or no solutions. Given the symmetry of firms, a symmetric equilibrium requires that the diagonal elements, g_1, of G are equal as are the off-diagonal, g_2, elements. If G is estimated subject to this restriction, all elements except the *i*th in the column vector on the right-hand side of (9) are equal and there exists a unique solution to (9).

More generally, suppose that the symmetry assumption is not used. For example, set i = 1 and let

$$p_1 = a_1(t) - \sum_{j=1}^{n+1} b_{ij} q_{jt}.$$

Then (9) can be rewritten as

$$\begin{bmatrix} 2b_{11} & b_{12} & \cdots & b_{1,n+1} \\ b_{12} & 0 & \cdots & 0 \\ \cdot & \cdot & & \cdot \\ \cdot & \cdot & & \cdot \\ \cdot & \cdot & & \cdot \\ b_{1,n+1} & 0 & \cdots & 0 \end{bmatrix} \begin{bmatrix} 1 \\ v_{11} \\ \cdot \\ \cdot \\ \cdot \\ v_{1n} \end{bmatrix} = \begin{bmatrix} y_{11} \\ y_{12} \\ \cdot \\ \cdot \\ \cdot \\ y_{1,n+1} \end{bmatrix} \delta_1,$$

where the y_{ij} are obtained from the right side of (9) and depend only on β and the elements of G. The existence of a solution to (9) requires

$$\frac{b_{ij}}{b_{ik}} = \frac{y_{ij}}{y_{ik}}, \qquad (10)$$

for all i and all j, k $\neq$ i, and gives $n^2 - 1$ restrictions involving the demand system, b_{ij}, and the feedback system, G. Under the symmetry assumption and G as in (8), these restrictions are satisfied. This approach is the simplest but not the most general way to satisfy (9).

If (9) is satisfied, δ_i can be uniquely estimated as $\delta_i = b_{ij}/y_{ij}$, j $\neq$ i. That is, given the estimated G matrix, δ_i is linear in the estimated demand slope coefficient(s). Using the previous equation to eliminate δ_i gives

$$\sum_{j \neq i} \frac{b_{ij}}{b_{ii}} v_{ij} = \frac{y_{ii} b_{ik}}{y_{ik} b_{ii}} - 2, \tag{11}$$

for all i, k $\neq$ i. There are n + 1 equations in (n + 1) n unknowns. An additional assumption, such as $v_{ij} = v_i \ \forall \ j \neq i$, is required. That is, for each firm, we can obtain an aggregate index (or aggregate conjectural variation), but we cannot distribute this index over a firm's rivals.

If we assume that $b_{ij} = b_{ik}, \ \forall \ j, k \neq i$ (a weaker condition than the symmetry assumption) and also that $v_{ij} = v_i$, then (10) and (11) simplify to

$$1 = \frac{y_{ij}}{y_{ik}} \qquad \forall \ j, k \neq i \tag{12}$$

and

$$v_i = \frac{(y_{ii} - 2y_{ik})}{ny_{ik}}. \tag{13}$$

Thus, it is possible to estimate the $(n + 1)^2$ elements of G subject to the $n^2 - 1$ restrictions of (12) and use (13) to infer v_i. This approach does not require estimation of the demand slope parameters, b_{ij} (b if the product is homogeneous). The slope parameters are necessary only to recover δ_i and, of course, to test whether the hypotheses $b_{ij} = b_{ik}$ (j, k $\neq$ i) are reasonable.

Feedback

To estimate v and δ in the feedback case, define the vectors

$$w_i = \left[I - \beta (G' \otimes G') \right]^{-1} \left[(G' \otimes G') \, (\text{vec } K_i) \right],$$

$$x_i = \left[I - \beta (G' \otimes G') \right]^{-1} \left[(G' \otimes G') - (I \otimes G') - (G' \otimes I) + I \right] [\text{vec}(e_i e_i')],$$

where vec(Z) stacks the columns of the matrix Z into a vector. Reversing the vec operation to obtain matrices, w_i and x_i are converted into the (n+1) $\times$ (n+1) matrices W_i and X_i. By inspection, W_i is linear in Firm i's demand coefficient(s) and X_i depends only on β and G. If agents use feedback strategies, v and δ must satisfy (see the Appendix).

$$\left[K_i + \beta W_i + (e_i e_i' + \beta X_i) \, \delta_i \right] v_i = G'^{-1} e_i \delta_i \equiv y_i^* \delta_i. \tag{14}$$

Given the complexity of W_i and X_i, it is difficult to analyze (14) for the general case. However, for the symmetric case, the left side of (14) is of rank 2. Under the assumption of symmetry, the estimate of v is independent of b. If we define the matrix A^i and B^i such that $bA^i \equiv K_i + \beta W_i$ and $B^i \equiv e_i \, e_i' + \beta X_i$, A^i and B^i depend only on β and G. To recover v and δ, we rewrite the *i*th and the *k*th (k $\neq$ i) equation of (14) as

$$b \left(A_{ii} + v \sum_{j \neq i} A_{ij} \right) + \left(B_{ii} + v \sum_{j \neq i} B_{ij} \right) \delta = y_{ii}^* \, \delta , \qquad (15)$$

and

$$b \left(A_{ki} + v \sum_{j \neq i} A_{kj} \right) + \left(B_{ki} + v \sum_{j \neq i} B_{kj} \right) \delta = y_{ik}^* \, \delta , \qquad (16)$$

where A_{ij}, B_{ij}, and y_{ii}^* are elements of A^i, B^i, and y^*. Solving (16) gives δ as a linear function of b and a nonlinear function of v. Substituting this function into (15) gives a quadratic in v that is independent of b. Hence, given symmetry, v can be estimated with knowledge of only β and G.

Although there are two solutions to (15) [or (14)], extensive simulation experiments show that one value is close to the open-loop value and that the other is implausible ($v \notin [-1/n, 1]$); therefore, in practice it is easy to choose the correct root. Using simulations, a comparison of the estimated v and δ under the open-loop and feedback models, maintaining the assumption of symmetry, shows that v and δ are larger under feedback.

Testing

It is possible for a value (or values in the asymmetric case) of v and a value of δ to satisfy (9) or (14) without the implied game being meaningful. Given the solution to (9) or (14), it is necessary to check that each player's second-order conditions are satisfied and that the underlying Ricatti difference equations are stable. For the open-loop equilibrium, these tests can be performed using the fact that the open-loop equilibrium can be generated by solving a control problem; it is simply necessary to check whether that control problem satisfies certain properties (Karp and Perloff, 1988). For the feedback game, it appears necessary to solve the game using the estimated δ and v; but this computation is straightforward.

We can test whether the equilibrium is open loop or feedback in addition to estimating the degree of competitiveness (v). If symmetry is assumed so that G is estimated as in (8), exactly the same restrictions are imposed in estimating the parameters of demand and the control rule under both open loop and feedback. In order to distinguish the two, an overidentifying restriction is needed. For example, given information on cost, it would be possible to estimate jointly a cost function involving δ and the demand function and control rule subject to (9) or (14). In principle, one could apply methods of nonnested hypothesis testing or, less formally, compare the values of the likelihood functions under the two sets of restrictions. Unfortunately, reliable cost data are rarely available. Most firm-specific cost data are constructed by allocating total cost to a set of categories which does not include "adjustment." It would be surprising if, using this data, one could obtain a reliable estimate of δ.

In the absence of cost data it is, in principle, possible to test open-loop vs. feedback behavior by dropping the symmetry assumption. The estimation of b_{ij}, G, v_i^k, and δ^k (k = o, f) subject to (9) or (14) will, in general, result not only in different estimates but also different values of the likelihood function. That is, in the absence of symmetry, the two sets of restrictions, (9) and (14), are not equivalent.

We can demonstrate this nonequivalence by means of an example. For n = 2, we chose an arbitrary G, b_{11}, and b_{12} and then chose b_{13} to satisfy (9); by construction, a unique estimate of v_1 and δ_1 satisfies the open-loop restrictions (9). However, for these values of G and b_{ij}, the feedback restrictions constitute three independent equations in δ_1 and v_1, and no solution exists. In order to make these equations consistent, we could, for example, change b_{13}. In that case the open-loop restrictions would cease to be consistent. Therefore, the value of the likelihood function may be either greater or less under open loop: The open-loop and feedback models are observationally distinct even without cost data. Unfortunately, imposing the restrictions in (9) and (14), other than by using symmetry, is computationally difficult.

As demonstrated in the previous example, the feedback model is capable of providing more information than the open loop. As mentioned above, it is possible to estimate only a single value v_i in the open-loop model. However, for the example we constructed, the feedback restrictions constitute three independent equations so it would have been possible to estimate v_{12} and v_{13}.

Empirical Applications

We discuss an application of our methodology to the rice export market (Karp and Perloff, 1989a) in detail and briefly mention another application to the international coffee export market (Karp and Perloff, forthcoming). In both cases, we concentrate on some of the largest exporting countries and assume that the other countries act as a competitive fringe and treat their output levels as exogenous.

The Rice Export Market

We concentrate on three of the largest exporters—China, Pakistan, and Thailand—who account for half of all exports. Thailand's share exceeded a third in recent years; China's share was that large a decade ago. China, Pakistan, Thailand, and the United States were responsible for two-thirds of all exports over the last couple decades.

We treat the United States as part of the fringe because, unlike the other large exporting countries, the United States does not have a single exporting agency. U. S. firms appear to act independently. Though the U. S. government intervenes often, the nature of the intervention varies substantially over time and appears not to be motivated by conditions in international markets.

We expect noncompetitive behavior by China, Pakistan, and Thailand for three reasons. First, these countries have large shares and central agencies that handle all their sales. Second, nontariff barriers affect 93% of world rice imports and 76% of exports. Third, the rice export market is thin. There are no large organized exchanges, so major brokerage houses in the United States, Europe, Singapore, and Hong Kong are able to charge fees as high as 5% to 10%. With other high transaction or search costs and high shipping costs, f. o. b. prices may not equalize rapidly.

The nominal rice export prices and quantities are from the United Nations Food and Agriculture Organization (various years). The price of wheat, a substitute, is from the International Wheat Council (various years). The world commodity wholesale price index and gross domestic product indexes are from the International Monetary Fund (various years).

Estimation Results

We used instrumental variables to estimate the linear demand curves, equation (1), including various substitutes prices and time trends; and we estimated adjustment equations (8). These two sets of equations can be used to solve for v.

In the demand curves, all prices were deflated by a world commodity wholesale price index. Because the demand slope coefficients are only necessary to estimate δ and not v, we do not discuss them further here.

The adjustment equations (8) were estimated using Zellner's seemingly unrelated equations method and are reported in Karp and Perloff (1989a, forthcoming). Each country's exports are regressed on its own lagged exports, the other country's lagged exports, a time trend, and dummies for adverse weather conditions. For coffee, a dummy was included for the major freeze in Brazil in 1977-78. For rice, a 1973 dummy reflected bad weather that reduced the rice crop substantially.

We imposed the cross-equation symmetry constraints that the coefficient on the own lagged exports was equal across equations as was the coefficient on the other country's lagged exports. That is $g_{11} = g_{22} \equiv g_1$, and $g_{12} = g_{21} \equiv g_2$ where g_{ij} is the (i, j) element of G. For neither commodity could we reject these equality restrictions at the 0.05 level. For both we could reject the hypotheses that $g_1 = 0$ or $g_2 = 0$.

Classical Estimates. We assume that $\beta = .95$ and then use the adjustment equations to infer v for both the open-loop and feedback models for rice (Table 6.1). The first set of estimates in Table 6.1 are based on standard (classical) regressions.

The classical (subscript c) estimates are $v_c^o = \phi_o(G) = -0.37$ and $v_c^f = \phi_f(G) = -0.32$, where $\phi_o(G)$ is a nonlinear function of G as shown in equation (9) and $\phi_f(G)$ is a nonlinear function shown in equation (14). These point estimates of v appear close to the price-taking level, $-1/n = -0.5$. Based on standard errors calculated using Taylor expansions, we cannot reject price-taking or Nash-Cournot (0) behavior, but we can reject collusive behavior (1).[9] The results for coffee are similar: we cannot reject price taking (-1) but can reject Nash-Cournot (0) or collusive behavior (1).

For this estimated dynamical system to "make sense," it must have three properties:

- The system is *stable*: $-1 < g_1 + ng_2 < 1$ and $-1 < g_1 - g_2 < 1$.
- The *market structure* lies between collusion and price taking: $1 > v^k = \phi_k(G) > -1/n$, k = o or f.
- The *adjustment* parameter in each of the models is positive: $\delta^k = \psi_k(G, b) > 0$, k = o or f.

Our classical point estimates of the elements of G and our estimates of v^k and δ^k meet these restrictions.

Bayesian Estimates. Rather than estimating the unconstrained system and hoping that the point estimates lie in the desired range as we did using the classical approach, we can impose these three sets of restrictions. Although it would be extremely difficult, if not impossible, to impose these inequality restrictions using a classical approach or to test them, Geweke (1986, 1988, 1989), Chalfant and White (1988), and Chalfant, Gray, and White (1991) show how such inequality restrictions can be imposed and tested using Bayesian techniques.

In this Bayesian approach, a prior is used that is the product of a conventional uninformative distribution and an indicator function that equals 1 where the inequality constraints are satisfied and 0 elsewhere. The posterior distribution is calculated using Monte Carlo numerical integration with importance sampling.[10]

To estimate the probability that the restrictions hold, we calculate the (importance weighted) proportion of Monte Carlo replications satisfying the restrictions. The odds that these restrictions hold are as shown in Table 6.1

TABLE 6.1 Classical and Bayesian Inequality Constrained Estimates for Rice

	v^o	v^f
Classical estimates		
v_c^k (unrestricted)	-0.37	-0.32
Standard deviation (Taylor approximation)	0.17	0.20
Bayesian inequality constrained estimates[a]		
Quadratic loss		
v_b^k (mean)	-0.26	-0.21
Standard deviation (σ)	0.26	0.27
Precision of the mean of v^k ($\sigma/\sqrt{T}$)	0.004	0.004
Absolute loss		
v_a^l (median)	-0.35	-0.30
Standard deviation	0.28	0.28
Reject because (%)		
Unstable	0.004	0.004
$\delta^k \leq 0$	8.2	7.8
$v_b^k < -1/n$	8.2	8.2
$v_b^k > 1$	3.2	3.2
Total rejections (1 - p)	11.4	11.4
Asymptotic standard error of p: $\sqrt{p(1 - p)/T}$	0.005	0.005

[a] 5,000 replications (T).

(based on 5,000 importance-sampling replications). The stability conditions are virtually always met. Because all three sets of conditions hold in approximately seven-eighths of the replications for rice (and three-quarters of the replications for coffee), imposing these restrictions is reasonable. Because the restriction that δ is positive holds for nearly 90% of the replications for rice (and three-quarters, for coffee), the data indicate there is dynamic adjustment. The odds in favor of a positive δ are 9 to 1 for rice (and 3 to 1 for coffee).

Given a quadratic (absolute difference) loss function, estimates of the parameters consistent with the restrictions are obtained by calculating the mean (median) of the coefficient estimates for all replications where the constraints are satisfied (Zellner, 1971, pp. 24-25). Indeed, we obtain the full posterior distributions of v_b^o and v_b^f.

TABLE 6.2 Distribution of v^k Based on Bayesian Estimates for Rice

Weight Between[a]		v^o	v^f
-1/n	0	88.1%	84.5%
0	1/2	9.0	12.4
1/2	1	2.9	3.1
-1/n	v_c^k	43.6	46.5
v_c^k	0	44.5	38.0
-1/n	v_b^k	66.5	64.5
v_b^k	0	21.6	19.9

[a]The classic estimate is v_c^k (k = o or f) and v_b^k is the Bayesian estimate based on a quadratic loss function. By definition, half the weight lies between -1/n and the absolute loss (median estimators), which are $v_a^o = -0.35$ and $v_a^f = -0.30$.

Table 6.1 summarizes the results of the classical and the Bayesian estimates. The v_b^k based on an absolute difference loss function (medians) are close to the classical point estimates. The v_b^k based on a quadratic loss function (means) are somewhat (.1 for rice and .2 for coffee) higher than the classical estimates. The standard deviations on the quadratic loss function v_b^k are only slightly greater than the Taylor approximations for the classical estimates.

The Bayesian estimates, which provide an entire posterior distribution for the market structure parameter, v_b^k, can be used to calculate the probability that v^k lies within a certain range. A histogram for both models for rice are shown in Figure 6.1. Some of the interesting ranges for both sets of Bayesian estimates are summarized in Table 6.2.

The probability that v_b^k lies between price taking and Nash-Cournot is nearly 90% for rice (and slightly higher for coffee). Two-thirds of both distributions lie below the mean (quadratic loss) estimates of v_b^k. The posterior odds ratio for the feedback model that the market structure lies between price taking and Nash-Cournot rather than between Nash-Cournot and collusive is 12.6 for rice and 12.9 for coffee. That is, it is nearly 13 times as likely that these markets are more competitive than Nash-Cournot than they are less competitive than Nash-Cournot.

Simulations

These estimates have implications for steady-state outputs, as we show using the rice estimates. We normalize so that, if the major exporting countries were price takers, each country's steady-state output would be 100. If the countries played open-loop Nash-Cournot, the steady-state output would be 75 (66.7 for coffee with two firms); and if they were a perfect cartel, their output would be 50 as shown in Table 6.3.

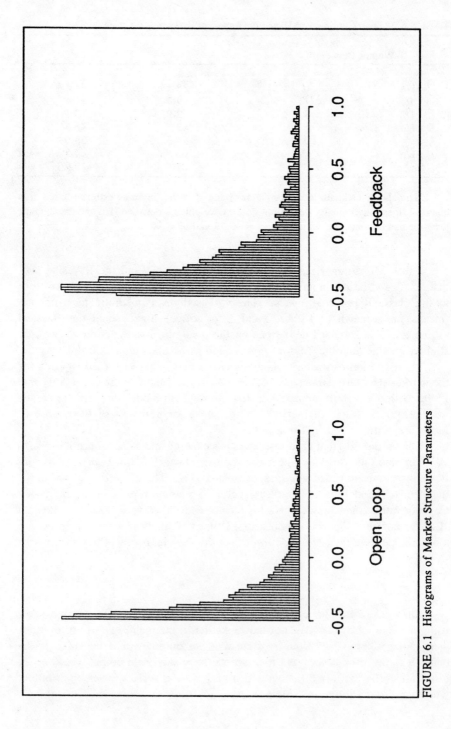

FIGURE 6.1 Histograms of Market Structure Parameters

TABLE 6.3 Rice Exports as a Percentage of the Price Taking Levels

Price-Taking	100.0
Feedback Steady State (classical $v_c^f = -0.32$)	94.2
Open-loop Steady State or Static (classical $v_c^o = -0.37$)	92.2
Static model (classical $v_c^f = -0.32$)	89.3
Feedback Steady State (Bayesian quadratic loss $v_b^f = -0.21$)	88.9
Open-loop Steady State or Static (Bayesian quadratic loss $v_b^o = -0.26$)	86.2
Static model (Bayesian quadratic loss $v_b^f = -0.21$)	83.8
Nash-Cournot Open-Loop Steady-State or Static	75.0
Collusive	50.0

Using the classical point estimates, the open-loop ($v_c^o = -0.37$) steady-state output is 92.2. The feedback ($v_c^f = -0.32$) steady-state output is 94.2. Using the Bayesian quadratic loss estimates, the open-loop ($v_b^o = -0.26$) output is 86.2 and the feedback ($v_b^f = -0.21$) output is 88.9. Thus, while the estimated market structure is "close" to price taking in the sense that v is close to -0.5, the steady-state outputs are below the price-taking levels by between 6 and 14 percent, depending on the model and the market structure estimate used for rice (and between 7 and 14 percent for coffee).

The cost of adjustment affects the steady state in the feedback model. In a static model with a v = -0.32 (classical v_c^f estimate), each firm produces 89.3; but in the dynamic feedback model, the steady-state output is 94.2. Similarly, with a $v_b^l = -0.21$ (Bayesian quadratic loss estimate of v^f), the static output is 83.8, whereas the feedback model's steady-state output is 88.9. Thus, where costs of adjustment are positive, more is produced in the steady state.

Summary and Conclusions

New families of open-loop and feedback models that are consistent with a range of market behaviors can be used to estimate the degree of competition among dynamic oligopolists and to test whether firms use open-loop or feedback strategies. This linear-quadratic estimation method, which is easy to implement, is applied to the rice and coffee export markets. Both classical and a Bayesian technique imposing inequality constraints are demonstrated. The Bayesian results are generally close to the classical ones. One advantage of the Bayesian approach is that an empirical approximation to the distribution of the behavior strategy variable is obtained, so the probability that market structure lies in a particular range can be easily calculated. For example, the steady-state coffee outputs may be between 7 and 14 percent lower than if Brazil and Colombia were pure price takers.

Notes

1. We thank Peter Berck, Diana Burton, James Chafant, Drew Fudenberg, George Judge, Max Leavitt, Wen-Ting Lu, and Ken White. All estimation and simulations were done on with SHAZAM.

2. In both our empirical applications, the agricultural exports in each country are controlled by a central agency. We refer to these agencies as "firms" and use the terms exports and outputs interchangeably.

3. Nonconstant marginal costs could be incorporated easily into the model; however, in neither of our applications do we have detailed enough cost data to estimate nonconstant marginal costs.

4. We treat this index, v, as a single parameter but, more generally, it might be a function of exogenous variables, as for example, in Gollop and Roberts (1979). Similarly, Iwata (1974) and Karp and Perloff (1989b) estimate v_i that vary over firms.

5. Our model is a generalization of early linear-quadratic cost-of-adjustment models that assume a competitive structure (Sargent, 1978; Hansen and Sargent, 1980; and Blanchard, 1983). Hansen, Epple, and Roberds (1985) use the dynamic linear quadratic model to study different open-loop markets as well as the open-loop and feedback Stackelberg models. Fershtman and Kamien (1987) and Reynolds (1987) compare the open-loop and feedback linear-quadratic Nash-Cournot models in theoretical models only.

6. There are at least two alternatives to the open-loop linear-quadratic model. One uses instrumental variables to estimate the game analog of the stochastic Euler equations (Hansen and Singleton, 1982; and Pindyck and Rotenberg, 1983). Similar methods could be used to estimate noncompetitive markets, but, because the Euler equations restrict the equilibria to be open loop, this approach is not pursued here. The second method uses dynamic duality (Epstein, 1981). Although, in principle, this method could be used to estimate both open-loop and feedback noncompetitive equilibria, it implies very complicated restrictions for the feedback case and may be of limited practical use. Roberts and Samuelson (1988) estimate a dynamic oligopoly model and test and reject the assumption of open-loop behavior. They do not, however, estimate the model under the assumption that firms use fully rational Markov strategies because of the complexity of the restrictions such a model implies when general functional forms are used.

7. Fershtman and Kamien (1987) and Reynolds (1987) show this result for the Nash-Cournot equilibria; we have generalized it for other v and describe the entire path, not just the steady state.

8. We could impose or test the rational expectations hypothesis by including an exogenous state vector of current information in the feedback game (see Chow, 1981). A large literature (see the cites in footnote 4) shows how to do the same in an open-loop model.

9. Because the exporters are not equal in size, v = 1 leads to the collusive solution only in the long run where their exports become equal.

10. Geweke (1986) explains this approach for a single equation. The multiequation generalization used here is based on Chalfant and White (1988) and Chalfant, Gray, and White (1991). These papers and Kloek and Van Dijk (1978) discuss Monte Carlo integration using importance sampling. Geweke briefly discusses our problem of lagged endogenous variables.

References

Appelbaum, E. 1982. The Estimation of the Degree of Oligopoly Power. *Journal of Econometrics* 19(2/3):287-299.

Blanchard, O. J. 1983. The Production and Inventory Behavior of the American Automobile Industry. *Journal of Political Economy* 91(3):365-400.

Blinder, A. S. 1982. Inventories and Sticky Prices: More on the Microfoundations of Macroeconomics. *American Economic Review* 72(3):334-349.

Chalfant, J. A. and White, K. J. 1988. Estimation and Testing in Demand Systems with Concavity Constraints. University of California, Department of Agricultural and Resource Economics, Berkeley, Working Paper No. 454.

Chalfant, J. A., Gray, R. S., and White, K. J. 1991. Evaluating Prior Beliefs in a Demand System: The Case of Meats Demand in Canada. *American Journal of Agricultural Economics* forthcoming.

Chow, G. C. 1981. *Econometric Analysis by Control Methods.* New York: John Wiley and Sons, Inc.

Epstein, L. G. 1981. Duality Theory and Functional Forms for Dynamic Factor Demands. *Review of Economic Studies* 48(1):81-95.

Fershtman, C. and M. R. Kamien. 1987. Dynamic Duopolistic Competition with Sticky Prices. *Econometrica* 55(5):1151-1164.

Geweke, J. 1986. Exact Inference in the Inequality Constrained Normal Linear-Regression Model. *Journal of Applied Econometrics* 1(2):127-41.

_____. 1988. Antithetic Acceleration of Monte Carlo Integration in Bayesian Inference. *Journal of Econometrics* 38(1/2):73-89.

_____. 1989. Bayesian Inference in Econometric Models Using Monte Carlo Integration. *Econometrica* 57(6):1317-1339.

Gollop, F. M., and J. M. Roberts. 1979. Firm Interdependence in Oligopoly Markets, *Journal of Econometrics* 10(3):313-331.

Hansen, L. P., D. Epple, and W. Roberds. 1985. Linear-Quadratic Duopoly Models of Resource Depletion. In *Energy Foresight and Strategy*, ed. T. J. Sargent, Resources for the Future, Washington, D. C.

Hansen, L. P., and T. J. Sargent. 1980. Formulating and Estimating Dynamic Linear Rational Expectations Models. *Journal of Economic Dynamics and Control* 2(1):7-46.

Hansen, L. P., and K. J. Singleton. 1982. Generalized Instrumental Variables Estimation of Nonlinear Rational Expectations Models, *Econometrica* 50(5):1269-1286.

Iwata, G. 1974. Measurement of Conjectural Variations in Oligopoly, *Econometrica* 42(5):947-966.

Karp, L. S. and J. M. Perloff. 1988. Open-Loop and Feedback Models in Dynamic Oligopoly, unpublished manuscript.

_____. 1989a. Dynamic Oligopoly in the Rice Export Market. *Review of Economics and Statistics* 71(3):462-470.

_____. 1989b. Estimating Market Structure and Tax Incidence: The Japanese Television Market. *Journal of Industrial Economics* 37(3):225-239.

_____. forthcoming. A Dynamic Model of Oligopoly in the Coffee Export Market. *American Journal of Agricultural Economics.*

Kloek, T. and H. K. Van Dijk. 1978. Bayesian Estimates of Equation System Parameters: An Application of Integration by Monte Carlo, *Econometrica* 46(1):1-19.

Pindyck, R. S., and J. J. Rotenberg. 1983. Dynamic Factor Demands and the Effects of Energy Price Shocks, *American Economic Review* 73(5):106-1079.

Reynolds, S. S. 1987. Capacity Investment, Preemption and Commitment in an Infinite Horizon Model, *International Economic Review* 28(1):69-88.

Roberts, M. J. and L. Samuelson. 1988. An Empirical Analysis of Dynamic, Nonprice Competition in an Oligopolistic Industry. *Rand Journal of Economics* 19(2):200-220.

Sargent, T. J. 1978. Estimation of Dynamic Labor Demand Schedules under Rational Expectations, *Journal of Political Economy* 86(6):1009-1044.

Starr, A. W., and Y. C. Ho. 1969. Nonzero Sum Differential Games, *Journal of Optimization Theory and Applications* 3(3):184-206.

Sumner, D. 1981. Measurement of Monopoly Behavior: An Application to the Cigarette Industry, *Journal of Political Economy* 89(5):1010-1019.

Zellner, A. 1971. *An Introduction to Bayesian Inference in Econometrics.* New York: John Wiley and Sons, Inc.

Appendix: Derivation of Restrictions

Because we are only interested in imposing restrictions on the demand slopes and the coefficients on q in the control rule and not on the intercepts of the demand and control systems, we can restrict our attention to the quadratic part of the problems.

The Open-Loop Restrictions

The Lagrangian for the ith player is

$$Q_i = \sum_{\tau=i}^{T} \beta^{\tau-i} \left[-\frac{1}{2} q_\tau' K_i q_\tau - \frac{1}{2} u_\tau' S_i u_\tau + \lambda_{i\tau}'(q_{\tau-1} + u_\tau - q_\tau) \right].$$

The first-order conditions for q_τ and $u_{i\tau}$ are

$$-K_i q_\tau - \lambda_{i\tau} + \beta \lambda_{i,\tau+1} = 0 \qquad (A.1a)$$

and

$$-v_i' S_i u_\tau + v_i' \lambda_{i\tau} + 0. \qquad (A.1b)$$

We can show that, at time T, λ_{iT} is a linear function of q and that, if λ_{it} is a linear function of q, then so is $\lambda_{i,t-1}$. Thus, by induction $\lambda_{i\tau} = H_{i,\tau}q_\tau$, for some matrix $H_{i,\tau}$. Letting $T \to \infty$ so that $H_{i\tau} \to H_i$, (A.1b) becomes

$$v_i' H_i q_\tau = \delta_i u_\tau, \qquad i = 1, \ldots, n + 1.$$

Stack these conditions to obtain $Eq_\tau = Su_\tau$, where the ith row of E is $v_i'H_i$ and the ith row of S is $\delta_i e_i'$. As a result, $Eq_\tau = S(q_\tau - q_{\tau-1})$ or $q_\tau = Gq_{\tau-1}$ where $G \equiv (S - E)^{-1}S$.

Use the previous definitions to rewrite (A.1a) as

$$0 = -K_i q_\tau - H_i q_\tau + \beta H_i q_{\tau+1} = (-K_i - H_i + \beta H_i G)q_\tau = 0$$

so

$$H_i(1 - \beta G) = -K_i$$

and

$$H_i = -K_i(1 - \beta G)^{-1}.$$

From the definition of G, we have

$$E = S(1 - G^{-1}).$$

Premultiply both sides by e_i' and use the definition of E and the previous expression for H_i to obtain

$$e_i'E = v_i'H_i = -v_i'K_i(I - \beta G)^{-1} = e_i'S(I - G^{-1}) = \delta_i\,e_i'(I - G^{-1}),$$

so that

$$-v_i'K_i = \delta_i\,e_i'(I - G^{-1})\,(I - \beta G)$$

and

$$K_i v_i = -\left[(I - G^{-1})\,(I - \beta G)\right]'e_i\,\delta_i.$$

Factoring out $G^{-1'}$ gives (7).

The Feedback Restrictions

The stationary dynamic programming equation is

$$-\frac{1}{2}q_{t-1}'H_i q_{t-1} = \max_{q_t} -\frac{1}{2}q_t'(K_i + S_i + \beta H_i)q_t + q_t'S_i q_{t-1} - \frac{1}{2}q_{t-1}'S_i q_{t-1}. \quad (A.2)$$

The first-order condition is

$$-v_i'(K_i + S_i + H_i)q_t + v_i'S_i q_{t-1}.$$

Stack the $n + 1$ first-order conditions to obtain

$$Eq_t = Sq_{t-1},$$

where the ith row of E is $v_i'(K_i + S_i + H_i)$ and the ith row of S is $\delta_i e_i'$. Rewrite this as $q_t = G q_{t-1}$ where $G \equiv E^{-1}S$. Substitute this into the maximized value of (A.2) to obtain

$$H_i = G'(K_i + S_i + \beta H_i)G - G'S_i - S_i G + S_i.$$

Apply the vec operation and simplify to obtain vec $H_i = w_i + x_i\delta_i$ where w_i and x_i are defined in the text. Convert the vectors back into a matrix to obtain $H_i = W_i + X_i\delta$. Take the ith row of $E\,G = S$ and use the definition of E to obtain

$$v_i'\Big[K_i + e_i\,e_i'\,\delta_i + \beta(W_i + X_i\,\delta_i)\Big]\,G = e_i'\,\delta_i.$$

Rearranging this equation gives (14).

PART FOUR

Market Structure-Strategy Studies

7

Generic Advertising as a Nonprice Marketing Strategy

John E. Lenz and Olan D. Forker

Introduction

Generic advertising has become a very important component of the marketing strategies of farmers worldwide. In the U.S., farmers provide almost $1 billion annually to collectively advertise and promote products produced from the commodities they market. Historically, economists have questioned the social desirability of advertising in general, and of generic advertising specifically. More recently, however, economists have begun to develop a rationale, and supporting empirical evidence, for viewing advertising as a productive economic endeavor which contributes to social and economic welfare. In recent years, some agricultural economists have included generic advertising variables in demand models; the results generally support the notion that generic advertising can be an effective component of nonprice marketing strategies.

The purpose of this paper is to describe generic commodity advertising, and examine its potentially important role as a nonprice marketing strategy for farmers. This will be done, after an introductory review of the status of generic commodity promotion in the U.S., by discussing alternative views of advertising's value to society, presenting the theoretical underpinnings of generic advertising, and summarizing the empirical evidence concerning the impacts of generic advertising.

Producer-funded generic commodity promotion has been undertaken for a variety of agricultural commodities over the past half-century. Among the oldest extant programs in the United States are the National Dairy Council, formed in 1915 (and several affiliated State Dairy Council units), and the Florida citrus program which began in the 1930s (Frank, 1985). The early programs were modest by current standards; over the years the number, variety, and size of programs have increased substantially. Rather than being legislatively authorized, the early programs were simply cooperative arrangements in which groups of individual producers with common marketing needs voluntarily contributed money to a promotion fund. The free-rider problem motivated producers to begin securing first state and then federal legislative authority for mandatory checkoffs.[1] Enabling legislation for such programs typically includes provisions allowing the funds to be used for both commodity promotion and various types of commodity-related research. Early legislation typically included provisions under which producers could request and receive refunds of the monies they paid into the promotion funds; concerns about producer equity have provided the impetus for recent movements toward mandatory programs with no refund provisions.

In contrast to brand advertising, where a primary goal is to increase a firm's market share within a product category, generic advertising of agricultural commodities is a cooperative effort undertaken by a group of producers, and processors in some instances, to increase total demand for their commodity. Such demand expansion can be brought about directly by increasing demand for the raw commodity, or indirectly by increasing demand for processed products in which the raw commodity is an input. Farmers do not generally retain title to their commodity as it moves through the marketing system, and the number of farmers (sellers) is usually large. Thus, farmers are usually unable to exert much individual influence on retail prices or consumer demand, and have instead resorted to various collective marketing strategies including generic advertising directed at consumers. These approaches to marketing are attempts to increase total revenues from commodity sales. While retail prices may increase if generic advertising activities are successful, the extent to which such price increases will be passed on to producers depends on the degree to which the increases are captured by retailers, processors, and distributors during the price transmission process. As does any positive demand shift, advertising-induced increases in retail-level demand for commodity-based products hold the potential for expanding the market for all participants. The extent to which producers benefit depends on the degree of competition in the both the retailing and the processing/distribution sector.

Regardless of economists' reservations, large numbers of farmers have joined together to collectively advertise and promote the commodities they produce. The results of a mail survey conducted during the summer of 1990 showed 52 different commodities promoted by the 116 commodity promotion organizations (Table 7.1) that responded (Lenz, *et al.*, 1991). Fruit and nut growers and dairy farmers support the two largest combined programs, accounting for over half of all reported expenditures (Table 7.2). At least 39 state, regional and national organizations are involved in promoting fluid milk and manufactured dairy products. Dairy promotion organizations reported a combined budget of $209 million, which is 28 percent of the total expenditures reported by all respondents. The 30 responding organizations involved in

TABLE 7.1. Classifications Used to Construct Commodity Groups

GRAINS & OILSEEDS	VEGETABLES
Corn	Artichokes
Dry Beans	Asparagus
Grain Sorghum	Avocados
Rice	Lettuce
Soybeans and Soy products	Olives
Wheat and Wheat products	Onions
	Potatoes
	Tomatoes

FRUITS & NUTS	MEAT, POULTRY, SEAFOOD, & EGGS
Almonds	Beef and Beef Products
Apples and Apple Products	Eggs and Egg Products
Apricots	Fish and Seafood
Tart Cherries	Lamb
Dates	Pork
Dried Figs	Turkey
Grapes and Grape Products	
Hazelnuts	**DAIRY**
Melons	Fluid Milk
Nectarines	Manufactured Dairy Products
Orange Juice	
Papayas	**FIBERS**
Peaches	Cotton
Peanuts	Mohair
Pears	Wool
Pistachios	
Plums	**OTHER**
Prunes	All Wyoming Agricultural Products
Raisins	German Agricultural Products
Strawberries	Honey
Walnuts	Indoor Tropical Plants
Watermelons	Nursery Stock
Wine	Sugar

promotion of fruits and nuts invested $218 million, or 29 percent of the total promotion expenditures by all responding commodity promotion organizations. For dairy, as well as some other commodities, the check-off that generates promotion funds typically represents just over 1 percent of the farmer's gross receipts from sales of the commodity to which it applies.

TABLE 7.2. Number of Commodity Promotion Organizations, Total Staff, and Total Budgets, All Survey Respondents by Commodity Category, United States, 1990

COMMODITY CATEGORY	NUMBER OF RESPONDENTS	TOTAL STAFF	TOTAL BUDGET ($THOUSANDS)
GRAINS & OILSEEDS	14	262	$48,047
DAIRY	39	603	208,856
FRUITS & NUTS	30	410	217,903
MEAT, POULTRY, SEAFOOD, & EGGS	12	254	193,854
VEGETABLES	11	71	23,867
FIBERS	4	316	56,379
OTHER	6	101	3,070
TOTAL	116	2,017	$751,976

Alternate Views of Advertising

As with most economic issues, the debate pertaining to the economics of advertising has proved quite lively. Although the opposing views are probably not as distinct or as polarized as presentations in the literature frequently are, it is useful to consider two schools of thought regarding advertising and its economic consequences. These schools exist under a variety of rubrics, for present purposes the labels "advertising-as-market-power" and "advertising-as-information" will suffice. Historically, both agricultural economists and economists have found the market power line of reasoning most persuasive. Recently, however, some have begun to articulate various information-based arguments, especially Albion and Farris (1981), and Ekelund and Saurman (1988).

Those espousing the advertising-as-market-power line of reasoning, hold that advertising, through its persuasive powers, alters consumers' tastes and preferences and contributes to greater product differentiation in an industry. Consumers become less price sensitive—demand becomes less price-elastic—and firms are able to exploit this situation by charging higher prices. To the extent that altered tastes and preferences lead to increased brand loyalty, entry barriers are heightened. With heightened entry barriers, existing firms are sheltered from competition by potential rivals; some argue that in such a sheltered industry large firms are able to consolidate their positions at the expense of smaller existing firms, with the industry becoming more concentrated. Under such conditions, remaining firms have more discretion as to the prices they charge and, it is argued, are less likely to compete on the basis of either price or quality. In addition they may have little incentive to be innovative in their product offerings. Under this view advertising is self-perpetuating; the high prices and "excessive" profits enjoyed by advertisers create incentives for further advertising. The market outcome is elevated prices and restricted output, in other words, an economic welfare loss to society. The advertising-as-market-power line of reasoning is, generally most readily applicable to brand advertising.

Those espousing the advertising-as-information position view the effects of advertising as being more benign and pro-competitive. Here, advertising is seen as providing product-related information—regarding such things as product existence, availability, price, quality, and other attributes—not as something capable of altering consumers' tastes and preferences. By expanding the consumer's information set, advertising contributes to increased price sensitivity (Wittink, 1977). Rather than erecting an entry barrier, advertising is a means for new, or expanding, firms to convey information about their products and thus attain a tenable market position. No argument is made regarding industry concentration; while competitive forces, including advertising, may induce the exit of inefficient firms' from the industry, others are free to enter and employ advertising as a facet of their strategy to secure a workable market position. Under this view of advertising it is the provision of information which is self-perpetuating. Better informed consumers create incentives for product innovations, information about which may be most efficiently presented via further advertising. The market outcome is lower prices for the industry with a larger, more diverse output. Though competition is enhanced, no clear-cut argument as to the effects on profits is posited. But one could infer from the arguments that something approximating normal profits would be the case, though this likely depends mostly on competitive factors other than advertising. The advertising-as-information view is applicable to both brand and generic advertising.

Undoubtedly, some admixture of these polar positions comes closest to providing an accurate characterization of most existing markets for advertised products. The empirical evidence generated by studies of a variety of generic advertising programs has generally suggested that these programs have resulted in greater demand for the commodities than would have existed in the absence of the advertising. Although this observation does not necessarily support either school of thought, it does seem to indicate that consumers value the information conveyed. Otherwise, being sovereign consumers, they would not respond positively to the advertising effort.

The Theory of Generic Advertising

It is generally stated that the purpose of generic commodity advertising is to expand the demand for the commodity or product category[2] being advertised. Generic advertising is not brand specific, rather, the advertising focus is on product or commodity characteristics common to all product items and brands produced from the commodity. Thus the purpose is to increase overall demand. If successful, the result will be a higher price, or an increased sales volume, or both, depending on the nature of the supply response function.

The theory behind this is that generic advertising conveys information and changes consumer perceptions about the products within a category. This, in turn, will alter their purchase behavior relative to the category being advertised as well as their behavior toward other products. Viewed in terms of traditional consumer demand theory, advertising is a demand shifter. The information conveyed by the advertising may change consumers' perceptions about the products within the category, reduce their search times, or perhaps even alter the nature of their utility functions. Regardless of the motivation that consumers have to alter behavior, the practical result is a shift of the demand function to the right.

In theory, generic advertising[3] can have a positive, negative or neutral effect on market shares of the various brands that make up the product category. If the characteristics of a particular brand are different than those of other brands, and those differences are being advertised, then it is possible that the generic advertising could increase the market share of that particular brand at the expense of others. Indeed, there may actually be a complementarity between generic and brand advertising if both are focussed on the same product characteristics. It is also conceptually possible that generic advertising could erode brand preference by highlighting the fact that branded products within a category share a common set of desirable characteristics. This could result in a shift in the market shares of particular brands, with those brands which are attempting counterfactual product differentiation seeing an erosion of their brand franchises. If the generic advertising increases aggregate demand without negatively affecting any particular brand, then every firm could realize a gain in sales in proportion to its market share.

Using consumer demand theory, it is possible to derive a functional relationship between sales and advertising. Several functional forms are feasible, though, as usual, theory is moot on this point. One that is sometimes posited is that depicted in Figure 7.1. With this relationship, the theory holds that a threshold level of advertising must be achieved before any change in sales will occur; expenditures from a_o to a_1 will have no effect on sales. Beyond the threshold level (a_1), the advertising effort has a positive effect up to the point of diminishing returns (a_3) where additional advertising efforts result in sales losses. Many different functional relationships can be envisioned over the

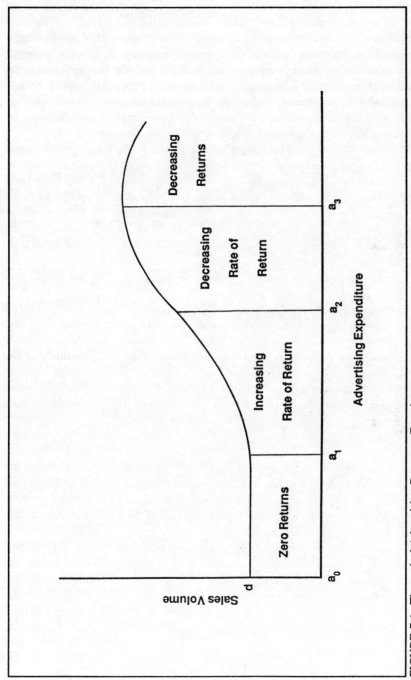

FIGURE 7.1 Theoretical Advertising Response Function

effective and experienced range of advertising expenditures. The impact might be immediate and linear or curvilinear. The curvilinear relationship could be the traditional "S" curve shape (a_1 through a_3 in Figure 7.1), continuously increasing at an increasing rate or continually increasing at a decreasing rate. *A priori*, there is no theory to predict the exact shape nor the magnitude; the impact could be anything from small to substantial. Determining the shape and magnitude of this functional relationship is an empirical issue.

The Empirical Evidence

Over the past half-century, a considerable body of literature has grown out of research conducted on various issues related to generic promotion of agricultural commodities. In the main, the reported research has dealt with evaluation of specific commodity programs. The studies tend to be fairly straightforward applications of standard demand theory, though some rather innovative approaches appear from time to time.

During the past several years Hurst and Forker have collected and annotated over 100 research reports that involved the study of generic (collective) advertising efforts. Most of the studies use traditional techniques to estimate demand functions, with advertising expenditures as an included variable. Some involve the estimation of single demand equations, others a system of equations. One study involved the simultaneous estimation of supply and demand equations for several products and several levels of trade (Liu, *et al.*, 1989). The one thing these studies have in common is that advertising expenditures are included as an explanatory variable to provide an estimate of the extent to which generic advertising shifts aggregate commodity demand.

The advertising response function depicted in Figure 7.1 in conjunction with the results of these studies provides a basis for considering the hypothesis that generic advertising has a positive impact on aggregate demand. In general, the positive advertising elasticities in all these studies lie somewhere within the theoretical range of effects depicted in Figure 7.1. The larger the elasticity measure, the larger the relative impact. However, magnitude alone does not provide a useful indication of the shape of the entire advertising response function, nor does it indicate the extent to which the expenditures generate a positive return to those providing the funds.

Of the studies reviewed, 37 provided estimates of advertising elasticities and 24 provided estimates of return on investment. The advertising elasticities are mostly point elasticities and, where necessary, are estimated at the average expenditure level.[4] All of the reported elasticities are positive, although the magnitude spread is quite large (Table 7.3). These studies constitute substantial evidence to support the general hypothesis that generic, or collective, advertising efforts by a commodity group can have a positive impact on demand.

TABLE 7.3 Generic Advertising Elasticities, Selected Studies

Product/Market	Elasticity	Source
Fluid Milk:		
California	0.275	Thompson (1974)
New York City	0.047	Thompson (1978a)
New York City	0.029	Thompson (1978b)
New York City	0.041	Kinnucan (1981)
New York City	0.051	Kinnucan (1986)
New York City	0.042	Forker & Liu (1986)
New York City	0.011	Liu & Forker (1989)
Rochester, NY	0.015	Thompson (1979)
Buffalo, NY	0.121	Kinnucan (1983)
Syracuse, NY	0.022	Liu & Forker (1989)
Albany, NY	0.007	Liu & Forker (1989)
10 Regions U.S.	0.009	Ward & McDonald (1986)
12 Regions U.S.	0.003	NDB (1986)
12 Regions U.S.	0.010	NDB (1987)
12 Regions U.S.	(0.017-0.046)	Warman & Stief (1990)
Ontario, Canada	0.004	Goddard & Tielu (1988)
Ontario, Canada	0.044	Kinnucan & Belleza (1989)
Ontario, Canada	(0.044-0.060)	Venkateswaran & Kinnucan (1990)
England & Wales	0.023	Ball & McGee (1970)
United Kingdom	0.036	Strak & Gill (1983)
Cream:		
England & Wales	0.009	Bryant (1984)
England & Wales	0.006	Yau (1990)
United Kingdom	0.029	Strak & Gill (1983)
Cheese:		
New York City	(0.035-0.088)	Kinnucan & Fearon (1984)
New York City	0.059	Kinnucan (1986)
United Kingdom	(0.030-0.133)	Strak & Gill (1983)
Butter:		
Canada	0.010	Goddard & Amuah (1989)
Canada	0.023	Chang & Kinnucan (1990)
England & Wales	0.061	Homatenos (1982)
Beef:		
Australia	0.037	Ball & Dewbre (1989)
Pork:		
Australia	0.029	Ball & Dewbre (1989)
Lamb:		
Australia	0.010	Ball & Dewbre (1989)
Eggs:		
United Kingdom	0.010	Strak & Ness (1978)
Apples:		
Ontario, Canada	0.008	Goddard (1990)
Citrus:		
United States	0.240	Nerlove & Waugh (1961)
United States	0.027	Ward (1988)
Potatoes:		
United States	(0.054-0.071)	Jones & Ward (1989)

Table 7.3 contains advertising elasticity estimates from 37 studies. The large numbers of studies available on dairy and citrus advertising programs are due primarily to the fact that producers of these two commodities have been funding generic advertising for longer and at higher expenditure levels than have producers of other commodities. To some extent, at least for particular commodities, the variability in these estimates can be attributed to the studies being conducted for different time periods, different markets, and different media campaigns, as well as to the use of a variety of analytical methods. It appears that factors such as these are more important than commodity type in explaining inter-study variations. Advertising elasticities that are between 0 and 1 indicate that the advertising expenditure level is on the "decreasing rate of return" segment of the function in Figure 7.1. All of the elasticities reported in Table 7.3 are consistent with theory, and thus provide some evidence as to the rationality of advertising investments, when viewed from the producer's standpoint.

The relatively small magnitude of most of the estimated elasticities for fluid milk advertising may at first lead to a conclusion that this advertising is of limited effectiveness. However, the estimated elasticities are all positive and the effects, in terms of returns at the farm level, are magnified by the classified pricing of raw milk. Of the studies which have included a supply response component, most have shown quite modest supply responses to the demand-stimulating effects of advertising at the retail level. Since milk for fluid use is priced at a premium over that used for manufactured products, any increase in demand involves shifting milk from manufacturing uses to higher-valued fluid uses. With such a shift comes an increase in the weighted average price which dairy farmers receive for the raw product. Additionally, for fluid milk and other commodities, it should be noted here that investments in generic advertising are typically only on the order of one percent of gross sales value. Thus even a very small elasticity can be indicative of a substantial demand response, in absolute dollar terms.

Those funding generic advertising are at least once-removed from the retailing segments of most commodity markets, thus it is necessary to consider evidence in addition to retail-level advertising elasticities when evaluating generic commodity advertising programs. Estimated farm-level returns to advertising are useful in this regard as they provide an indication of what portion of the retail gains find their way back to the farm level.

As do the estimated advertising elasticities, estimates of farm-level returns to fluid milk and dairy product advertising exhibit a good deal of inter-study variation. A portion of Table 7.4 contains reported estimates of farm-level returns, a mixture of average and marginal returns, from several studies of fluid milk and dairy product advertising programs. Only one of these studies (Thompson and Eiler, 1975) obtained an estimate of less than a one-for-one marginal return to advertising expenditure, this for the Syracuse, New York

TABLE 7.4 Return on Investment in Generic Advertising, Selected Studies

Product/Market	Rate of Return[a]	Source
	($)	
Fluid Milk:		
New York City	1.11 A	Thompson & Elier (1975)
New York City	2.47 M	Thompson (1978a)
New York City	6.07 A	Kinnucan (1986)
New York City	1.40 A	Forker & Liu (1986)
New York City	1.50 A	Liu & Forker (1988)
Albany, NY	1.61 M	Thompson & Eiler (1975)
Syracuse, NY	0.40 M	Thompson & Eiler (1975)
Rochester, NY	1.47 A	Thompson (1979)
Buffalo, NY	17.00-22.00 M	Kinnucan (1983)
10 Fed. Orders	1.85 M	Ward & McDonald (1986)
United States[b]	7.04 A	Liu et al. (1989)
Ontario, Canada	8.00 M	Goddard & Tielu (1988)
Ontario, Canada	10.00-24.00 A	Venkateswaran & Kinnucan (1990)
Fluid Milk & Manufactured Dairy Products:		
United States[c]	4.77 A	Liu et al. (1989)
Fluid Milk & Cheese:		
New York City[d]	11.29 M	Kinnucan & Forker (1988)
Butter:		
Canada	1.11 A	Goddard & Amuah (1989)
Beef:		
12 Regions U.S.	5.25-12.72 A	Ward (1990)
Eggs:		
United Kingdom[e]	£10 A	Strak & Ness (1978)
Apples:		
Ontario, Canada	12.00 A	Goddard (1990)
Citrus:		
United States	10.44 A	Lee (1981)
United States	2.28 M	Lee & Fairchild (1988)
Europe	1.00-11.50 A	Lee et al. (1979)
Europe	5.51 A	Lee & Brown (1986)
Soybeans:		
Global	14.20 A	Williams (1985)
Wool:		
United States	1.94 A	Bureau of Agricultural Economics (1987)

[a]A = average return at actual expenditure level, M = marginal return at actual expenditure level.

[b]Return at actual expenditure level for fluid milk.

[c]Return at actual expenditure level for all products.

[d]Expenditure allocation of 60% for fluid milk, 40% for cheese.

[e]Average return of £10 per £1 invested in advertising.

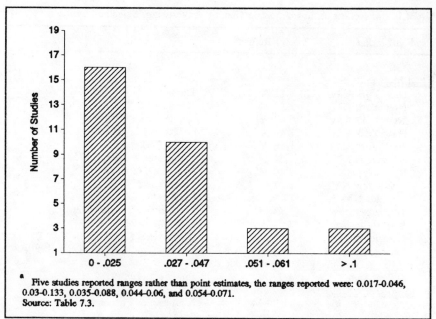

FIGURE 7.2. Distribution of Generic Advertising Elasticities, from 32 Studiesᵃ

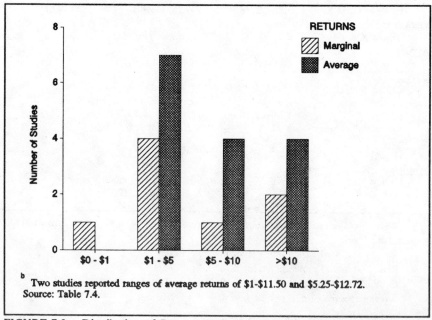

FIGURE 7.3. Distribution of Returns to Generic Advertising Investments, from 21
Studiesᵇ

market over the period 1971-74. More recent evidence for this same market has indicated a greatly changed situation, with fluid milk advertising being estimated as only at one-third its optimal level for the period 1984-87 (Liu and Forker).

Other than the Syracuse study, the other dairy-related studies reporting marginal returns all provide evidence that advertising expenditures fell short of the economic optimum for the periods under study; in some cases, the shortfall was substantial. Average returns alone do not provide an adequate basis for judging where expenditures are in relation to the optimum. The most that can be said is that, with the reported average returns all being greater than unity, the total returns to advertising exceeded the amounts expended on it. Although it does seem reasonable to conclude that where relatively large average returns were found the expenditure levels could have been profitably increased.

In general, our review of the literature pertaining to evaluation of fluid milk and dairy product advertising indicates that investments in such programs by dairy farmers have been beneficial. Though the differences in the many existing studies preclude any strict comparisons of results, the differences themselves do illustrate some interesting points. The most striking, perhaps, is the dynamic nature of fluid milk markets. This is probably best illustrated by considering the results obtained in the various studies of the New York City market, both in terms of advertising elasticities and farm-level returns to advertising.

With the non-dairy commodities a similar situation exists, in regard to elasticity estimates, as exists for the estimated elasticities for fluid milk and manufactured dairy products. There is still a good bit of variation in the estimates, though over a somewhat narrower range. The positive signs of all the reported elasticities indicate that in each instance advertising had a positive impact on consumer demand. The reported estimates of return on investment for the non-dairy commodities have quite a wide range. Again, the reported average returns are inadequate for making any judgement as to where the advertising expenditures lie in relation to their optimal level. Figures 7.2 and 7.3 provide a graphical summary of the number of studies reporting generic advertising elasticities and returns on generic advertising investments in various ranges.

Unanswered Questions

Evidence from a variety of studies supports the general conclusion that generic advertising does, or can, increase demand. The nature of the studies provides some insight into the shape of the advertising response function. In most studies the functions are nonlinear and increasing at a decreasing rate over the range of the observed expenditure levels. Thus with most programs operating in the area of decreasing rate of returns (Figure 7.1), the programs appear to be economically rational from the perspective of producers.

Yet there are many unanswered questions. Chief among these is "What is the shape of the entire sales response to advertising function?" The limited range of data available for empirical studies precludes estimation of the entire function. At best, researchers are left to extrapolate their estimated functions, within the confines imposed by the chosen functional form, in attempting to determine what the optimal level of advertising may be. In a related vein, a question exists as to whether or not there is some threshold level of advertising investment, below which it simply doesn't pay to advertise. If the function is indeed as postulated in Figure 7.1, then this threshold level is represented by the point where the function initially turns up. Answers to these two questions will be difficult to obtain in the absence of experiments wherein either distinct, but comparable, markets are subjected to varying levels of advertising, including levels outside the normal ranges, or a single market is subjected to such variations over different time periods.

Another important issue is "What is the most appropriate response to changing market conditions?" The dynamic nature of the many forces affecting both the supply of and demand for agricultural commodities necessitates ongoing studies of commodity markets. When market conditions change new program approaches may be needed. If commodity promotion monies are to be used in the most efficient and effective manner, the organizations expending them must continually monitor their markets to be in the best position to make appropriate program alterations.

In order to develop an appropriate response to the question "What is the farm-level return on the funds invested in retail-level advertising and promotion?" an understanding of the market forces and marketing functions operating between the two ends of the marketing channels must be developed. The varying nature of price transmission in different commodity sectors needs to be quantified so that retail-level responses to advertising can be translated into farm-level returns. The study by Liu *et al.* (1989) is based on a model that includes a retail to farm linkage, and takes an important first step along the path to a fuller understanding of these issues.

Given the proliferation of generic promotion programs in recent years, another important question is "How do the advertising and promotion activities of one group affect the activities of other groups?" The data requirements for such analyses are considerably more far-ranging than those for any single-commodity study. Chang and Green (1989) have done some preliminary work in this area. Their results indicate positive own-advertising elasticities for some commodity groups. The results on cross-advertising elasticities are inconclusive. Conceptually, it should be possible to determine, from the perspectives of various commodity promotion groups, an optimum level of generic advertising for each that would maximize overall producer welfare. However, in the absence of further studies in this direction researchers are unable to do much more than hazard speculative, though informed, guesses as to what the overall effects of the myriad of commodity promotion programs are on producer, consumer, and social welfare. Any definitive conclusion as to the social desirability, or lack thereof, of commodity promotion programs awaits further research on the cross-commodity effects of the many and varied programs in operation.

Concluding Comments

Based on our research, and our study of the results of other work in the area of generic commodity advertising and promotion, we are led to an admittedly tentative conclusion that much of the direct investment in the generic advertising of agricultural commodities has been profitable for the producers who fund the programs. Reported research results generally show positive, and statistically significant, advertising elasticities. When reported, returns have generally been estimated to be greater than advertising expenditures. However, a mixture of average and marginal return estimates have been reported. While the reported average returns typically show that total returns exceed total investments, the question of where expenditures are in relation to optimal advertising levels can only be adequately addressed by an estimate of marginal returns to advertising over the range of possible expenditure levels.

As the previous discussion of unanswered questions indicates, there are a variety of interesting and important issues related to commodity promotion which are deserving of further research effort. Definitive conclusions as to the overall efficacy and desirability of generic commodity promotion programs await the results of research cast in a much broader and all-encompassing framework.

Notes

1. In the checkoff mode of funding generic promotion, a small percentage is deducted from producers' checks for each unit of commodity they market commercially.

2. Commodity and product category will be used interchangeably in this paper to convey the idea that a commodity can be used to produce an array of products under a variety of brand labels.

3. Generic advertising can also be referred to as collective advertising. This refers to the fact that a group of firms pool their funds and collectively conduct an advertising effort. It is called generic because it is not brand or item specific.

4. The traditional definition of elasticity applies. That is, it is the percent change in sales associated with a one per cent change in advertising expenditures.

References

Albion, M. S. and P. W. Farris. 1981. *The Advertising Controversy: Evidence on the Economic Effects of Advertising*. Boston: Auburn House Publishing Company.

Ball, K. and J. Dewbre. 1989. An Analysis of the Returns to Generic Advertising of Beef, Lamb, and Pork. *Discussion Paper 89.4*. Australian Bureau of Ag. and Resource Econ. Canberra. 37 pp.

Ball, R. J. and J. McGee. 1970. An Econometric Analysis of the Demand for Milk. Economic Models Ltd. Unpublished.

Blaylock, J. R. and W. N. Blisard. 1988. Effects of Advertising on the Demand for Cheese. USDA, ERS, Technical Bulletin 1752. Washington DC. 33 pp.

Bryant, K. 1984. An Analysis of Cream Advertising. Staff Paper, Milk Marketing Board, U.K. Unpublished.

Bureau of Agricultural Economics. 1987. Returns from Wool Promotion in the United States. *Occasional Paper 100*. Australian Government Publishing Service. Canberra. 33 pp.

Chang, H. S., and R. Green. 1989. The Effects of Advertising on Food Demand Elasticities. *Can. J. Ag. Econ.* 37(3):481-94.

Chang, H. S. and H. W. Kinnucan. 1990. Advertising and Structural Change in the Demand for Butter in Canada. *Can. J. Ag. Econ.* 38(2):295-308.

Conner, J. M. and R. W. Ward, eds. 1983. *Advertising and the Food System*. N.C.Project 117, Mono 14. Research Div, Col. of Agr. and Life Sci., Univ. of Wisconsin-Madison.

Ekelund, R. B., Jr. and D. S. Saurman. 1988. *Advertising and the Market Process: A Modern Economic View*. San Francisco: Pacific Research Institute for Public Policy.

Forker, O. D. and D. J. Liu. 1986. An Empirical Evaluation of the Effectiveness of Generic Advertising: The Case of Fluid Milk in New York City. *A.E. Research 86-12*. Dept. of Agr. Econ., Cornell Univ., Ithaca, NY.

Frank, G. 1985. Generic Agricultural Promotion and Advertising: An Overview. In *Research on Effectiveness of Agricultural Commodity Promotion*, ed. W. Armbruster and L.H. Myers. Oak Brook, IL: Farm Foundation.

Goddard, E. W., and Tielu. 1988. Assessing the Effectiveness of Fluid Milk Advertising in Ontario. *Can. J. Ag. Econ.* 36(2):261-78.

Goddard, E. W., and A. K. Amuah. 1989. The Demand for Canadian Fats and Oils: A Case Study of Advertising Effectiveness. *Am. J. Ag. Econ.* 71(3):741-49.

Goddard, E. W. 1990. Demand for Fruit in Ontario: A Case Study of Apple Advertising Effectiveness. *Working Paper WP90/24*. Dept. of Ag. Econ. & Bus. U. of Guelph. Guelph, Ontario. 15 pp.

Homatenos, D. 1982. The Effectiveness of Advertising for Butter. Staff Paper, Milk Marketing Board, U.K. Unpublished.

Hurst, S., and O. D. Forker. 1991. Annotated Bibliography of Generic Commodity Promotion Research (Revised). *A.E. Research 91-7*. Dept. of Agr. Econ., Cornell University, Ithaca, NY.

Jones, E. and R. W. Ward. 1989. Effectiveness of Generic and Brand Advertising on Fresh and Processed Potato Products. *Agribusiness* 5(5):523-536.

Kinnucan, H. W. 1981. Performance of Shiller Lag Estimators: Some Additional Evidence. *A.E. Research 81-8*. Dept. of Agr. Econ., Cornell Univ., Ithaca, NY.

_____. 1983. Media Advertising Effects on Milk Demand: The Case of the Buffalo, New York Market. *A.E. Research 83-13*. Dept. of Agr. Econ., Cornell U., Ithaca, NY.

_____. 1986. Demographic Versus Media Advertising Effects on Milk Demand: The Case of the New York City Market. *Northeastern J. Ag. & Resource Econ.* 15(1):66-74.

_____. 1987. Effect of Canadian Advertising on Milk Demand: The Case of the Buffalo, New York Market. *Canadian Journal of Agricultural Economics* 35(1):181-196.

Kinnucan, H. W. and D. Fearon. 1986. Effects of Generic and Brand Advertising of Cheese in New York City with Implications for Allocation of Funds. *N. Central J. Ag. Econ.* 8(1):93-107.

Kinnucan, H. W. and O. D. Forker. 1988. Allocation of Generic Advertising Funds Among Products: A Sales Maximization Approach. *Northeastern J. Ag. & Resource Economics* 17(1):64-71.

Kinnucan, H. W. and E. Belleza. 1989. Measurement Error and Advertising Evaluation: The Case of Fluid Milk in Ontario. *Working Paper 89-5* Dept. of Agr. Econ. and Rur. Soc., Auburn Univ., Auburn, AL.

Lee, J. Y., L. H. Myers, and F. Forsee. 1979. Economic Effectiveness of the Brand Advertising Programs of Florida Orange Juice in European Markets. FL Dept. Citrus. *ERD Rep. 79-1*. Gainesville, FL. 46 pp.

Lee, J. Y. 1981. Generic Advertising, FOB Price Promotion, and FOB Revenues: A Case Study of the Florida Grapefruit Juice Industry. *South. J. Ag. Econ.* 13(2):69-78.

Lee, J. Y. and M. G. Brown. 1986. Economic Effectiveness of Brand Advertising Programs for United States Orange Juice in the European Market: An Error Components Analysis. *J. Ag. Econ.* 37(3):385-94.

Lee, J. Y. and G. F. Fairchild. 1988. Commodity Advertising, Imports, and the Free Rider Problem. *J. Food Dist. Res.* 19(2):36-42.

Lenz, J. E., O. D. Forker, and S. Hurst. 1991. U.S. Commodity Promotion Organizations: Objectives, Activities, and Evaluation Methods. *A.E. Research 91-4*. Dept. of Agr. Econ., Cornell Univ., Ithaca, NY.

Liu, D. J., and O. D. Forker. 1988. Generic Fluid Milk Advertising, Demand Expansion, and Supply Response: The Case of New York City. *Am. J. Ag. Econ.* 70(2):229-36.

_____. 1989. Optimal Fluid Milk Advertising in New York State: A Control Model. *A.E. Working Paper 89-4*. Dept. of Agric. Econ., Cornell Univ., Ithaca, NY.

Liu, D. J., H.M. Kaiser, O. D. Forker, and T. D. Mount. 1989. The Economic Implications of the U.S. Generic Dairy Advertising Program: An Industry Model Approach. *A.E. Research 89-22*. Dept. of Agric. Econ., Cornell Univ., Ithaca, NY.

National Dairy Promotion & Research Board. 1986. In Report to Congress on the Dairy Promotion Program. USDA, Washington, DC.

_____. 1987. In Report to Congress on the Dairy Promotion Program. USDA, Washington, DC.

Nerlove, M. and F. V. Waugh. 1961. Advertising Without Supply Control: Some Implications of a Study of the Advertising of Oranges. *J. Farm Econ.* 43:813-37.

Sheth, J. N., ed. 1975. *Models of Buyer Behavior: Conceptual, Quantitative, and Empirical.* New York: Harper & Row.

Strak, J. 1983. Optimal Advertising Decisions for Farmers and Food Processors. *J. of Agri. Econ.* 34(3):303-315.

Strak, J. and M. Ness. 1978. A Study of the Generic Advertising in the U.K. Egg Industry, 1971-1976. *Bulletin 165/EC69.* Dept. of Ag. Econ. U. Manchester, England.

Strak, J. and L. Gill. 1983. An Economic and Statistical Analysis of Advertising in the Market for Milk and Dairy Products in the UK. *J. Ag. Econ.* 34(September).

Thompson, S. R. 1974. Sales Response to Generic Promotion Efforts and Some Implication of Milk Advertising on Economic Surplus. *J. Northeastern Ag. Econ. Council* 3(2):78-90.

Thompson, S. and D. A. Eiler. 1975. Producer Returns from Increased Milk Advertising. *Am. J. Ag. Econ.* 57(3):505-8.

Thompson, S. 1978a. The Response of Milk Sales to Generic Advertising and Producer Returns in the New York City Market Revisited. *A.E. Staff Paper 78-8.* Dept. of Agr. Econ., Cornell Univ., Ithaca, NY.

_____. 1978b. An Analysis of the Effectiveness of Generic Fluid Milk Advertising Investment in New York State. *A.E. Research 78-17.* Dept. of Agr. Econ., Cornell Univ., Ithaca, NY.

_____. 1979. The Response of Milk Sales to Generic Advertising and Producer Returns in the Rochester, New York Market. *A.E. Staff Paper 79-26.* Dept. of Agr. Econ., Cornell Univ., Ithaca, NY.

Venkateswaran, M. and H. W. Kinnucan. 1990. Evaluating Fluid Milk Advertising in Ontario: The Importance of Functional Form. *Can. J. Ag. Econ.* 38(2):471-488.

Ward, R. W., and W. F. McDonald. 1986. Effectiveness of Generic Milk Advertising: A Ten Region Study. *Agribusiness* 2(1):77-89.

Ward, R. W. 1990. Economic Impact of the Beef Checkoff Programs. Unpublished, Univ. of Florida, Gainesville, FL. 39 pp.

Warman, M. and M. Stief. 1990. Evaluation of Fluid Milk Advertising. Market Research Branch, CSSD, ASMS, USDA, Washington, DC. 25 pp.

Williams, G. W. 1985. Returns to U.S. Soybean Export Market Development. *Agribusiness* 1(3):243-263.

Wittink, D. R. 1977 Advertising Increases Sensitivity to Price. *J. Advertising Res.* 17(2):39-42.

Yau, C. 1990. A Quantitative Analysis of Household Consumption of Cream. Staff Paper, Milk Marketing Board, U.K. Unpublished.

8

Branded Product Marketing Strategies in the Cottage Cheese Market: Cooperative Versus Proprietary Firms

Lawrence E. Haller

Introduction

Marketing cooperatives have several strategies available to maximize their farmer-members' returns. The most basic is to organize horizontally into farmgate level commodity supply and bargaining associations. Cooperatives can integrate forward into food processing, producing intermediate inputs (e.g. butter powder) or private label products (e.g. store brand cottage cheese). Cooperatives can also develop and market their own differentiated brands. This paper examines the behavior of cooperatives that market their own brands of a single product (cottage cheese) and compares their pricing and marketing strategies with those of investor-owned firms (IOFs).

Cooperative theory developed by Helmberger (1964), Cotterill (1987) and others suggests that a cooperative will behave differently when facing the same market conditions as an IOF. Under some conditions cooperatives may price lower than IOFs in branded product markets, while under other conditions they may lead the market towards higher prices. In late 1988, the chairman of the FTC claimed that cooperatives "do business just like other large food companies and should be subject to the same statutory obligations as their competitors" (Food Institute Report (1988), p. 10). How cooperatives actually behave is an important question. In a study of the competitive impacts of cooperatives, Petraglia and Rogers (1991) found that the percentage of a market's shipments held by the largest cooperatives was negatively related to the market's price-cost margin, especially in concentrated markets. Wills (1985) has also examined this question using national data from 1979 and 1980 on 145 products, with about half of the categories containing at least one brand marketed by a co-op. Wills' results indicate that cooperatives tend to price lower than IOFs, *ceteris paribus*.

This paper extends Wills' work by examining cooperatives' influence on the price of a specific product at the local market level. The cottage cheese industry is an attractive choice for analysis because there are many local and regional brands marketed both by cooperatives and by IOFs, as well as several national brands. In addition, cottage cheese has no close substitutes and minimal quality differences from brand to brand[1].

The following section describes the data used in the analysis and examines the descriptive statistics for several subgroups of brands in some detail. The third section presents the results of a regression analysis of structural and strategic factors influencing average local market price at the brand level. The fourth section draws conclusions from the foregoing analysis, and the appendix contains several descriptive tables.

The Data

The data used in this study cover cottage cheese sales in 47 markets for the fourth quarter of 1988 and were obtained from Information Resources, Inc. (IRI). The markets represented are listed in appendix Table 8.A.1. Not all brands and manufacturers of cottage cheese are included in the IRI market-level data base because some are sold in areas not included in IRI's 47 markets where the data are collected. The data include all observations provided by IRI with a local market share of at least 0.5% and a price no greater than $2.00 per pound[2]. There are 104 brands marketed by 74 manufacturers included in this study[3]. On average, a brand is sold in 3.125 of the 47 markets. The most widely distributed brand (Breakstone) is found in 30 of the markets, yet there are 63 brands found in only a single market. Each manufacturer produces an average of 1.4 brands. Two manufacturers market 5 brands each while the majority sell only 1. All but one of the co-ops represented in the study market a single brand, and one sells two brands. On average, a manufacturer sells its brands in 3.03 of the 47 markets. The most widely distributed manufacturer can be found in 37 of the markets, the second most widely distributed in 22, while there are 44 that can be found in one each. Complete descriptive statistics for the brand- and manufacturer-level data are given in appendix Tables 8.A.4 and 8.A.5.

The following section examines the descriptive statistics more closely. Data on brands are examined in groups: those brands ranked nationally[4] in the top 4 group, those ranked from 5 to 10, those ranked from 11 to 20, those ranked 21 to 50, those ranked 51 to 100, and those ranked greater than 100. Co-ops are also considered as a group. Additionally, data on private label sales are included as a group. Care is needed when interpreting the private label data, however. Each market contains a single observation for private label sales, representing all store-brand sales for that market.

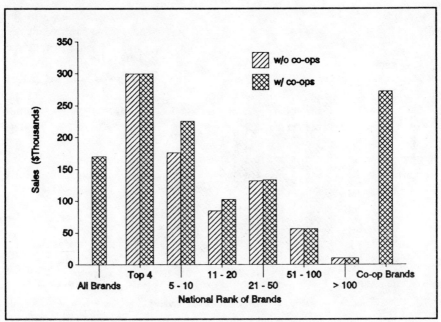

FIGURE 8.1 Average Local Sales, Cottage Cheese, 4th Quarter, 1988

Local sales for all brands averaged $169,900 (Figure 8.1). Sales per market follow the expected trend of declining as the national rank declines, with the exception of a smaller than expected average local sales figure for the group 11 - 20. The top four nationally ranked brands (Top 4) had average sales of $299,360 in each of the markets where they were sold. The six brands ranked 5 - 10 sold an average of $225,200 per market when the single co-op brand in the group is included, and an average of $176,310 when it is not, indicating that the co-op's average local sales is significantly higher. As mentioned, there is a dip in average local sales for brands ranked 11 - 20. Two of the ten brands in this group are "lite" varieties (Weight Watchers and Lite Line) with broad distribution but relatively small local market shares. Without these brands, the average local sales increases to $104,900 excluding co-op brands, or $155,130 when co-op brands are included in the group. Groups 21 - 50 and 51 - 100 show almost no variation in average sales when co-ops are excluded. The group of brands ranked greater than 100 contains no co-op brands. Taken as a group, brands marketed by cooperatives had average sales of $271,690 per market, higher than any group of brands except the Top 4. There are fewer co-op brands in the higher nationally ranked groups because co-op brands are sold in fewer markets than investor-owned brands. Co-op brands average 2.1 markets per brand, compared to 19.3 markets per brand for the Top 4 brands and 3.1

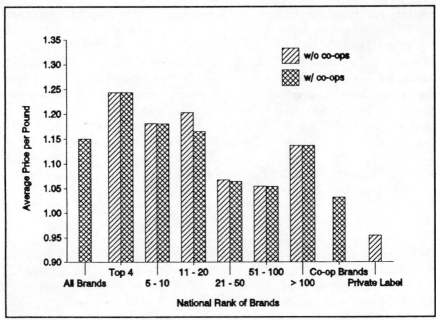

FIGURE 8.2   Average Local Price, Cottage Cheese, 4th Quarter 1988

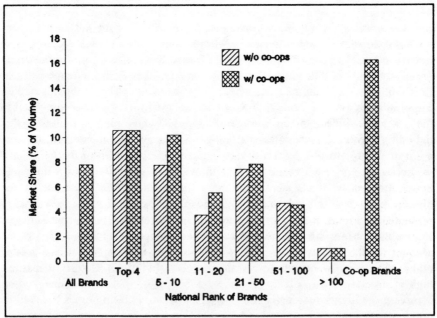

FIGURE 8.3   Average Local Market Share, Cottage Cheese, 4th Quarter 1988

markets per brand for all non co-op brands. Private label sales averaged $689,720 per market.

The (nonweighted) average local price for all brands is $1.15 per pound (Figure 8.2). Average price also tends to decline as the national rank declines. There is a spike for non-co-op brands ranked 11 - 20, and also a higher price for brands ranked greater than 100. Perhaps some of the brands in the latter group are specialty or flavored brands, filling a relatively small niche and commanding a premium price. Co-ops taken together are significantly lower priced than IOF brands with an average price of about $1.03 per pound. Private label brands on the average are even lower priced than the co-op brands with an average price of just over $.95 per pound.

The local market share by volume for all brands averages 7.8% (Figure 8.3). This chart is similar to the chart presenting average local sales, except that in the market share chart there is less spread between the groups. This is because the higher market share brands also have higher prices, accentuating the differences in local volume sales. Co-ops as a group again show a marked difference in performance. Their average market share, 16.3%, is more than 1.5 times higher than the next closest group, the Top 4, whose average share is 10.6%. Co-ops are clearly the branded product volume leaders in the markets in which they are sold. While accounting for less than 9% of the brand-level observations in the data set, co-ops hold the number one volume position in over 21% of the markets (10 out of 47). Private label cottage cheese accounts for 44.6% of sales in the markets included here. It should be remembered that this represents the shares of several store brands, not a single retailer's product.

The largest selling national brands are also the most widely distributed brands in local markets (Figure 8.4). Consumer acceptance of these brands is obviously high. The percentage of stores the brand is carried in within a market falls with a brand's national ranking. On the whole, a given brand could be found in stores that account for about half the total grocery sales[5] in a market (51.5%). Co-ops' brands' mean distribution is below the average for all brands; only 44.7% of retailers in a typical market sold a given co-op brand. Almost 4 out of 5 (79%) stores in the markets in this study offer a private label brand.

Next we look at a brand's market share in only those stores selling that brand, rather than the market-level share (Figure 8.5). This measure is not directly available, but can be computed by dividing the market-level share presented in Figure 8.3 by the average market-level distribution presented in Figure 8.4. Here we find that the brands ranked highest nationally do not account for the largest shares; the distribution of in-store share varies across ranks. The groups with the lowest unit price (21 - 50 and 51 - 100) have the highest store-level shares. Co-op brands do appear to outsell their brand-name competitors in the stores in which they are carried, indicating that consumers prefer co-op brands when they are available, perhaps because of their lower price. This chart presents a good opportunity to make a comparison between

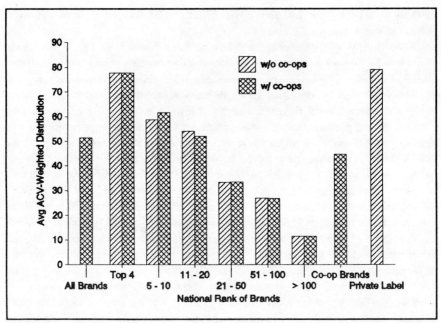

FIGURE 8.4   Percent of Stores Carrying Brand, Cottage Cheese, 4th Quarter 1988

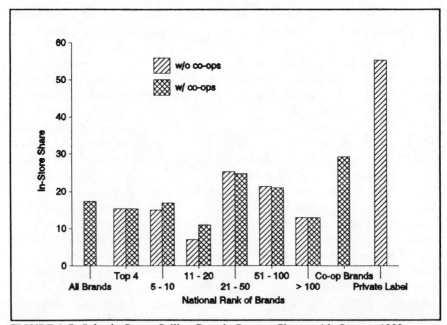

FIGURE 8.5   Sales in Stores Selling Brand, Cottage Cheese, 4th Quarter 1988

sales of private label and of brands on more equal footing, since this is a per store share figure. Private label continues to outsell its branded competition, but the gap is not as great as it appears in the market-level figures.

Most brands make use of some form of merchandising[6], selling 15% to 20% of volume with the aid of merchandising tools (Figure 8.6). The high level of merchandising (25%) for the group ranked 5 - 10 appears to be an exception—these numbers are driven by the merchandising of two of the brands (Sealtest and Borden). Without these, the group would fall within the range of the other groups. Co-ops as a group are slightly more aggressive in the use of merchandising, and stores promoting their own brands use the most merchandising. Stores promoting private label brands are the leaders in the use of feature (newspaper) advertising. Co-ops and private label brands used price promotions most heavily, followed closely by the 5 - 10 group. In-store displays were used most extensively by the group 5 - 10, followed by the group of brands ranked greater than 100. This reinforces the speculation that this group contains niche or specialty brands, since in-store promotion is often used to entice consumers to try out-of-the-ordinary products.

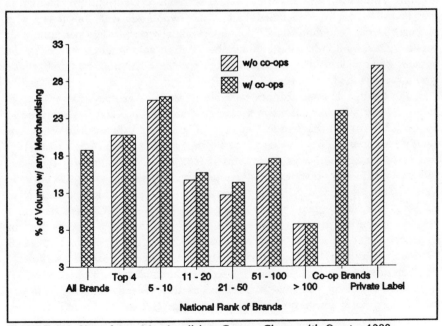

FIGURE 8.6 Use of Any Merchandising, Cottage Cheese, 4th Quarter 1988

Regression Results

Average price per pound at the brand level was modelled as a function of several structural and firm-specific variables using OLS regression techniques. The unit of observation is a brand within a market. There were 595 observations in the original data set. After eliminating observations with a local market share less than 0.5% and those with a price above $2.00 per pound, as explained above, and the 47 private label observations, 325 valid observations remain. Complete descriptive statistics can be found in appendix Table 8.A.4. **Volume Market Share** is the percentage of cottage cheese sold in a particular market by a given brand, calculated on a volume basis (pounds of cottage cheese), rather than on a sales dollar basis. The Volume Market Share can be broken into two components: Average Weighted Distribution and In-Store Share. The **Average Weighted Distribution** measures how widely distributed within a market area a brand is. Wide distribution may be a sign of consumer acceptance and so allow a brand to command a higher price, and therefore is expected to be positively related to average price. **In-Store Share** is the share of sales a brand enjoys in only the stores carrying that brand, rather than in all stores in a market[7]. In-Store Share is expected to be negatively related to average price, since to sell more within a store, *ceteris paribus*, price must be lowered. In-Store Share is not directly available, but can be calculated by dividing Volume Market Share by Average Weighted Distribution. The **Units per Pound** variable is included to test the hypothesis that consumers who purchase larger "economy sized" containers actually do pay less per pound. It is constructed by dividing the total number of units sold within a market by the number of pounds sold[8]. A positive sign for the coefficient of this variable would support the hypothesis.

The **Co-ops Present Binary** variable is set on a market-wide basis. Its value is 1 for all observations in a market if any co-op brand has achieved at least a 0.5% market share in that market. The **Co-op Volume Market Share** variable contains the same value as the Volume Market Share if the observation is a co-op brand, and a value of zero if it is not a co-op brand. If co-ops behave no differently than their IOF competitors, the coefficient of this variable should not be significantly different from zero. If co-ops are less apt to exercise market power, this variable should be negative. The **Market CR₄** is a structural variable measuring the sales of the top 4 chains in a market as a percentage of all grocery sales. If a higher concentration ratio is indicative of conditions that facilitate collusion by retailers, this variable should have a positive coefficient. The **Population** of the market area is used to examine the effect of market size. A positive coefficient on this variable means that consumers living in larger markets pay more; a negative coefficient would indicate that there are increasing economies of scale in the range of the size of markets in this sample. The **Percent of Volume with any Merchandising** looks at the price effect of promotional activities. The **Private Label Price per Pound** can be looked at as a floor price. A higher private label price allows manufacturers to sell their brands at a higher price. If private label price is a "competitive" price, reflecting the costs of selling the product in a market (Connor and Peterson 1992), then this variable is a proxy for differential costs across markets. A simple correlation matrix is presented in appendix Table 8.A.6.

Regression results are summarized in Table 8.1. Equation 1 models the basic price-market share relationship, using **Volume Market Share** and **Units per Pound**. It finds a positive relationship between price and share that is significant at the 5% level. Units per Pound is positive and highly significant, indicating that larger sizes are indeed less expensive on a per volume basis. Equation 2 adds the **Co-ops Present** binary variable and the **Co-op Volume Market Share**. They are both negative and significant at the 1% level, and their addition improves the significance of the Market Share variable to the 1% level. The coefficient of the Co-op Volume Market Share is as large as the coefficient on the Volume Market Share and combining the two share effects shows that co-ops exercise little or no share-based market power. Equation 3 replaces the Volume Market Share with the **Average Weighted Distribution** and **In-Store Share**[9]. Their inclusion improves the explanatory power of the model greatly, increasing the R^2 by 50 percent. The coefficients of both variables are highly significant and of the hypothesized signs. The Co-ops Present binary variable continues to be negative and significant at the 1% level.

Equation 4 introduces the **Market CR₄** and the **Population** structural variables to the model in Equation 2, and Equation 5 introduces them to the model in Equation 3. The market share variables retain the same signs and levels of significance they held in Equations 2 and 3. The Co-ops Present and Co-op Volume Market Share variables continue to be negative in Equations 4 and 5 at the same significance level they held in Equations 2 and 3. Market CR₄ is positive and significant at the 10% level in Equation 4 and positive but not significant in Equation 5. Taken together, this provides only weak support for the hypothesis that consumers pay more for cottage cheese in more concentrated markets. The Population variable is positive and significant at the 1% level. This suggests that consumers in larger market areas face higher prices, a result worthy of additional research.

Equations 6 and 7 introduce the **Percent Volume with any Merchandising** (PctVolMerch) and **Private Label Price per Pound** (PLPrice) variables into Equations 4 and 5. PctVolMerch is significant at the 1% level and negative, as expected. The coefficient of PLPrice is positive and significant at the 1% level, indicating that brand name prices rise as the price of store brands goes up. Population is highly correlated with Private Label Price per Pound ($r = 0.43$), and its coefficient falls to insignificance in these equations.

Table 8.2 presents the effects of changes in the explanatory variables on the price of cottage cheese. The effects caused by co-ops are the most notable.

TABLE 8.1 Regression Results for the Cottage Cheese Market, 4th Quarter 1988

Dependent Variable is Average Price per Pound

Eq.	Volume Market Share	Average Wtd. Distrib.	In-store Share	Units per Pound	Co-ops Present Binary	Co-op Volume Market Share	Retail Market CR$_4$	Pop.	Percent Volume w/ any Merch.	Private Label Price/ Pound	Constant	R^2	Adj. R^2
1	0.00208 (2.080)			0.605 (10.120)							0.599 (11.055)	0.2508	0.2461
2	0.00480 (3.832)			0.595 (10.303)	-0.0676 (3.436)	-0.00511 (3.035)					0.626 (11.667)	0.3078	0.2991
3		0.00233 (9.527)	-0.00167 (3.082)	0.454 (8.533)	-0.0737 (4.348)						0.693 (13.521)	0.4630	0.4563
4	0.00438 (3.500)			0.553 (9.351)	-0.0602 (3.055)	-0.00466 (2.781)	0.00127 (1.701)	9.94 E-9 (2.630)			0.557 (8.467)	0.3252	0.3126
5		0.00226 (9.380)	-0.00195 (3.610)	0.407 (7.527)	-0.0632 (3.735)		0.000818 (1.248)	11.9 E-9 (3.606)			0.654 (10.766)	0.4844	0.4747
6	0.00381 (3.275)			0.495 (9.029)	-0.0361 (1.965)	-0.00365 (2.353)	0.000974 (1.414)	0.790 E-9 (0.210)	-0.00236 (4.675)	0.590 (6.458)	0.132 (1.366)	0.4348	0.4205
7		0.00235 (10.708)	-0.00170 (3.535)	0.350 (7.236)	-0.0384 (2.511)		0.000548 (0.939)	4.22 E-9 (1.310)	-0.00305 (7.082)	0.486 (6.274)	0.321 (3.760)	0.5960	0.5858

Note: Significance levels for a two-tailed test are: $t > 1.645$, 10%, $t > 1.960$, 5%, $t > 2.576$, 1%, (t statistics in parentheses).

TABLE 8.2 Change in Price Resulting From a Change in Explanatory Variable

Explanitory Variable	Magnitude of Change in Explanatory Variable	Resulting Change in Price
Volume Market Share[a]	10 percentage points	$0.044
Average Weighted Distribution[b]	10 percentage points	0.024
In-Store Share[b]	10 percentage points	-0.017
Units per Pound[a]	0.083[c]	0.046
Co-ops Present Binary[a]	1	-0.060
Co-op Volume Market Share[a]	10 percentage points	-0.047
Market CR₄[a]	10 percentage points	0.013
Population[a]	1,336,450[c]	0.013
Percent Volume with any Merchandising[b]	8.79%[c]	-0.027
Private Label Price per Pound[b]	$0.055[c]	0.027

[a] Calculated from Equation 4, Table 8.1.
[b] Calculated from Equation 7, Table 8.1.
[c] Represents one half of one standard deviation.

Co-ops exercise little or no share-based market power. Calculating the effect of a 10% change in a co-op-owned brand's market share, we see that the price actually declines about 0.3 cents (a 4.4 cent rise from Volume Market Share and a 4.7 cent decline from the Co-op Volume Market Share variable). Additionally, the prices of all brands in a market are affected by the presence of a co-op brand. Brands in markets with at least one co-op are 6 cents cheaper than similarly positioned brands where no co-ops compete.

Conclusions

We find evidence supporting the premise that market power is being exercised in the cottage cheese market. Price rises with an increase in market share. We also find weak evidence that price rises as the retail market-level four firm grocery concentration ratio (Market CR₄) rises, indicating that retailers may raise prices in more concentrated markets. We find that one of the strongest influences of the price of cottage cheese is the extent of market penetration of the brand. Brands with a higher average weighted distribution have higher prices. This may be because these brands have won high consumer acceptance and so are able to charge more, indicating the exercise of market power on the manufacturer level. Brands' prices rise with an increase in the price of private label cottage cheese.

We also find strong evidence that cooperatives are not exercising market power or "unduly enhancing price." Cooperatives charge a lower price than their IOF competition under the same conditions and, unlike their IOF competition, co-ops do not capitalize on higher share to raise price. Moreover, the presence of cooperatives in a market brings the price of competing brands down. Cooperatives do not differ greatly from IOFs in the use of merchandising tools, using them as much or slightly more in most cases. If cooperatives wish to increase sales of their brands, gaining wider distribution in their current markets should allow them to boost their sales without requiring a price reduction.

Notes

1. *Consumer Reports* (1986) tested 24 national and store brands of cottage cheese for taste and quality. Brands were sampled in each of three locations across the country (California, Texas, and New York), yielding a total of 54 observations. Of these, 50 were ranked as "excellent" or "very good", and 3 as "good". Only one brand was ranked "fair".

2. Observations with a volume share less than 0.5% were dropped at the suggestion of IRI. Observations with a price greater than $2.00 per pound were dropped because these were suspected of having errors in data collection or misclassification (e.g. ricotta cheese classified as cottage cheese). Two dollars per pound is more than 4 standard deviations from the mean price in the study.

3. A complete list of all brands used in this study and their national ranks can be found in appendix Table 8.A.2. Table 8.A.2 also lists the total number of markets the brand was found in, whether it is marketed by a cooperative, and a summary of the market positions held. Table 8.A.3 lists similar information at the manufacturer level.

4. In addition to market-level data, IRI also provides information on total U.S. sales. The total U.S. figures were used to assign national rankings. Several regional brands were assigned ranks based on their national sales, but were not sold in any of the 47 markets included in the data set and do not appear in Tables 8.A.2 and 8.A.3. Thus, the 104 brands included in the study are ranked from 1 to 122.

5. Total grocery sales are also referred to as "All Commodity Volume", or ACV. The average distribution of a brand is weighted by the sales volume (ACV) of the stores carrying that brand. As an example, assume there are three stores serving a market. Store A sells half of all food purchased in the market, while store B sells 30% and store C sells the remaining 20%. If a brand of cottage cheese is carried only by store B, its average weighted distribution is 30%. If it is carried by stores A and B, its average weighted distribution is 80%.

6. Merchandising as used here is the use of feature (newspaper) advertising, price discounts, and in-store promotional displays. It does not include the use of coupons, TV or national magazine advertising.

7. An example may be helpful. Let us assume equal-sized stores to avoid complicating the example. A brand which is carried in all stores in a market and has a 10% share of cottage cheese sales in each of those stores has a 10% market share (100%

distribution * 10% in-store share). A brand which is carried in half the stores in a market but has a 20% share of cottage cheese sales in each of those stores also has a 10% market share (50% * 20%). Claiming a 10% share for each may mask important competitive differences between the two brands.

8. Thus, this variable would have a value of 2 for a brand sold exclusively in 8 oz. containers, and a value of 0.5 for a brand sold exclusively in the larger two-pound economy size.

9. The variable Co-op Volume Market Share is not included in this and subsequent equations which include Average Weighted Distribution and In-Store Share. Its interpretation is unclear in these equations. As an alternative, Co-op Average Weighted Distribution and Co-op In-Store Share were tried, but proved unsatisfactory.

References

Connor, J. M. and E. B. Peterson. 1992. Market-Structure Determinants of National Brand-Private Label Price Differences of Manufactured Food Products, *Journal of Industrial Economics*, forthcoming.

Consumer Reports. 1986. Cottage Cheese. 51(3).

Cotterill, R. W. 1987. Agricultural Cooperatives: A Unified Theory of Pricing, Finance, and Investment. in *Cooperative Theory: New Approaches*, ACS Service Report Number 18, U.S. Department of Agriculture, Washington, D.C.

Food Institute Report. 1988. American Institute of Food Distribution, Fair Lawn, N.J., December 3, 1988.

Helmberger, P. G. 1964. Cooperative Enterprise as a Structural Dimension of Farm Markets. *Journal of Farm Economics*. 46(3): 603-17.

Petraglia, L. M. and R. T. Rogers. 1991. The Impact of Agricultural Marketing Cooperatives on Market Performance in U.S. Food Manufacturing Industries for 1982, NE-165 Research Report No. 12, University of Connecticut.

Wills, R. L. 1985. Evaluating Price Enhancement by Processing Cooperatives. *American Journal of Agricultural Economics*. 67(2): 183-92.

Appendix

TABLE 8.A.1 IRI Geographic Markets

Los Angeles	Oklahoma City
New York	Sacramento
Chicago	San Diego
Memphis	Portland, Or.
Houston	Salt Lake City
Pittsburgh	Phoenix / Tucson
Seattle / Tacoma	Miami / Ft. Lauderdale
Detroit	Nashville
Cleveland	Raleigh / Greensboro
St. Louis	Albany
Dallas / Ft. Worth	Baltimore / Washington
Kansas City	Milwaukee
Birmingham	New Orleans / Mobile
Boston	Buffalo / Rochester
San Francisco / Oakland	Hartford / Springfield
Tampa / St. Petersburg	Jacksonville
Minneapolis / St. Paul	Louisville
Denver	San Antonio
Philadelphia	Columbus
Atlanta	Omaha
Providence	Grand Rapids
Cincinnati / Dayton	Little Rock
Indianapolis	
Wichita	
Orlando	

TABLE 8.A.2 Listing of Brands Showing Frequencies of Share Rankings, 4th Quarter, 1988

Rank	Co-op	Brand	Manufacturer	#Mkts	#1	#2	#3	#4	>4
1		Light N Lively	Philip Morris Co. Inc	29	12	7	5	2	3
2		Knudsen	Philip Morris Co. Inc	5	3	1	1	0	0
3		Breakstone	Philip Morris Co. Inc	30	0	6	9	5	10
4		Kemps Slim Trim	Quality Chekd Dry Prdts Assn	13	1	2	3	2	5
5	C	Hood	Agway, Inc	5	4	0	1	0	0
6		Sealtest	Philip Morris Co. Inc	27	0	5	10	3	9
7		Borden	Borden, Inc	18	3	4	1	6	4
8		Friendly Farmer	Friendship Food Products Inc	5	1	1	1	1	1
9		Deans	Dean Foods Co	7	2	2	1	0	2
10		Borden Viva	Borden Inc	5	2	2	0	0	3
11	C	Darigold	Darigold, Inc	2	2	0	0	0	0
12		Crowley	Crowley Foods, Inc	6	0	1	0	3	2
13		Anderson Erickson	Anderson Erickson Dairy Co	2	0	0	0	0	2
14		Weight Watchers	H J Heinz Co	21	0	2	4	3	12
15	C	Cream O Weber	Intermountain Milk Producers	2	1	0	0	1	0
16	C	Prairie Farms	Prairie Farms Dairy, Inc	4	1	0	1	0	2
17		Nordica	Nordica International, Inc	2	1	0	0	0	1
18		Michigan	General Mills	6	2	0	1	1	2
19		Axelrod	Crowley Foods, Inc	5	0	0	0	2	3
20		Lite Line	Borden Inc	12	0	3	0	1	8
21		Meadow Gold	Borden Inc	6	0	0	2	0	4
22		Carnation	Nestle Co	3	0	2	1	0	0
23		Old Home	Old Home Foods, Inc	1	1	0	0	0	0
24		Shamrock	Shamrock Foods Co	1	1	0	0	0	0
25		Crystal	Crystal Cream & Butter Co	2	1	0	0	1	0
26		Quality Chekd	Quality Chekd Dry Prdts Assn	4	0	0	1	1	3

(continues)

TABLE 8.A.2 (continued)

Rank	Co-op	Brand	Manufacturer	#Mkts	#1	#2	#3	#4	>4
27		Friendship	Friendship Food Products Inc	4	0	0	0	0	4
28	C	Bison	Bison Foods Co	1	1	0	0	0	0
29		Driggs Farms	Driggs Dairy Farms, Inc	1	0	0	0	0	1
30	C	Golden Guernsey	Golden Guernsey Dairy	1	1	0	0	0	0
31		Pet	Whitman Corp	2	0	0	0	0	2
32		Hillside	Hillside Old Meadow Dairy	4	0	0	0	1	3
33		Berkeley Farms	Berkeley Farms, Inc	1	0	1	0	0	0
34		Purity	Philip Morris Co. Inc	1	1	0	0	0	0
35		Farm Fresh	Farm Fresh Dairy, Inc	2	1	1	0	0	0
36		Reiter	Reiter Foods, Inc	3	0	1	0	0	2
37		Vitamilk Dairy	Vitamilk Dairy, Inc	1	0	1	0	0	0
38		Hoosier	East Side Jersey Dairy, Inc	1	1	0	0	0	0
39		Louis Trauth	Louis Trauth Dairy, Inc	1	1	0	0	0	0
40		Kemps Lite	Marigold Foods, Inc	3	0	0	0	0	3
41		Gilt Edge Farms	Gilt Edge Farms, Inc	2	0	0	1	1	0
42	C	Roberts	Roberts Dairy Co	2	1	0	0	1	0
43		Pevely	Pevely Dairy Co	1	0	1	0	0	0
44	C	Swiss Valley Farms	Swiss Valley Farms Co	1	0	0	0	0	1
45		Bestever	East Side Jersey Dairy, Inc	1	0	0	1	0	0
46		Alta Dena	Bongrain S A	2	0	0	0	0	2
47	C	Cabot	Cabot Farmers' Coop Creamery	3	0	0	0	0	3
48	C	Flav O Rich	Flav-O-Rich, Inc	3	0	1	0	1	1
49		Country Side	Finevest Foods, Inc	1	0	0	0	0	1
50	C	Zarda	Zarda Brothers Dairy, Inc	2	0	1	0	0	1
51		Fairmont	Utotem, Inc	2	1	0	0	0	1
52		Sani Dairy	Johnstown Sanitary Dairy	1	0	0	0	0	1
53		Melody Farms St Fr	Melody Farms Dairy Co	1	0	0	0	1	0
54		Dellwood	Arabian Investment Banking	1	0	0	0	0	1

No.	C	Brand	Company					
55		Steffen	Steffen Dairy Foods Co, Inc	1	0	0	0	0
56		Slendrella	Old Home Foods, Inc	1	0	0	0	1
57		Fieldcrest	Dean Foods Co	2	0	0	1	1
58		Dairy Fresh Superior Dairy	Fresh Milk Co	1	0	0	1	0
59		Barber	Barber Dairy, Inc	2	0	0	0	1
60		Baremans	Bareman's Dairy, Inc	1	0	0	0	0
61		Bluebunny	Wells Dairy, Inc	1	1	0	0	0
62		Bay View Farms	Quality Chekd Dry Prdts Assn	1	0	0	1	0
63		Kemps	Marigold Foods, Inc	1	0	0	0	0
64		Farmcrest	Pevely Dairy Co	1	0	0	1	1
65		Londons	London's Farm Dairy, Inc	1	0	0	0	1
66		Schroeder	Schroeder Milk Co, Inc	1	0	0	0	0
67	C	Land O Lakes	Land O Lakes, Inc	1	0	0	0	1
68		Pevely Delightfully Light	Pevely Dairy Co	1	0	0	1	1
69		Fikes	R Bruce Fike & Sons Dairy	1	0	0	0	0
70		Jersey Farms	Pevely Dairy Co	1	0	0	1	1
73		Quality Chekd	McDonald Coop Dairy Co	1	1	0	0	0
74		Quality Chekd	Coleman Dairy, Inc	1	0	0	1	1
75		Trauth Dairy	Louis Trauth Dairy, Inc	1	0	0	0	0
77		Swiss	Swiss Dairy	1	0	0	1	1
78	C	Shenandoah Pride	Valley Of Virginia Cooperative	1	0	0	0	0
79		Country Girl	Farm Fresh Dairy, Inc	2	0	1	0	1
80		Yoders	Delphos Frozen Specialties	1	0	0	0	1
81		Great Scott	Borden, Inc	1	0	0	0	1
82		Melody Farms	Melody Farms Dairy Co	1	0	0	0	1
83		Alpen Rose	Alpenrose Dairy, Inc	1	0	0	0	1
84		McColls Trim	All Star Dairy Assn, Inc	1	0	1	0	0
85		McColls	All Star Dairy Assn, Inc	1	0	0	0	1
86		Smiths	Smith Dairy Products Co	2	0	0	0	2
87	C	Norris	Norris Creameries, Inc	1	0	0	0	1

(continues)

172
172

TABLE 8.A.2 (continued)

Rank Co-op	Brand	Manufacturer	#Mkts	#1	#2	#3	#4	>4
88	Johnsons	Johnson Creamery Co	1	0	0	0	0	1
89	Kleinpeter	Kleinpeter Farms Dairy, Inc	1	0	0	0	1	0
90	Premier	Gazelle, Inc	1	0	0	0	0	1
91	Meyer	H. Meyer Dairy Co	1	0	0	0	0	1
93	Oak Farms	Morningstar Foods, Inc	1	0	0	0	1	0
96	Clover Stornetta	Clover Brand Dairy Products	1	0	0	0	0	1
98	Vitalite	Vitamilk Dairy, Inc	1	0	0	0	0	1
99	Bowman	Dean Foods Co	1	0	0	0	0	1
100	Curlys	Quality Chekd Dry Prdts Assn	1	0	0	0	0	1
101	Gerlands	Gazelle, Inc	1	0	0	0	0	1
102	Cabells	Morningstar Foods, Inc	1	0	0	0	0	1
103	Pine State	Pine State Creamery Co	1	0	0	0	1	0
104	Oak Grove	Oak Grove Dairy	1	0	0	0	0	1
106	Schepps	Schepps Ice Cream Co, Inc	1	0	0	0	0	1
108	Verifine	Dean Foods Co	1	0	0	0	0	1
110	Foremost	McKesson Corp	1	0	0	1	0	0
116	Gillette	Gillette Dairy, Inc	1	0	0	0	0	0
117	Lite A Rite	Gillette Dairy, Inc	1	0	0	0	1	0
120	Holland	Holland Dairy, Inc	1	0	0	0	0	1
122	Atlanta Dairies	Land-O-Sun Dairies	1	0	0	0	0	1

Note: Some brands were not marketed in the markets in this study; they are not listed in this table. They were, however, assigned a national rank based on their total sales.

TABLE 8.A.3 Listing of Manufacturers Showing Frequencies of Share Rankings, 4th Quarter, 1988

Rank	Co-op	Manufacturer	#Mkts	#1	#2	#3	#4	>4
1		Philip Morris Co. Inc	37	18	12	4	1	2
2		Borden, Inc	22	5	8	4	4	1
3		Quality Chekd Dairy Products Assn	14	1	2	4	1	6
4	C	Agway, Inc	5	4	0	0	1	0
5		Friendship Food Products Inc	5	0	3	1	1	0
6		Dean Foods Co	8	1	4	0	0	3
7		Crowley Foods, Inc	10	0	3	4	1	2
8	C	Darigold, Inc	2	0	0	0	0	2
9	C	Anderson Erickson Dairy Co	2	2	0	0	0	2
10		H J Heinz Co	21	0	3	3	6	9
11	C	Prairie Farms Dairy, Inc	4	0	0	2	1	1
12	C	Intermountain Milk Producers	2	1	0	0	1	0
13		Nordica International, Inc	2	0	1	0	0	1
14		General Mills	6	1	0	2	1	2
15		Old Home Foods, Inc	1	1	0	0	0	0
16		Nestle Co	3	0	2	1	0	0
17		Shamrock Foods Co	1	1	0	0	0	0
18		Crystal Cream & Butter Co	2	1	0	0	1	0
19	C	Bison Foods Co	1	1	0	0	0	0
20		Whitman Corp	2	0	0	1	0	1
21		Driggs Dairy Farms, Inc	1	0	0	0	0	1
22		Pevely Dairy Co	1	1	0	0	0	0
23	C	Golden Guernsey Dairy	1	1	0	0	0	0
24		East Side Jersey Dairy, Inc	1	1	0	0	0	0
25		Hillside Old Meadow Dairy	4	0	0	2	0	2
26		Louis Trauth Dairy, Inc	1	1	0	0	0	0

(continues)

TABLE 8.A.3 *(continued)*

Rank	Co-op	Manufacturer	#Mkts	#1	#2	#3	#4	>4
27		Farm Fresh Dairy, Inc	2	1	1	0	0	0
28		Berkeley Farms, Inc	1	0	1	0	0	0
29		Marigold Foods, Inc	3	0	0	1	2	2
30		Reiter Foods, Inc	3	0	0	0	2	1
31		Vitamilk Dairy, Inc	1	0	1	0	0	0
32		Gilt Edge Farms, Inc	2	0	0	1	1	0
33	C	Roberts Dairy Co	2	1	0	0	1	0
34	C	Swiss Valley Farms Co	1	0	0	0	1	0
35		Bongrain S A	2	0	1	1	0	0
36	C	Cabot Farmers' Cooperative Crmy	3	0	1	2	0	1
37		Melody Farms Dairy Co	1	0	0	0	1	1
38	C	Flav-O-Rich, Inc	3	0	1	0	1	0
39		Finevest Foods, Inc	1	0	1	0	1	0
40	C	Zarda Brothers Dairy, Inc	2	0	1	0	0	1
41		Utotem, Inc	2	1	0	0	0	1
42		Johnstown Sanitary Dairy	1	0	0	0	1	0
43	C	Land O Lakes, Inc	1	0	0	0	0	1
44		Arabian Investment Banking	1	0	0	0	1	0
45		Steffen Dairy Foods Co, Inc	1	1	0	0	0	0
46		Superior Dairy Fresh Milk Co	1	0	0	0	1	1
47		Barber Dairy, Inc	2	1	0	0	0	0
48		All Star Dairy Assn, Inc	1	0	0	0	1	1
49		Bareman's Dairy, Inc	1	0	0	0	0	1
50		Wells Dairy, Inc	1	0	1	0	0	1
51		London's Farm Dairy, Inc	1	0	0	0	0	1
52		Schroeder Milk Co, Inc	1	0	0	1	0	1
53		R Bruce Fike & Sons Dairy	1	0	0	0	0	0
54		McDonald Coop Dairy Co	1	0	0	0	0	1

57		Morningstar Foods, Inc	2	0	0	1	1	0
58		Coleman Dairy, Inc	1	1	0	0	0	0
60	C	Swiss Dairy	1	0	0	0	1	0
61		Valley Of Virginia Cooperative	1	0	0	1	0	0
62		Delphos Frozen Specialties	1	0	0	0	0	1
63		Alpenrose Dairy, Inc	1	0	0	1	0	0
64		Gazelle, Inc	1	0	0	0	0	0
65	C	Smith Dairy Products Co	2	0	0	1	0	2
66		Norris Creameries, Inc	1	0	0	0	0	1
67		Johnson Creamery Co	1	0	0	0	0	1
68		Kleinpeter Farms Dairy, Inc	1	0	0	1	0	0
70		H. Meyer Dairy Co	1	0	0	0	0	1
72		Clover Brand Dairy Products	1	0	0	0	0	1
73		McKesson Corp	1	0	0	0	0	1
74		Gillette Dairy, Inc	1	0	0	0	0	1
75		Pine State Creamery Co	1	0	1	1	0	0
76		Oak Grove Dairy	1	0	0	0	0	1
78		Schepps Ice Cream Co, Inc	1	0	0	0	0	1
84		Holland Dairy, Inc	1	0	0	0	1	1
85		T G Lee Foods, Inc	1	0	0	0	1	0
87		Land-O-Sun Dairies	1	0	0	0	1	0

Note: Some manufacturers did not market their cottage cheese in the markets in this study; they are not listed in this table. They were, however, assigned a national rank based on their total sales.

TABLE 8.A.4 Descriptive Statistics for Brand-level Data

Variable Name	Mean	St. Dev.	Min.	Max.
Average Price per Pound	1.1504	0.2063	0.67	1.751
Volume Market Share	7.812	9.934	0.5	67.187
Average ACV-Weighted Distribution	51.531	35.817	1.078	100
In-Store Share	17.41	16.290	1.09	82.995
Percent Volume with any Merchandising	18.761	17.571	0	77.419
Percent Volume with Feature A B Only	7.405	10.165	0	67.792
Average Percent Price Reduction	15.55	10.896	0	48.942
Percent Volume Sold with Display	0.8185	3.146	0	34.39
Units per Pound	0.8839	0.166	0.500	2.000
Co-ops Present Binary	0.4708	0.4999	0	1
Retail Market CR_4	61.231	13.149	23.9	84.1
Supermarket Grocery Sales Ratio	80.235	6.846	67.2	95.3
Population (x000)	2,856.8	2,672.9	603.49	15,582
Private Label Price per Pound	0.943	0.110	0.753	1.305

Note: There are 325 observations for each variable.

TABLE 8.A.5 Descriptive Statistics for Manufacturer-level Data

Variable Name	Mean	St. Dev.	Min.	Max.
Average Price per Pound	1.137	0.210	0.670	1.934
Volume Market Share	11.229	14.131	0.505	67.187
Average ACV-Weighted Distribution	47.939	35.868	1.078	100.000
Percent Volume with any Merchandising	17.104	16.513	0.0	71.123
Percent Volume with Feature A B Only	6.689	9.185	0.0	51.666
Average Percent Price Reduction	15.538	11.108	0.0	48.942
Percent Volume with Display	0.642	2.345	0.0	18.119

Note: There are 229 observations for each variable.

TABLE 8.A.6 Correlation Table for Data Used in Price Regressions

	Average Price per Pound	Volume Market Share	Average Weighted Distrib.	Sales per $Million ACV	Units per Pound	Co-ops Present Binary	Co-op Volume Market Share	Retail Market CR4	Pop.	Percent Volume w/ any Merch.	Private Label Price per Pound
Average Price per Pound	1.000										
Volume Market Share	0.112	1.000									
Average Weighted Distribution	0.507	0.542	1.000								
Sales per $Mil ACV	-0.274	0.462	-0.195	1.000							
Units per Pound	0.491	0.0233	0.208	-0.243	1.000						
Coops Present Binary	-0.206	0.0749	-0.0048	0.0744	-0.0448	1.000					
Co-op Volume Market Share	-0.0684	0.635	0.158	0.337	0.0102	0.204	1.000				
Retail Market CR4	0.126	0.0199	0.116	-0.131	0.138	-0.00052	-0.0138	1.000			
Population	0.242	0.0615	0.0553	0.0760	0.187	-0.166	-0.0442	-0.179	1.000		
Percent Volume with any Merch.	-0.234	0.0911	0.150	-0.0124	-0.102	0.101	0.0428	-0.0087	-0.0058	1.000	
Private Label Price / Pound	0.428	0.0848	0.176	-0.170	0.172	-0.209	-0.0891	0.0114	0.433	0.0197	1.000

9

Strategic Groups, Competition, and Retail Food Prices

Bruce W. Marion, Keith Heimforth and Wiltse Bailey

Background

The concept of minimal service, low-price grocery retailing dates back to the advent of the supermarket in the 1930s. Independent grocers in selected areas have successfully operated warehouse stores for decades. However, it was not until the economically turbulent 1970s that the warehouse store began making significant inroads in a number of urban grocery markets. Early warehouse efforts by chains such as Penn Fruit, Thriftmart, Acme and A&P were primarily last-ditch attempts to save failing stores (Progressive Grocer 1972). Although many of those conversions ultimately failed, successful adaptation of the format throughout the 1970s and 1980s resulted in warehouse and super warehouse stores' capturing 30% or more of sales in some markets.

The latest addition to the economy type supermarkets are "hypermarkets," which are giant food and nonfood stores requiring a buying market of about 500,000 people within 20 miles. There were only an estimated 12 hypermarkets in operation in 1990 (Gilbert 1990). Throughout this paper, warehouse stores, super warehouse stores, and hypermarkets will be considered as different store formats and will be referred to separately by those names, and collectively under the rubric "depot stores." In total, "depot stores" accounted for 15 percent of U.S. supermarket sales in 1989. (Progressive Grocer, 1990).

In this article, we focus on the competitive impact of depot stores during the period 1977-1987. These stores represent new retail "formats" with substantially lower costs and prices than traditional supermarkets (Progressive Grocer 1985; Grocer's Spotlight 1985; Supermarket News 1986).

The significantly lower prices in depot stores has led to aggressive responses by some conventional supermarkets when depot stores invaded their territory. A recent study of the Washington, D.C. market is a case in point. At Giant and Safeway stores *not* located near a warehouse store, prices were 13 percent higher than Shoppers Food Warehouse stores and nearly 20 percent above Food Lion, a new discount price entrant (Swisher 1990). However Giant and Safeway stores located near a depot store had prices that were 7 to 13 percent lower than their remaining stores. Zone pricing, as this type of geographic price discrimination is called in food retailing, has been used in response to depot stores in Indianapolis, Chicago, San Antonio, Atlanta and numerous other markets. In other markets, conventional supermarkets have avoided a price response and have countered depot stores by emphasizing quality and service attributes.

We believe that in most markets which depot stores have successfully entered, their presence will provide a downward pressure on market prices. This is the primary hypothesis that this study tested. Additional hypotheses tested are that when other factors are held constant, change in market concentration will be positively related to change in food prices, and denovo entry will be negatively related to change in prices.

Strategic Groups in Retail Food Markets

Porter (1979) comments: "An industry can . . . be viewed as composed of clusters or groups of firms, where each group consists of firms following similar strategies in terms of key decision variables. . . . I define such groups as *strategic groups*" (p. 215).

Marion (1984) has suggested that each of the food store formats illustrated in Figure 9.1 is a strategic group. Each format offers a unique mix of price, service, and products. The eight strategic groups in Figure 9.1 fall into at least two relevant markets: those on or above the horizontal axis compete with each other for the major shopping trips of consumers; the remaining three groups largely compete for fill-in shopping. The major shopping market can be labelled the "supermarket market". The stores in this market account for about 75% of food store sales and have the greatest effect on prices charged to consumers.

To have a positive effect on supermarket rivalry, depot stores must capture or be expected to capture a significant share of the "supermarket market." If they never capture more than 1 or 2 percent of a market's sales, depot stores would be part of the fringe that the major competitors can ignore. Some observers have estimated the market share potential of depot stores at 15-20% (Supermarket News 1983, 1984). Depot stores have already far exceeded this share in a few markets. Thus, these stores hold the potential of increasing the number of strategic groups and adding a significant competitor in many metropolitan areas. Those incumbent strategic groups that are strategically closest to the new group are expected to be affected the most and to respond the most aggressively.

New strategic groups may provide a means of circumventing entry barriers, at least in the short-run. Because depot stores represent a significant innovation

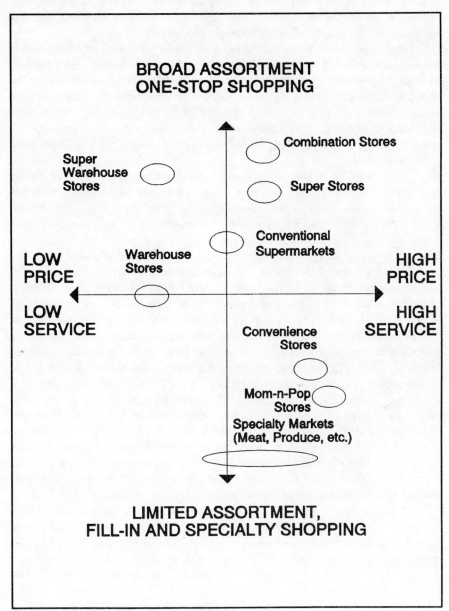

FIGURE 9.1 Retail Food Store Formats

in the retailing of food with particular appeal to price-oriented customers, they provide a means of overcoming the significant barriers to entry into the supermarket sub-market of many metropolitan areas.[1]

Thus, depot stores may enhance price rivalry in the supermarket sub-market for several reasons: (1) they represent a means of circumventing entry barriers so that denovo entry by depot stores is more likely than denovo entry by conventional supermarkets or superstores, (2) they represent new strategic groups which are likely to reduce the ability of incumbent supermarkets to coordinate their competitive actions, and (3) because depot stores emphasize low prices, their presence is likely to stimulate price rivalry with at least some incumbent retailers.

The first factor—lower entry barriers for depot stores—will affect market rivalry primarily during the period of entry and market penetration. Although the existence of warehouse store companies "waiting in the wings" could also affect competition in markets they have not yet entered, we hypothesize that the prices of incumbent superstores and conventional supermarkets will mainly be affected by actual entry of depot stores as compared to potential entry.

Significant denovo entry is normally expected to affect market rivalry simply because of the addition of capacity and the resulting displacement effect of the new entrant. The entry of depot stores is expected to have an additional impact on market prices because of the substantially lower prices of these firms.

Whether depot stores have a negative influence on market prices beyond the entry and market penetration period remains to be seen. If Porter is correct, oligopolistic coordination will be more difficult where depot store strategic groups exist; rivalry will be greater and prices lower. However, this is a hypothesis that remains to be tested. In addition, depot stores are likely to reduce average market prices vis a vis markets in which there are no depot stores because of the lower prices in depot stores. Even if market segmentation in a new equilibrium results in conventional supermarket and superstore prices returning to levels similar to markets in which depot stores are not present, depot stores are expected to continue to charge lower prices. Their effect on average market prices would depend upon their market share.

There are few markets (if any) where depot stores have achieved their market share potential and a new equilibrium has emerged. Thus, for the period examined in this study, we will primarily be examining the price effects of depot store introduction and expansion.

Assessing the Impact of Warehouse and Super Warehouse Stores on Retail Food Markets

When warehouse or super warehouse stores or hypermarkets enter a market, we expect them to reduce the prices paid by consumers in two ways.

First, since these stores carry significantly lower prices than conventional supermarkets, consumers who shop at depot stores receive directly the benefits of their lower prices. In addition, we expect the entry of depot stores to trigger price reductions by some incumbent retailers. Consumers shopping at the latter stores receive the "indirect" benefits of depot stores. Both sources of price reduction must be combined to understand the full impact of depot stores on market prices.

Unfortunately, data appropriate to assess both effects are unavailable. First, the most accurate and readily available price data, the Consumer Price Index (CPI) published by the U.S. Department of Labor, Bureau of Labor Statistics (BLS), do not allow meaningful comparisons of price *levels* across metropolitan areas. (U.S. Department of Labor 1984, p. 6) Thus, it is necessary to restrict interarea analyses to comparisons of price *changes* over time. Second, the CPI does not capture the full effects of depot stores on area prices because BLS methodology intentionally obscures the "direct effect" of new outlets. Since BLS procedures are not widely understood and have a major influence on the meaning of the CPI for food consumed at home, we digress briefly to examine BLS methodology.

The Food At Home Component of the CPI. The primary objective of BLS in calculating the CPI is to measure changes in price for a fixed basket of goods and services, not to measure the prices consumers actually pay in the market. Stated differently, the intent is to create a price index, not a cost of living index. Thus, to maintain comparability of the CPI over time, BLS attempts to "filter out" what it perceives as qualitative changes in the goods and services being priced.

To obtain a representative sample of food and other items with which to monitor price changes in an area, BLS combines information from two surveys, the Consumer Expenditure Survey (CE) and the Continuing Point of Purchase Survey (CPOPS). The CE has been conducted approximately once every ten years, and provides information on the absolute and relative amounts spent by consumers on various categories of goods and services in different BLS regions. The CPOPS has been conducted since 1977, and is intended to provide data for periodically updating the sample of outlets at which CPI items are priced; each year, one-fifth of the 87 metropolitan areas regularly surveyed by BLS are updated.

When the CPOPS updates the sample of outlets (and, hence, the specific items priced), it is necessary for BLS to "blend in" the price data obtained based upon a new CPOPS sample. To do so, in one month the BLS obtains prices for both the old and the new samples of outlets and items for the area being updated. Published CPI figures for that month are based on the price relative from the preceding to the current month in the old sample only; in the following month the CPI is based on the price relative in the new sample only. Using this procedure, BLS is able to blend in the price data from the new sample without directly comparing the price levels of the two different samples. A stylized example is:

	April	May	June
Market Basket Cost			
Old Sample w/o Depot Stores	$110	$121	-
New Sample w/ Depot Stores	-	$100	$110
CPI	100	110	121

Continuity of the CPI is maintained since BLS computes the CPI by a "chaining process" using price relatives, not price levels. In this process, the price relatives from the new sample are multiplied by the estimates (based on the old sample) of the previous month's expenditures in the stratum. Thus, in the above example, the CPI change from April to June is 21 percent.

In general, then, published CPI figures conceal reductions in area price levels that directly result from depot store entry. In the above example the published CPI figures would accurately reflect the monthly 10% increase in prices, but they would hide the lower *level* of prices in the new sample. Furthermore, the BLS holds weights given to different outlets constant between CPOPS surveys. So, even though depot stores might be represented in the BLS sample, the CPI would conceal any effect on area price levels due to increases in the market share of depot stores.

Note, however, that the CPI does capture the extent to which depot stores affect the prices charged by incumbent supermarkets—what we refer to as the indirect effect.[2] This price effect is the main one we will examine since data don't exist on the CPI of depot stores alone.

The Model

Our analysis focuses on changes in food prices across markets. We limit our attention to large, urban metropolitan areas for which BLS data are available.

The model empirically tested is of the form:

$$Y = \alpha + \beta_i X + \beta_j Z + e$$

where the dependent variable (Y) is the *percentage change* in the BLS Food-at-Home Price Index, X represents a set of binary variables reflecting depot store activity, and Z are control variables suggested by economic theory that affect changes in retail food prices. The analyses are performed on two data sets. One is a set that pools annual percentage price change observations from 1977-1987 across 25 Metropolitan Statistical Areas (n = 249). The second set contains the percentage price change during the ten year period, 1977-1987, for 25 *MSAs* (n = 25).

The Variables and Their Measurement in the Pooled Data Set

Change In Food Prices. The annual percentage change in the BLS CPI for Food—at—Home for each MSA was calculated from the average annual price index published for each of 25 MSAs in the CPI detailed reports. These data are based on a sample of all food stores, and thus include a number of store types (e.g., convenience stores, meat markets, produce markets) that ideally would be excluded for purposes of this analysis. However, the prices of small stores and specialty markets receive relatively little weight in nearly all MSAs.

Depot Share Variables. Our hypothesis is that the emergence of a strong depot store strategic group(s) in a market results in lower food prices in that market. Since the lower prices charged by depot stores is not directly captured by BLS data, and since we are restricted to analyzing food price changes (not levels), our testable hypothesis is that markets with strong depot groups experience a reduction in food prices or a smaller increase in food prices compared to markets without a significant depot store presence.

To test this hypothesis, five binary "dummy" variables were constructed to reflect the depot share[3] in a market in year t:

$$0\% < D_1 \leq 5\%$$
$$5\% < D_2 \leq 10\%$$
$$10\% < D_3 \leq 20\%$$
$$20\% < D_4 \leq 30\%$$
$$30\% < D_5$$

Change in Concentration. The linkage between market concentration and market prices has been well established by over 75 studies in a wide variety of industries (Weiss, 1989). The relationship between *change* in concentration and change in price has received much less attention. However, at least one study has found evidence of a long run relationship between *changes* in concentration and *changes* in prices in the food and tobacco manufacturing industries (Kelton 1982). That evidence, plus the logical extension of the concentration level-price level relationship, suggests that a change in concentration variable should be tested in this model. Four-firm concentration ratios were calculated from shares reported by Metro Market Studies, with some adjustments.[4]

Change in Per Capita Income. Per capita income may influence store prices in at least two ways. First, demand is more inelastic in high income markets. This means that the monopoly price is higher in high income compared to low income markets. Second, as income per capita in an MSA increases, consumers are expected to shift their purchases towards higher priced products and towards stores with higher levels of service. The long-run effects of these shifts would be an increase in food store sales that stem in-part from higher prices and in-part from changes in products sold.[5]

The percentage change in personal disposable income was calculated using Sales and Marketing Management's (S&MM) estimate of "Effective Buying Income," as reported in the Annual Survey of Buying Power.

Change in Population. Change in population is a measure of growth in the MSAs examined. A faster rate of growth in population would be expected to expand demand and increase prices, all else the same. Rapid growth is likely to result in high capacity utilization. Conversely, slow or negative growth is more likely to encounter over capacity, greater rivalry and lower prices. Thus a positive sign is expected. We measure this variable by calculating the percentage change in population as estimated in S&MM's survey of buying power.

Change in Operating Costs. The combination of utility and rental costs account for 10-12% of grocery store operating costs and roughly 2% of grocery stores sales. As a proxy for the cost of these inputs to store owners, CPI data for fuels and for residential rent were combined to form an "energy plus rent" price index. The annual percentage change in this index was then calculated. The "energy plus rent" index was calculated as a simple average of the two price indices since grocery industry data show that utilities (excluding telephone service) accounted for roughly the same percentage of operating costs as did rent payments during 1978-1982 (Progressive Grocer, April 1983 p. 96). We expect that price *changes* for these inputs are similar for consumers and commercial purchasers.

Labor Costs. By far the largest single operating expense for food stores is payroll, accounting for nearly 60% of operating expenses (Progressive Grocer 1985). Lamm and Wescott (1981) found changes in wages of food and grocery store employees to be a large and significant factor contributing to national changes in grocery prices over time. Such wage changes differ considerably across markets due to, inter alia, differing levels of unionization and union success in affecting wages (Lamm 1982; Harp 1979), and differing levels of unemployment.

The effects of changing wage rates on prices depends upon the extent to which productivity also changes. Thus, payroll per dollar of sales is preferred to wage rates as a measure of labor costs since it incorporates changes in wages, productivity and the mix of employees (part-time v. full-time). Change in labor costs were measured as the annual percentage change in payroll as a percent of sales:[6]

$$\Delta \ PAYRAT \ = \ \left[\frac{(Payroll_t \ /Sales_t)}{(Payroll_{t-1} \ /Sales_{t-1})} \right] \ -1.0$$

Entry. In some markets, depot stores have been developed by supermarket firms already in the market. In other cases, depot stores were introduced by new entrants. In the latter cases, a new competitor is added to the market whereas in the former cases, only a new strategic group is added. We hypothesize that the response of the market will be different in the two situations. An entry dummy variable is included in most models to distinguish between the two methods of introducing depot stores and to also capture the effects of denovo entry by major supermarket firms. The dummy variable has a value of one for the year of entry and the following year. A negative sign is expected.

Variables in Cross-Sectional Analysis of Average Price Changes for 1977-1987 Period.

Whereas the previous data set is used in pooled time series-cross sectional analysis in which the dependent variable is year to year percentage changes in prices, the cross sectional analysis examines average price changes over the period 1977-1987. The dependent variable is the simple average of year to year changes in the Food at Home CPI over the ten-year period based upon BLS data.

The market share of depot stores changed considerably in some MSAs over the ten years studied. Thus, the presence of depot stores was measured by the average annual change in depot store shares over the ten year period. Three binary variables were used to designate three rates of growth:

$$0 \ < \ D_1 \ \le \ 1\%$$
$$1\% \ < \ D_2 \ \le \ 2\%$$
$$2\% \ < \ D_3$$

Average annual change in concentration used the data discussed above and were calculated as the simple average of the ten annual changes. The percentage changes in per capita income and labor costs were calculated for the entire ten year period in a similar manner as the dependent variable.

Variables analogous to the population change and energy plus rent cost change for the ten year period were constructed and included in preliminary analysis. However, collinearity problems and the limited degrees of freedom in this data set resulted in our omitting these variables from subsequent analyses.

Empirical Results

Pooled Data Set

The pooled data set consisted of 249 observations: ten annual observations for 24 geographic markets for the years 1977-1987, plus nine annual observations for the Miami MSA (data for 1977 were unavailable). The average annual percentage changes in the Food-at-Home CPI for observations grouped by depot market share are shown in Table 9.1. On average, the percentage change in CPI drops as depot store shares increase up to 20 to 30 percent of the market.

The regression model estimated using the pooled data set was:

$$FOODP_{it} = \alpha + \beta_1 D1_{it} + \beta_2 D2_{it} + \beta_3 D3_{it} + \beta_4 D4_{it} + \beta_5 D5_{it}$$
$$+ \beta_6 \Delta CR4_{it} + \beta_7 E_{it} + \beta_8 \Delta INC_{it} + \beta_9 \Delta POP_{it}$$
$$+ \beta_{10} \Delta FRP_{it} + \beta_{11} \Delta PAYRAT_{it} + \beta_j Tj_t + e_{it}$$

where: i,t = area, year subscripts;

	Expected Sign
FOODP = percentage change in BLS Food-at-Home price index expressed in decimals	
D0 = 1 if depot store share is = 0	
D1 = 1 if depot store share is $0 < D_1 \le 5\%$, 0 otherwise	<0
D2 = 1 if depot store share is $5 < D_2 \le 10\%$, 0 otherwise	<0
D3 = 1 if depot store share is $10 < D_3 \le 20\%$, 0 otherwise	<0
D4 = 1 if depot store share is $20 < D_4 \le 30\%$, 0 otherwise	<0
D5 = 1 if depot store share is $>30\%$, 0 otherwise	<0
$\Delta CR4$ = change in CR4 expressed in decimals	>0
Entry dummy = 1 if denovo entry by depot store or major supermarket firm occurred during year or during previous year	<0
ΔINC = percentage change in per capita disposable income expressed in decimals	>0
ΔPOP = percentage change in population expressed in decimals	>0
ΔFRP = percentage change in price index of energy plus rent in decimals	>0

TABLE 9.1 Descriptive Statistics, Pooled Data Set

Depot Shares	D0 0	D1 $0 < D_1 \leq 5\%$	D2 $5\% < D_2 \leq 10\%$	D3 $10\% < D_3 \leq 20\%$	D4 $20\% < D_4 \leq 30\%$	D5 $>30\%$
Average Annual Percent Δ MSA CPI	6.57%	5.48%	4.33%	4.00%	1.87%	3.30%
Average Annual Percent CPI- Δ U.S. CPI	.182	-.016	-.003	-.412	-.776	+.330
Average Annual Percent Δ CR4	1.16%	-.41%	1.27%	1.40%	-.47%	-1.08%
Number of Observations	113	37	49	30	10	10
Number of Areas	18	15	16	10	4	3

ΔPAYRAT = percentage change in payroll/sales expressed in
     decimals > 0
Tj$_t$ = 1 when year = t, 0 otherwise. An alternative variable, the
     change in the U.S. Food-At-Home CPI, was used in some
     models instead of the time dummies $\neq 0$

The set of time-related dummy variables was introduced to control for the impact of other trend-related factors on food price such as inflationary expectations, recession, and worldwide price fluctuations. Without a time trend, these factors would be partially captured by the warehouse dummies since warehouse market share generally increases with time. The percentage change in the U.S. food-at-home CPI was used instead of time dummies in some models. The annual change in this index ranged from slightly over 10% to 1% during 1977-87.

Tests on OLS residuals showed a strong positive correlation between the logarithm of the squared residuals and the logarithm on the concentration ratio so that we may assume that Var (e^2) = ß$_o$CR4$_{\text{ß}}$. Table 9.2 shows the OLS and GLS regression results for the pooled data set.

Analysis. Regression results strongly support our hypothesis that the introduction of depot stores and the increase in their market share affected food prices negatively. Whenever depot stores made significant inroads into a market, capturing 10 to 20% (D3) or 20 to 30% (D4), a marked slowdown in food price increases was apparent. We note, however, that the coefficients on D1 and D2 are not statistically significant, indicating that these levels of entry were insufficient to significantly affect prices market-wide.

For those few MSAs in which depot share exceeded 30% (D5), the annual percentage change in BLS priced became positive. Perhaps by this point, the market reaction to depot stores had played itself out. Since the coefficient on D5 is not significant, we will not spend much space interpreting.

From a theoretical perspective, one of the more interesting results is the strong positive impact of an *increase* in the four-firm concentration ratio on food price increases. Of all the variables except the time dummies, this variable has the largest "t" value in the GLS results. A 10 percentage point increase in CR4 was associated, on average, with an increase in food prices of .66%.

The entry variable also had the expected sign and was marginally significant. Thus, when depot stores were introduced into an MSA by denovo entry, the prices of incumbent supermarkets were negatively affected even more than when the depot stores were introduced by an incumbent retailer. Based upon the GLS coefficients, the entry effect on prices was roughly equal to depot stores capturing 10 to 30% of the market.

Of the remaining control variables, only the change in per capita disposable income was significant in the GLS results (and positive as hypothesized). The two variables used as proxies for cost increases were never significant. Change

TABLE 9.2 Regression Analysis Explaining Annual Change in BLS Food-At-Home Prices in 25 Metropolitan Areas, 1977-1987 [a] (Pooled Time-Series/Cross Section Set, n=249)

	OLS		EGLS	
	Coeff.	(t-stat)	Coeff.	(t-stat)
Intercept	.0376	(10.07)[c]	.0362	(9.97)[c]
D1	-.0002	(-0.06)	-.0008	(-0.30)
D2	-.0011	(-0.44)	-.0012	(-0.50)
D3	-.0054	(-1.87)[b]	-.0060	(-2.09)[b]
D4	-.0083	(-1.76)[b]	-.0060	(-1.39)[d]
D5	.0029	(0.61)	.0042	(1.02)
ΔCR4	.0470	(2.49)[c]	.0664	(3.80)[c]
ΔINC	.0286	(0.92)	.0505	(1.64)[d]
ΔPOP	.0481	(1.04)	.0604	(1.30)
ΔFRP	.0071	(0.36)	-.0051	(-0.25)
ΔPAYRAT	.0029	(0.26)	.0091	(0.88)
Entry	-.0121	(-3.01)[c]	-.0074	(-1.57)[d]
T78	.0638	(14.82)[c]	.0635	(14.54)[c]
T79	.0693	(15.01)[c]	.0693	(14.76)[c]
T80	.0340	(6.90)[c]	.0366	(7.34)[c]
T81	.0323	(7.07)[c]	.0325	(7.11)[c]
T82	-.0027	(-0.64)	-.0021	(-0.48)
T83	-.0211	(-5.19)[c]	-.0215	(-5.31)[c]
T84	-.0055	(-1.34)	-.0116	(-2.84)[c]
T85	-.0218	(-5.65)[c]	-.0228	(-5.87)[c]
T86	-.0084	(-2.15)[b]	-.0103	(-2.60)[c]
R^2	.8721			
R^2	.8609			
F	77.74		271.15	

[a] 1978-1987 for Miami.
Level of Significance: [b] 1 Percent.
[c] 5 Percent.
[d] 10 Percent.

in population was positively linked to change in food prices but was not quite significant at the 10% level.

Cross Section Data Set

The cross section data set consisted of 25 observations. The percentage changes in the Food-at-Home CPI for areas grouped by average annual depot share growth were:

	Δ share=0	D_1 0 < Δ share ≤ 1	D_2 1 < Δ share ≤ 2	D_3 Δ share > 2%
Annual Δ CPI, '77-'87	5.58%	5.28%	5.17%	5.21%
Annual Δ CR4, '77-'87	1.20	1.26	0.96	-1.49
# Observations	8	7	7	3

The regression model estimated using the cross-section data was:

$$AFOODP_i = \alpha + ß_1 D1_i + ß_2 D2_i + ß_3 D3i + ß_4 ACR4_i + ß_5 AINC_i + ß_6 APOP_i + ß_7 APAYRAT_i + e_i$$

Expected
Sign

$AFOODP_i$ = annual average percentage change in BLS Food-At-Home price index for MSA i over the period 1977 to 1987.

D1 = low growth binary variable, value = 1 if average annual change in depot share is greater than 0 and less than or equal to 1, other wise 0.;   <0

D2 = medium growth binary variable, value = 1 if average annual change in depot share is greater than 1 and less than or equal to 2, otherwise = 0;   <0

D3 = high growth binary variable, value = 1 if average annual change in depot share is greater than 2, otherwise = 0;   <0

ACR4 = annual average percentage point change in concentration ratio during the period 1977 to 1987.   >0

AINC = annual average percentage point change in per capita disposable income for period 1977 to 1987.   >0

APOP = annual average percentage point change in population during 1977 to 1987.   >0

APAYRAT = annual average percentage point change in payroll/sales ratio during 1977 to 1987.   >0

Whereas the pooled data analysis allowed us to examine whether a certain degree of market penetration by depot store was necessary to affect prices, the cross-section analysis examines the effect of varying rates of growth of depot stores on market prices.

The regression results on the cross-section data indicate that a medium growth rate in market share by depot stores seems to have the strongest negative effect on market price (Table 9.3). Although all three depot store dummies have a negative sign only D2 (annual share growth of 1 to 2 percent) is significant in the GLS results.

TABLE 9.3 Impact of Depot Market Share on Area Food Prices, 1977-1987[a], (Cross-Section Data Set, n=25)

	OLS		E GLS	
	Coeff.	(t-Stat)	Coeff.	(t-Stat)
Intercept	.0489	(6.55)	.0563	(12.10)
ACR4	.0639	(1.22)	.0710	(2.40)[b]
APOP	.0610	(0.87)	.1200	(1.95)[b]
ADISPOP	.0678	(0.74)	-.0422	(-0.71)
APAYRAT	.0347	(0.59)	-.0034	(-0.06)
D1	-.0031	(-1.47)[b]	-.0018	(-1.08)
D2	-.0050	(-2.25)[b]	-.0040	(-1.94)[b]
D3	-.0021	(-0.65)	-.0011	(-0.58)
R^2	.369			
R^2	.109			
F	1.42		1743.5	

[a] 1978-87 for Miami.
[b] Significant at the 99% level.

    Change in CR4 is once again positive and significantly related to change in market prices. The coefficient on ΔCR4 is very similar for the cross section and pooled data. In the cross section analysis, change in population is also significant and positively related to change in food prices.

    The lack of significance of D3—rapid growth of depot store share—is somewhat puzzling. Three MSAs account for all the observations here. And these MSAs are also the only ones in which depot store share exceed 30% (D5=1 in pooled analysis). Tables 9.2 and 9.4 indicate a positive coefficient for D5—though not significant. However, it appears from these results that the negative influence of depot stores on price changes in these three MSAs has run its course. Figure 9.2 supports this conclusion. For the three MSAs in which depot stores have had the greatest penetration, the negative effect on market prices was largest when the market share of depot stores was 10 to 20 percent. Figure 9.2 shows the change in food prices at the MSA level *relative* to the U.S. change in food prices. Thus, after adjusting for the change in U.S. CPI for food-consumed-at-home, the prices in these three MSAs dropped by nearly 1 percent during each year in which depot stores had 10-20 percent of the market. By the time depot stores had captured over 30 percent of the market, food prices in these MSAs were increasing slightly faster than the national average.

TABLE 9.4 OLS Regression Analysis Explaining Annual Change in BLS Food-at-home Prices in 25 MSAs, 1977-1987, Without Time Dummies

	Full Sample		Slow Growth		Medium Growth	
	Coeff.	"t" Stat.	Coeff.	"t" Stat.	Coeff.	"t" Stat.
Intercept	.0021	0.75	.0109	1.67	.0104	1.17
D1	-.0001	-0.05	-.0026	-0.63	.0050	0.71
D2	-.0018	-0.72	-.0071	-1.57c	.0006	0.09
D3	-.0062	-2.19b			-.0055	-0.85
D4	-.0087	-1.86b			-.0202	-1.97b
D5	.0010	0.21				
ΔCR4	.0379	2.06b	.0050	0.15	.1497	2.43a
ΔINC	.0335	1.13	-.0171	-0.34	.0140	0.19
ΔPOP	.0323	0.70	.0794	1.10	-.0531	-0.48
ΔFRP	.0042	0.28	-.0299	-0.80	.0010	0.04
ΔPayrat	-.0019	-0.18	.0286	1.51	-.0050	-0.24
ENTRY	-.0119	-3.02a	-.0106	-2.22b	-.0321	-3.33a
ΔU.S. CPI	.9320	28.29a	.9228	15.52a	.8260	9.87a
n	249		70		69	
R²	.858		.896		.790	
F	125.4		67.3		24.2	

Level of Significance: [a] 1 Percent
[b] 5 Percent
[c] 10 Percent.

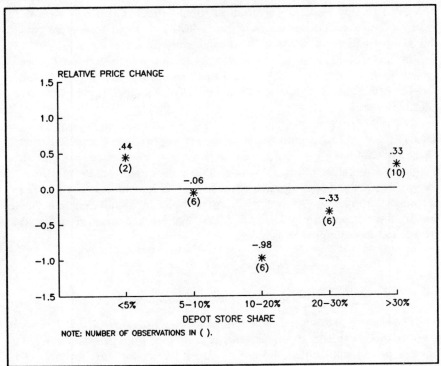

FIGURE 9.2 Relative Price Change in MSA s with Rapid Depot Share Growth, 1977-1987

Different Model Specifications

Table 9.4 provides regression results in which the time dummies are replaced with a variable measuring the change in U.S. CPI for food-consumed-at-home. With this model specification, neither serial correlation nor heteroscedasticity are a problem. Thus, the results in Table 9.4 are OLS.

Table 9.4 results using the full sample are similar to the OLS results in Table 9.2. Depot stores have a significant negative effect on the rate of MSA price increase when depot store share is between 10 and 30 percent. Entry also has a significant negative impact on price increases. Change in four-firm concentration has a significant positive effect as in the Table 9.2 models. The change in U.S. CPI for food-consumed-at-home has the expected strong positive relation to price changes in MSAs.

Table 9.4 also examines two subsets of the sample, one in which depot stores were present but grew slowly during 1977-87, and one in which depot stores experienced medium growth during that period. Omitted from both of these data sets are 29 observations in three MSAs with rapid depot store growth and 81 observations in MSAs with no depot store presence.

In the MSAs with slow growth of depot store share a negative influence is indicated when depot stores achieve a 5 to 10 percent market share. The entry variable is negative and significant but the coefficient on change in CR4 drops to insignificance.

In the MSAs with medium growth of depot stores, a significant negative impact is not indicated until depot stores achieve a 20 to 30 percent share. Change in CR4 has a significant positive effect on price increases and entry has a significant negative effect. Although D3, a depot store share of 10 to 20 percent, has a significant negative effect in the full sample, it is not significant in the smaller samples. Apparently the significance in the full sample is due to the six observations in rapid growth MSAs that are not included in the slow growth and medium growth samples.

Summary and Conclusions

The analysis in this paper indicates that metropolitan areas with a significant presence by depot stores experienced lower retail food price increases during the ten year period studied than did areas without warehouse store activity. The point at which this negative influence is exerted seems to vary some for different MSAs. No significant effect was found where depot stores had less than a 5 percent market share. In some MSAs, a negative effect was found when depot stores had 5 to 10 percent of the market. For all MSAs, however, the negative influence of depot stores became statistically significant when the depot store share was between 10 and 30 percent. In the three MSAs in which depot shares exceeded 30 percent, food prices increased faster than average. Thus, there is some evidence that the negative influence of depot stores eventually runs out of steam. A new equilibrium may emerge at that point.

The results are consistent with the hypothesis that depot stores constitute a strategic group that is sufficiently interdependent with other supermarket formats to be an important competitive force that increases rivalry and leads to substantial consumer benefits. The results provide at least modest support for Porter's hypotheses concerning the rivalry effects of strategic groups. Additional insights into the competitive behavior of retail food markets can be gained by considering the structure and conduct characteristics of strategic groups.

The results also support the hypothesis concerning the price effects of entry.

Entry of a "significant competitor" had a consistent negative influence on food price increases in the two years following entry.

The results also support the hypothesis that retail food prices are positively related to MSA concentration. Although in this case, change in retail prices was related to change in concentration, the main theoretical basis for testing such a relationship is that concentration and price are expected to be positively linked.

This finding takes on added significance given the "no relationship" findings of the recent USDA study (Kaufman and Handy, 1989) and the agnostic review of food retailing structure—performance studies by Anderson (1990). Because of the way the analysis was done, this study sheds some light on two of Anderson's concerns. Anderson contended that price comparison studies aimed at measuring market power must in some way deal with the possibility that, (1) costs may differ across markets and/or (2) that quality and service may vary across markets. While we disagree with Anderson's contention that these have been inadequately accounted for in previous studies, the present study avoids this criticism. Here we are examining the change in prices of the same market over time. BLS makes sure that quality/service are held constant over time. Two changes in cost variables were included in our models. Both were insignificant at even the 10% level. Thus, our results find a positive linkage between concentration and prices even after holding costs and quality/service constant.

The results of this study are consistent with six other studies that found a significant positive relationship between grocery store prices and the concentration of sales in local markets (Marion *et al.* 1979; Hall, Schmitz and Cothern 1979; Lamm 1981; Meyer 1983; Cotterill 1983; Cotterill 1986). The results are also consistent with Weiss's conclusions after his massive review of concentration—price studies in a wide variety of industries (Weiss 1989).

One implication of these results is the importance of ensuring that markets are free of any artificial barriers to the entry of new competitors—especially depot stores. Depot store entrants are vulnerable to predatory pricing by incumbent chains (Mueller 1984).

Finally, the study has raised a question regarding the extent to which published CPI data accurately reflect changes in food prices facing consumers in areas with a significant depot store presence. This is a matter of which the BLS is well aware and which they face in any consumer goods market in which discount outlets emerge. The changes in prices analyzed in this study reflect primarily the effects of warehouse stores on the prices of competing stores. Had the lower prices of depot stores been included in the data set (i.e., to constitute a cost of living index rather than a price index), the effects of warehouse stores would have been even greater. For example, if warehouse stores charge prices that average 7% less than conventional supermarkets and superstores, and if they achieve 30% of an MSAs sales, the actual cost of food to consumers in that market would be approximately 2% less than the CPI.

Notes

1. The development of a new strategic group in a market requires denovo entry and capacity expansion in many cases. This has been true with depot stores in some MSA's, although in the majority of MSA's examined, depot stores were first introduced by supermarket firms already in the MSA.

2. Since the relative weights given different outlets are held constant between CPOPS surveys, if warehouse stores are eroding the market position of incumbent supermarkets the indirect effect of depot stores on changes in market prices may be somewhat overstated.

3. Depot shares were estimated using a variety of trade sources, including Metro Market Studies, Supermarket News, Market Scope, and others. The imprecise nature of the estimates obtained from such sources prevented us from constructing a reliable continuous measure of depot store shares.

4. Since publication of Metro data lags collection of that data by one to two years, $CR4_t$ was first calculated as the average of Metro $CR4_{t+1}$ + Metro $CR4_{t+2}$. The resulting figure was then adjusted by the ratio for that market: (1977 Census CR4)/ (1977 Metro CR4), or 1982 Census CR4/1982 Metro CR4, or a combination of these ratios.

The following table shows the weights assigned to each coefficient of adjustment for all the years:

Year:	77	78	79	80	81	82	1983 to 1987
Coef1: (C77/M77)	(1)	(.8)	(.6)	(.4)	(.2)	(.0)	(.0)
Coef2: (C82/M82)	(.0)	(.2)	(.4)	(.6)	(.8)	(1)	(1)

5. Given the relatively short period of this empirical study and BLS methodology, it may not be possible to detect the expected positive relationship between price change and income change. For the five year periods between CPOPS surveys, the products and outlets on which BLS collects prices and the weights used are held constant. If consumers shift towards higher service stores as incomes increase, that will only be detected by BLS when a new CPOPS survey is conducted. Then--because of the bridging procedure followed, the shift in consumer patronage to high service-high priced stores (or low price stores) is not allowed to influence the change in CPI. Thus the nature of the BLS data set conceals certain effects of changes in income.

6. Payroll data were calculated from Bureau of Census County Business Pattern data which reports gross wages and other compensation for civilian employees of food stores. The geographic areas for these data were generally identical to BLS areas. Sales data for each MSA and year were from S&MM Survey of Buying Power, which estimates sales of stores selling food primarily for home consumption.

References

Anderson, K. 1990. A Review of Structure-Performance Studies in Grocery Retailing. Bureau of Economics, Federal Trade Commission, Washington D.C., June.

Cotterill, R. W. 1983. The Food Retailing Industry in Arkansas: A Study of Price and Service Levels. Unpublished report submitted to the Honorable Steve Clark, Attorney General, State of Arkansas, January 10.

_____. 1986. Market Power in the Retail Food Industry: Evidence from Vermont. *Review of Economics and Statistics* 68 (August): 379-86.

Gilbert, L. 1990. Smaller Format Seen For U.S. Hypermarkets. *Supermarket News*. (June): 1.

Grocer's Spotlight. 1985. Warehouse Stores - Format of the 80s. (August): 33.

Hall, L., A. Schmitz, and J. Cothern. 1979. Beef Wholesale-Retail Marketing Margins and Concentration. *Economica* 46:295-300.

Harp, H. 1979. Labor Costs in Food Marketing. *National Food Review* 8 (Fall): 12-14.

Kaufman, P., and C. Handy. 1989. Supermarket Prices and Price Differences: City, Firm and Store-Level Determinants. Economic Research Service Tech Bulletin 1776, USDA, December.

Kelton, C. M. L. 1982. The Effect of Market Structure on Price Change for Food Manufacturing Product Classes. Working Paper no. 60. Madison, WI: NC 117.

Lamm, R. 1981. Prices and Concentration in the Food Retailing Industry. *Journal of Industrial Economics* 30:67-78.

_____. 1982. Unionism and Prices in the Food Retailing Industry. *Journal of Labor Research* 3 (Winter): 69-79.

Lamm, R., and P. C. Westcott. 1981. The Effects of Changing Input Costs on Food Prices. *American Journal of Agricultural Economics* 63 (May):187-196.

Marion, B. W. 1984. Strategic Groups, Entry Barriers and Competitive Behavior in Grocery Retailing. Working Paper 81. Madison, WI: NC 117.

_____., W. F. Mueller, R.W. Cotterill, F.E. Geithman, and J.R. Schmelzer. 1979. *The Food Retailing Industry: Market Structure, Profits and Prices*. New York: Praeger.

Meyer, P. J. 1983. Concentration and Performance in Local Retail Markets. in John Craven, ed., *Industrial Organization, Antitrust, and Public Policy*. Boston: Kluwer-Nijhoff Publishers.

Mueller, W. F. 1984. Alleged Predatory Conduct in Food Retailing. Working Paper 78. Madison, WI: NC 117.

Porter, M. 1979. The Structure Within Industries and Companies' Performance. *Review of Economics and Statistics*. 61 (May): 214-227.

Progressive Grocer. 1972. What Makes Warehouse Markets Go. (August): 61.

_____. 1983. Composite Chain Report. (April): 96.

_____. 1985. Composite Chain Report. (May): 96.

_____. 1990. 57th Annual Report of the Grocery Industry, 1989. April.

Supermarket News. 1983b. Store Opening Puts Lucky in Super Warehouse Field. (December 12): 1.

_____. 1984. Cub Foods Out Front in Green Bay Market. (January 9): 1, 21.

_____. 1985. Town and Country Opens Cub to Stem Competition. (November 11): 8A.

_____. 1986. Says Big Depot Stores Are Losing Their Advantage. (April 14): 1, 22.

Swisher, K. 1990. Survey Reports Prices Not Uniform at Giants. *Washington Post*, Business Section (May 30): 1.

United States Department of Labor, Bureau of Labor Statistics. 1984. *BLS Handbook of Methods, Volume II, The Consumer Price Index*. Washington, D.C.: GPO.

Weiss, L., ed. 1989. *Concentration and Price*, Cambridge: MIT Press.

PART FIVE

The Food Retailing Debate

10

Structure-Performance Studies of Grocery Retailing: A Review

Keith B. Anderson[1]

In the past fifteen years or so, several studies have attempted to estimate the relationship between concentration in local food retailing markets and the level of prices charged or the profits earned by retailers operating in those markets.[2] With the exception of the recent USDA study conducted by Kaufman and Handy, these studies have found evidence of monopoly power in concentrated markets; and their results have been used by some to advocate a restrictive policy toward grocery-retailing mergers.

During roughly the same time period, economists generally have become skeptical of statistical cross-industry structure-performance studies that found a positive relationship between profits and concentration and were similarly interpreted as providing evidence of monopoly power. Questions have been raised about the theory underlying the studies. Critics such as Harold Demsetz asked why new entry did not undercut the higher profits apparently earned in industries with higher levels of concentration. As an alternative explanation of the apparent relationship between profitability and concentration, it was suggested that leading firms in more concentrated industries were more efficient than their smaller rivals and earned higher profits to reflect this greater efficiency.[3]

The results of more recent studies using cross-industry data have generally been consistent with greater efficiency, rather than monopoly power, as the explanation of high profits. Profitability has been found to be higher, ceteris paribus, for firms with higher market share but not in industries with higher levels of concentration. Indeed, several of the more recent papers find concentration and profitability to be negatively related.[4] In summarizing the newer findings, Scherer wrote:

> Profitability rises strongly with larger market share; belonging to a more highly concentrated industry adds little or nothing. Let me make the point

more strongly: At least for the United States, the many studies that found a positive association between aggregated industry profits and concentration were almost surely spurious, the victims of aggregation bias.[5]

The purpose of this paper is to critically review the grocery retailing papers in order to determine whether they are well grounded in economic theory and use proper statistical methodology. If they are, the obvious question is why the results in grocery retailing differ from the general results. If not, some directions for future research may be suggested. In addition, such findings may suggest that policy toward such mergers should not be based on the findings of these studies alone.

Price and Concentration

Cross-industry studies of the structure-performance paradigm must use profit data because comparing prices across industries is infeasible. Studying markets, such as local grocery retailing markets, that differ not by the product sold but rather by geography permits the researcher to use price as the dependent variable. All of the recent studies of grocery retailing have, in whole or in part, examined prices, perhaps in the belief that the use of price rather than profit data would allow a cleaner test of the hypothesis that increased concentration gives rise to monopoly power.

However, explaining prices is actually little, if any, easier than explaining profits, and anyone seeking to explain cross-market price differences must be careful to avoid several pitfalls. Most of these difficulties follow directly from simple price theory. Here we focus on two types of problems: inter-market differences in the cost of doing business and intra-market differences in the services provided.

To be credible, a study must account for such differences. If not, higher prices found in more concentrated markets may be the result of higher costs, not a lack of competition. For example, consumers in some markets may demand a higher level of services with the result that retailing costs are higher in those markets. There is some evidence that larger firms are more likely to provide these additional services.[6] As a result, concentration may tend to be higher in more service-intensive markets. Prices will also tend to be higher in such markets because of the costs of providing the extra services, creating the possibility of a positive correlation between price and concentration. However, any such correlation would be the result of the demand for higher quality not the presence of monopoly power.

Since none of the studies reviewed here fully controls for such differences, their findings cannot be taken as convincing evidence that prices are higher in more concentrated markets because of the exercise of market power or that higher concentration is alone a reliable indicator of monopoly power. Indeed, the paper which comes closest to controlling for such differences—the recent paper by Kaufman and Handy—is the one paper that finds no such relationship.

Inter-Market Differences in Costs

Competition among grocery stores will result in prices that just cover the costs incurred. Competitive prices in each market will be just high enough to cover the costs of operating in that market.[7] Competition will not lead to equal prices in markets with different costs: prices will be higher where costs are higher. If prices were equal in face of cost differences, either the firms in the high-cost market would be losing money or those in the low-cost market would be earning super-competitive profits.

There are several potential sources of cost differences among local grocery retailing markets, including wholesale and labor costs. All of the studies, with the exception of Cotterill's analysis of prices in Vermont, include at least one variable to measure labor cost differences; and Cotterill suggests, quite reasonably, that wage differences are unlikely to be significant in his data. However, only the studies by Lamm and the one by Hall, Schmitz, and Cothern explicitly account for wholesale costs, which account for about 78 percent of the price of items sold in grocery stores.[8,9] Failure to account for these cost differences could create substantial problems in interpreting any comparison of prices across markets.[10]

Differences in real estate and utilities costs and in taxes are also frequently not reflected in these studies. The studies by Lamm and Marion, *et al.*, may be capturing part, but not all, of the differences in these items; other studies are unlikely to capture any such differences. Lamm's model includes regional variables that should capture inter-regional differences. Marion, *et al.*, control for the size of the market and therefore should capture differences that vary with the size of the market.

Markets may also differ in the level of efficiency achieved by the firms selling there, which could result in differences in costs and the level of competitive prices. In those markets where consumer preferences lead to larger firms and larger stores, the costs of operating supermarkets may be lower because firms are able to more fully exploit economies of scale and, as a result, the competitive price may be lower than in markets with smaller stores and/or smaller firms.[11]

Some of the studies include variables that may be capturing, at least partially, differences in operating efficiencies. As noted, Marion, *et al.*, include the size of the market in their regressions. Grinnell, Crawford, and Feaster include two variables that may be capturing differential efficiencies—stores per capita and grocery store sales per capita. Lamm includes the average size of stores in the market in his analyses. On the other hand, while Cotterill includes firm- or store-specific measures of size and sales per square foot, he does not include any variables measuring differences in inter-market average efficiency, and it is market-average values that are relevant here.[12]

Omitting variables measuring cost differences across markets may cause problems in a regression analysis. For example, concentration tends to be somewhat higher in smaller markets.[13] At the same time, costs may be higher in smaller markets, because of, for example, higher warehousing and advertising costs per unit of sales. This may give rise to a positive correlation between concentration and costs, which would result in a positive correlation between concentration and competitive prices. However, the higher level of prices in the smaller, more-concentrated markets would be the result of the higher costs of operating in smaller markets, not necessarily monopoly.

Supermarket Competition Involves More than Just Price

Even within the same competitively-functioning market, different supermarkets may charge different prices for the same items. Supermarkets differ in terms of the services they provide, in terms of the selection of products they offer, and in terms of the quality of the goods they sell, in addition to the prices they charge. One supermarket may be open longer hours than another or have a more convenient location. Another supermarket may have a service seafood department and offer a wider selection of fresh seafood products.

Offering better service or higher quality products is not costless. Keeping a store open all night is likely to increase costs per unit of sales. Similarly, costs are incurred in keeping a store cleaner or keeping additional check-out lines open. Providing higher quality in the produce department will also result in higher costs because wholesale costs will be higher and the store may have to throw away more produce in order to maintain the high quality. Similarly, offering a wider selection of fresh, rather than frozen, fish may result in increased spoilage of inventory.

Because of the added costs, higher quality products and services will only be offered if customers are willing to pay for them. As a result, if stores differ in terms of the quality and services provided, they will also differ in the prices charged. However, the higher prices at higher quality stores will be the result of cost differences, not noncompetitive market performance.

There is considerable evidence of service differences among supermarkets. Larger stores tend to carry a larger number of items and to be open longer hours.[14] Larger stores also seem more likely to have services such as pharmacies, service seafood and service deli counters.[15]

Thus, particularly if a study uses firm- or store-specific price data, a valid test of the hypothesis that prices are higher in more concentrated markets can only be conducted if the researcher controls for the important quality differences among stores.[16,17] However, the only study that comes close to controlling for differences in services and quality is the work of Kaufman and Handy, who include in their model an index based on the number of services offered and a variable differentiating warehouse stores from other kinds of supermarkets. This study finds no relationship between concentration and prices but finds that prices are significantly higher where more services are offered and are significantly lower in warehouse stores.[18]

Concentration and Profitability

Most of the criticism of statistical studies of the structure-performance paradigm have been directed at studies that have used profits as their dependent variable. Thus, there is a considerable literature on which to draw in examining the one study that has considered grocery-retailing profitability—the work of Marion, *et al.*

Several aspects of this work warrant comment. First, as with all studies of profitability, there are serious questions about the use of accounting data to represent economic profits. Second, regression results designed to test the "critical concentration" hypothesis appear inconsistent with monopolistic output restrictions. Third, using relative firm market share (RFMS) rather than just the firm market share (FMS) appears to be incorrect. Fourth, even assuming the nonlinear relationship between profitability and concentration, an examination of the total effects of changing concentration shows that profits rise much less with increasing concentration than is reported. Indeed, at higher levels of concentration, the estimated relationship implies that profits fall rather than rise with increases in concentration.[19,20]

The Use of Accounting Data

One of the most troubling criticisms of profitability-concentration studies in recent years has concerned the use of accounting data to measure monopoly power. Economists generally accept that the Lerner index, which is equal to the difference between price and marginal cost divided by price, provides a useful measure of monopoly power.[21] The profit margin is commonly used as a proxy for this index. However, the profit margin may not properly measure market power unless accounting profits are equal to economic profits and other conditions are met.

Using accounting data to measure economic profits poses several problems. First, as Fisher has shown, accounting data generally do not correctly capture the cost of capital.[22] Measuring the cost of capital is complicated because of questions of the allocation of depreciation over the useful life of any capital item. Second, accountants generally expense such long-lived investments as advertising and research and development in the year they are incurred rather than treating them as assets. Third, assets are not valued at replacement cost. Fourth, if a firm that possesses market power has been sold, the capitalized value of its monopoly rents will be included in the price for which the firm was sold; and the new owner will not appear to be earning more than a competitive return on its assets.[23]

Further, the Lerner index is the ratio of price minus *marginal economic* cost divided by price, while the profit-sales ratio can be shown to be equal to price minus *average accounting* cost divided by price. As a result, the profit-sales ratio can only be an accurate proxy for the Lerner index in those cases where there is no divergence between marginal and average cost—*i.e.*, where there are constant returns to scale—or if adjustments are made for this deviation.

Hypothetical examples in Fisher's papers raise the possibility that the errors resulting from the use of accounting data could be large. However, the ultimate question is whether the errors that arise with the data for real firms are significant and whether they are correlated with variables included in the regression models being estimated.[24] Unfortunately, only limited information is available to answer this question. Salamon has shown empirically that the errors in the accounting rate of return are correlated with the size of the firm.[25] Fisher has also provided some reasons to conclude that the errors in the accounting profit-sales ratio are likely to be correlated with some or all of the variables in which one is likely to be interested.[26] However, much remains unknown. As Scherer has noted, Fisher's argument that accounting data can tell us virtually nothing about industry performance:

> is certain to be a spectre that haunts nearly every analysis of market structure and profitability in the future. It is a peculiarly difficult spectre to exorcise, since everyone admits that accounting data are imperfect, and it is virtually impossible to prove the negative proposition that the problems are not so serious as to preclude valid inferences. Much seems to hinge on basic matters of faith, reminding one of efforts to silence Galileo.[27]

The data used by Marion, *et al.*, poses an additional problem. If a profit margin is to correctly measure economic profits, the cost of capital—both debt and equity—along with other costs, must be subtracted from price in computing the numerator. While it is unclear whether Marion, *et al.*, included any capital costs, it is fairly clear that they had no way of including the cost of equity capital.[28] Since the cost of capital is an economic cost incurred in offering retailing services, a profit margin that does not include these costs cannot provide an appropriate measure of monopoly power.[29]

The Nonlinear Concentration Measure Employed
Does Not Appear to Capture the Critical Concentration Concept

In some of their regressions, Marion, *et al.*, assume a linear relationship between concentration and profitability. However, their results with this specification are not very robust. In only one of five cases is the relationship between profitability and concentration statistically significant at the levels usually used by economists.[30]

In other regressions, Marion, *et al.*, assume a nonlinear relationship between profitability and concentration, to test the "critical concentration hypothesis."[31] A critical level of concentration is hypothesized below which markets function competitively and above which either tacit or overt collusion results in effectively monopolistic output levels and the higher prices and profits associated with such reductions in output.[32]

Based on the critical concentration hypothesis, one would have expected the estimated relationship between profitability and concentration to be S-shaped— below the critical level profitability rises very slowly, if at all, as concentration increases; in the critical range profitability rises rapidly; and above the critical level there is again little increase in profitability. However, for the levels of concentration found in these authors' data set (*i.e.*, for four-firm concentration levels between 30 and 80),[33] the estimated nonlinear relationship is everywhere concave from below. At least for the range of observed concentration levels, there is no critical level below which markets would appear to be functioning competitively.[34] By the time four-firm concentration reaches a level of 30 percent, profitability is rising rapidly. These results are not consistent with the critical concentration hypothesis. In addition, the finding that profits are already rising rapidly when concentration is in the area of 30 percent appears to be inconsistent with other studies of the critical concentration hypothesis which have almost universally found the upsurge in profit rates to occur at C4s between 45 and 59 percent.[35]

It is also instructive to note the substantial increase in estimated profitability at extremely low levels of concentration. Profits appear to rise more as the four-firm concentration ratio rises from 30 to 40 percent than when concentration rises from 40 to 100.[36] This could only be the result of monopolistic behavior if grocery retailing firms can successfully collude at very low levels of concentration—an unlikely event. Indeed, Marion, *et al.*, appear to accept that a market with a C4 of 40 percent or lower is likely to function competitively.[37]

Using Relative Firm Market Share
Rather than Market Share is Incorrect

Rather than include firm market share (FMS) in their profit regression, Marion, *et al.* use RFMS, which is defined as firm market share divided by the four-firm concentration ratio. This variable is used because the authors believe that "a firm's discretion in pricing and its cost advantage or disadvantage depend largely on its relative position in the market."[38] We have two reasons for believing that it is incorrect to use RFMS rather than FMS to identify firms that are more efficient.

Recent work in oligopoly theory has shown that firms which are more efficient in the sense of having lower unit costs will have higher market shares than their less efficient competitors. In addition, because the more efficient firm has lower costs, the market price will, *ceteris paribus*, exceed the efficient firm's costs by more than it will exceed the costs of smaller, less efficient firms.[39] This implies a positive relationship between market share and profitability.[40] These models say nothing about relative market share, and we are unaware of any models that do.[41]

RFMS also allegedly provides a more relevant measure of scale economies because it provides a measure of a firm's size relative to that of the largest firms in the market. However, this reasoning demonstrates a misunderstanding of a competitive equilibrium and of the profits resulting from increased efficiency. It is not the efficiency of a firm relative to the largest, presumably most efficient, firms in the market that determines whether a firm earns high profits because of its efficiency—what economists call efficiency rents. Rather, efficiency rents are determined by the firm's efficiency relative to the least efficient firm in the market, which in a competitive equilibrium will be earning zero profits. If, for example, a firm must have 10 percent of the sales in a market of a particular size to realize available economies of scale, then any firm with a 10 percent share in such a market should be more profitable than one with a one percent market share. This difference in profitability will be the same whether such a firm is the largest firm in the market or there are four firms each with 20 percent of the market.[42] However, the value of RFMS will differ significantly in these two cases: being at least 25 in the case where the largest firm has a share of 10 percent and equal to 12.5 in the latter case.

Using Relative Market Share Rather Than Firm Share Overstates the Effects of Increasing Concentration

Use of RFMS rather than FMS leads the results of Marion, *et al.*, to overstate any positive relationship between profitability and concentration, because a change in concentration, holding market share constant, changes RFMS. For example, with FMS equal to 10, an increase in C4 from 40 to 50 percent will cause RFMS to fall from 25 to 20. Since the estimated relationship between profits and RFMS is positive, the decline in RFMS will at least partially offset the apparent increase in profits associated with an increase in C4. Indeed, an examination of the total effect of changing C4 in the relationship estimated by Marion, *et al.*, suggests that profits fall rather than rise with increased concentration in more highly concentrated markets—the markets that are presumably of greatest concern to antitrust policy makers.

Table 10.1 illustrates both the direct effect of increasing concentration—the effect reported by Marion, *et al.*—and the total effect, for two different values of firm market share.[43] Part A shows the effect for a firm with a market share of 10 percent—approximately the average of firms in Marion, *et al.*'s data set.[44] Part B shows the effects of increasing concentration where a firm has a 25-percent market share.

The direct effect—the result of the positive coefficient on the concentration variable—is a steadily rising level of profits as concentration increases.[45] The increases in profitability appear to be quite substantial. For example, in the

TABLE 10.1 The Effects of Changing Concentration on Profitability (Reported Effects Versus Total Effects)

	A. Average Market Share Case (Market Share = 10%)	
C4	Effect Reported By Marion, *et al.*	Total Effect
40	.000	.000
45	.342	.167
50	.570	.255
55	.706	.277
60	.780	.255
65	.817	.211
70	.832	.157
75	.836	.101
80	.837	.049

	B. High Market Share Case (Market Share = 25%)	
C4	Effect Reported By Marion, *et al.*	Total Effect
40	.000	.000
45	.342	-.095
50	.570	-.218
55	.706	-.368
60	.780	-.532
65	.817	-.698
70	.832	-.856
75	.836	-1.001
80	.837	-1.132

average market share case, an increase in C4 from 40 percent to 70 percent results in an increase in the profit-sales ratio of 0.832 percent, more than 50 percent of the average profit-sales ratio for firms in the sample used to estimate the relationship.[46]

However, the total effect of a change in concentration, including the effect on RFMS, presents a very different picture. Profitability does not rise steadily with concentration. For FMS equal to 10 percent, estimated profitability rises until C4 reaches 55 percent and then falls as C4 increases further.[47] Further, an increase in C4 from 40 percent to 70 percent results in an increase in the profit-sales ratio of only 0.157 percent—less than 20 percent of the increase in the direct effect.

The difference between direct and total effects is even more dramatic at higher market shares. If FMS equals 25, the direct effect shows the same positive relationship between concentration and profits. However, the total effect shows profitability falling continuously as C4 increases from 40 and 80 percent.[48] If direct estimation of a more correctly specified model were to confirm this finding, the Marion, *et al.*, results would be consistent with some of the newer cross-industry studies—such as those of Ravenscraft (1983) and Kwoka and Ravenscraft (1986). Of course, these studies find that market share and not concentration is the important determinant of profitability; and this is not, I think, how Professor Marion and his co-authors interpret their results.[49,50]

Conclusions and Implications for Future Research

Because of substantial methodological and data questions as well as the inevitable shortcomings of this type of statistical study, the studies reviewed here do not provide an appropriate basis for prohibiting certain mergers or groups of mergers. While such studies may contribute in a general way to our understanding of how markets work, standing alone they are simply unable to capture all of the factors that are important in determining whether a particular merger poses competitive problems.[51] And, relying on rules that fail to consider all the relevant evidence is likely to lead to the sacrifice of potential efficiency gains in cases where market power problems are unlikely to arise.

Looking at implications for future research, the first, and most obvious, is that future researchers need to carefully control for differences in costs and the quality of the products and services offered. While I have focused on the importance of controlling for cost and quality differences in the context of a price equation, controlling for quality differences is also important in profitability equations. (Costs differences should, of course, not be a problem since they are part of profits.) Professor Scherer has argued that the primary cause of post-war increases in concentration in manufacturing industries was innovation in consumer products or in their marketing that allowed the innovating firm to increase its market share.[52] Since the sale of innovative products is likely to increase the profitability of the innovating firm as well as permitting the firm to charge a price premium, it is obviously important to control for such differences in any equation that hopes to explain differences in either profitability or prices.

The second point is equally important. One must differentiate between the average value for all firms in a market and the value for an individual firm or store. Market averages for cost variables appear more important in price equations, while the difference between the individual firm or store values and the average for all firms in the market would appear to be important in profitability equations. However, provided sufficient numbers of observations are available, it may be best to include both market average values and the deviation of individual firm or store values from those market averages in both types of equations. In either a profitability or a price model, the proper measure of quality would be that related to the individual firm or store, perhaps adjusted for the average level of quality in the market.

Notes

1. This paper is drawn from a longer paper published by the Federal Trade Commission, Anderson (1990). The interested reader is referred to that paper for additional detail. Some of the points raised here are also made by Newark (1989) which came to the author's attention as he was completing the earlier paper.

The views expressed here are those of the author and not necessarily those of the Federal Trade Commission, the U.S. International Trade Commission or any of the Commissioners of either Commission.

2. These studies include the work of Marion *et al.*, (1979a and 1979b); two studies by Cotterill (1983, 1984, and 1986); two studies by Lamm (1981 and 1982); an unpublished study by Grinnell, Crawford, and Feaster (1976); a study by Hall, Schmitz, and Cothern (1979); and a study by Kaufman and Handy (1988 and 1989).

3. See, *e.g.*, Demsetz (1973 and 1974).

4. See, *e.g.*, Ravenscraft (1983) and Kwoka and Ravenscraft (1986).

5. Scherer (1986), p. 8. However, like Shepherd and some others, Scherer appears to believe that the higher profits earned by firms with high market shares may reflect monopoly power rather than, or in addition to, greater efficiency. (See Scherer (1986), p. 9.)

6. Indeed, larger supermarkets tend to offer more products and to be open longer hours than smaller stores. (See discussion in and around note 13, below.)

7. Technically, this is true only for the marginal firm if firms in the same market have different costs. More efficient firms will earn additional profits—efficiency rents resulting from their higher level of efficiency.

8. Lamm (1981), p. 70.

9. Cotterill and Grinnell, Crawford, and Feaster include variables designed to capture differences in the costs of transporting goods to the retail market. Marion, *et al.*, apparently experimented with a transportation distance variable like that used by Grinnell, Crawford, and Feaster and found it not statistically significant. They also argue that differences in transportation costs should be the only way in which wholesale cost differences would affect the prices of goods included in their price indexes, since all of the goods they used in constructing their indexes were produced nationally; and therefore the wholesale costs of *these goods*, except for transportation, should be the same in all markets. (Marion, *et al.* (1979a), p. 112) Whether this demonstrates that retail prices of these items should be the same across markets depends on how competition works in grocery retailing. If competition forces the price of each good to equal its cost, the argument made by Marion, *et al.*, would appear to be correct. However, shoppers often purchase many items in a single store and select a store based on the price and quality of the whole range of items they purchase. If, as a result, competition among grocery stores only leads to prices that on average over a range of products are equal to costs, rather than being equal item by item, differences in costs for locally produced goods may be reflected in the prices of other goods.

10. It is informative to note that the coefficient on the wholesale cost variable in Lamm's work is highly significant and very close to the value one would expect given that wholesale costs account for 78 percent of price. (Lamm (1981), pp. 72-73)

11. The issue here is differences among markets not differences among firms competing in the same market. Intra-firm efficiency differences in the same local market will result in efficiency rents for some firms, but should not cause prices to differ.

12. See, *e.g.*, Cotterill (1983), p. 108. Similarly, Hall, Schmitz, and Cothern do not account for differences in efficiency.

13. While the average four-firm concentration ratio in cities with populations of one million or more was 52.0 percent in 1977, the average in cities of less than 150,000 was 61.2 percent. (Parker (1986), pp. 37-39)

14. In 1987, the average store with a volume of $2 to $4 million per year which was operated by a chain, stocked 10,460 items. The average store with a volume in excess of $12 million stocked 22,823 items. The smaller stores were open 101 hours per week on average, while the larger stores were open an average of 143 hours per week. Similarly, while only 10 percent of the smaller stores were open 24 hours per day at least one day a week, 60 percent of the larger stores were open 24 hours per day at least one day a week. ("55th Annual Report of the Grocery Industry," *Progressive Grocer*, April 1988, p. 29)

15. If larger stores tend to be operated by firms with higher market shares, these effects may be captured by the market share variables included in several of the studies. However, it would appear that the correct variable would be market share, not the relative firm market share used by Cotterill and Marion, *et al.* Use of relative share will cause the relationship between price and concentration to be overstated as is discussed in the review of the relationship between profits and concentration below. (See the discussion beginning on page 12.)

16. The problem posed by different levels of quality is not, of course, unique to studies of the prices of grocery products. Control of quality and service differences would be necessary in any study of prices. For reviews of other studies examining the

relationship between prices and concentration, see Geithman, Marvel, and Weiss (1981) and Weiss (1987).

17. There can even be problems in studies using market-average price data since average prices would be expected to be higher in markets with higher average levels of quality.

18. Next to Kaufman and Handy, the study that comes closest to meeting this condition is Cotterill's study of prices in Arkansas, which includes store format as a variable. (Cotterill (1983)) However, store format does not capture all of the significant quality differences among stores. (See Anderson (1990), p. 42, n.100.) Other studies only capture quality effects if they are related to the size of the store.

19. Our conclusions are based on an analysis of the equation estimated by Marion, *et al.*, rather than on estimates from a correctly-specified equation, since we do not have the data these authors used.

20. Two other issues deserve brief mention. First, their analysis covers 1970 to 1974. During much of this period, retail food stores, like much of the U.S. economy, was subject to government price controls. One wonders, therefore, how generalizable their results are to other time periods. Second, the authors single out markets in which A&P was a competitor as being different from normal markets. It is not clear to us that such treatment is justified.

21. Indeed, several researchers have recently constructed explicit oligopoly models in which the Lerner index of market power is shown to be the correct measure of profits. (See, *e.g.*, Cowling and Waterson (1976), Kwoka and Ravenscraft (1986), and Martin (1984).)

22. Fisher and McGowan (1983) and Fisher (1987).

23. Liebowitz (1987).

24. See Long and Ravenscraft (1984).

25. Salamon (1985). See also Buijink and Jegers (1989) and Salamon (1989).

26. Fisher (1987).

27. Scherer (1986), p. 9.

28. The authors define their measure of profits as net profits before taxes, which would seem to imply that interest costs had been deducted. However, they then state that they do not have any data on assets or stockholder's equity by market. (See, *e.g.*, Marion, *et al.* (1979a), p. 78.) Absent such data, it is not clear how one would allocate interest expenses across markets.

29. Failing to adjust for the cost of capital may appear to be less important where, as here, a study looks at performance in only a single industry. However, the cost of capital may still differ across observations if land costs are higher in one area than another or if some stores in some areas provide more services than others and additional capital is needed to supply these services.

30. Equation 1b, Table 3.5, is significant at the 5 percent level in a one-tailed test. (See Marion, *et al.* (1979a), p. 82.) One other equation is significant at the ten percent level in a one-tailed test; and the other three coefficients are insignificant.

31. The nonlinear relationship is introduced into a linear regression by using CC4, a nonlinear function of C4, in the regression. CC4 is defined as

$$CC4 = (C4 + a)^3 / [1 - 3(C4 + a) + 3(C4 + a)^2]$$

where C4 is expressed as a fraction. The function CC4 lies between zero and one and has an S-shape provided the constant "a" is restricted so that (C4 + a) lies between zero and one for all observations in the data set. For Marion, *et al.*, this means "a" must lie between -0.3 and 0.2. Subject to this restriction, the value of the parameter "a" is estimated along with the other parameters in the profitability equation, apparently using an iterative approach. The profitability equation was apparently estimated for each of several values of "a" within the acceptable range and the value selected to minimize the standard error of the regression. (Marion, *et al.* (1979b), p. 422)

32. For a review of studies examining this question, see Pautler (1983).

33. The range of C4 in the data set used by Marion, *et al.*, is from a low of 29 percent to a high somewhat less than 80. (Marion, *et al.*, (1979b), p. 422 and Cross-examination of Dr. Bruce Marion, in Federal Trade Commission v. National Tea Company, United States District Court, District of Minnesota, Fourth Division, Civ. 4-79-161, transcript at p. 296.)

34. If one is willing to extrapolate outside of the range of the data used to estimate the relationship, it could be argued that a critical concentration value exists. However, that critical value would be in the range where the four-largest firms controlled only 10 to 15 percent of a market. It is implausible that control over such a small share of the market would allow for collusive behavior.

35. Scherer (1980), p. 280.

36. The value of CC4 is 0.5 when C4 is 30 percent, 0.77 when C4 is 40 percent, and 1.00 when C4 is 100 percent.

37. Marion, *et al.* (1979a), p. 138.

38. Marion, *et al.* (1979a), p. 71. The authors also claim that RFMS provides a "superior measure of the degree of enterprise differentiation among firms in a market."

39. See Kwoka and Ravenscraft (1986), p. 352.

40. Clarke, Davies, and Waterson (1984) also demonstrate that where there are economies of scale a positive relationship between market share and the measured profit-sales ratio may exist because of the difference between this ratio and the correct measure of monopoly power, the Lerner index. Because profit-sales ratios are equal to price minus average cost divided by price whereas the numerator of the Lerner Index is equal to price minus marginal cost, measured profit-sales ratios understate monopoly power to the extent that marginal cost is less than average cost. The difference between marginal and average cost is smaller for large firms that more fully exploit available economies of scale. Thus, the failure to account for the divergence between average and marginal cost may create a relationship between market share and the profit-sales ratio that is not an indicator of firm efficiency.

41. Indeed, correct empirical estimation of one of these oligopoly models, the one developed by Clarke and Davies, requires the inclusion of both market share and market share multiplied by concentration—rather than market share divided by concentration—as variables. (See Kwoka and Ravenscraft (1986), p. 352.)

42. Alone, FMS cannot capture the ability of firms to realize available economies of scale in different sized markets. It is necessary to include market size in the equation as well.

43. These illustrations are based on equation 1c from Marion, *et al.* (1979a), Table 3.5, p. 82.

44. Calculated from data in Marion, *et al.* (1979a), Table 3.3, p. 64.

45. The relevant derivative for evaluating this relationship is

$$d(P/S)/dC4 = 3.661 \, d\{(C4 + 0.2)^3/(1 - 3 (C4 + 0.2) + 3 (C4 + 0.2)^2)\}/dC4$$

where: P/S is the profit-sales ratio expressed as a percentage, and
C4 is expressed as a fraction.

46. As the average profit-sales ratio, we use 1.45 percent, the unweighted average across years of the annual averages reported in Marion, *et al.* (1979a), Table 3.2, p. 62.

47. The relevant equation for evaluating the total effect of changes in C4 is

$$d(P/S)/dC4 = 3.661 \, d\{(C4 + 0.2)^3/(1 - 3(C4 + 0.2) + 3(C4 + 0.2)^2)\}/dC4 + 0.063 \, d(100 \times FMS / C4)/dC4$$

where both FMS and C4 are measured as fractions.

48. With a higher value of FMS, the reduction in RFMS that accompanies any increase in concentration is greater. As a result, the total effect of increasing concentration holding FMS constant is smaller. For example, if FMS is equal to 10, an increase in C4 from 50 to 60 will cause RFMS to fall by 3.3 percentage points from 20 to 16.7. When FMS equals 20, the same increase in C4 will result in a decline of 6.7 percentage points in RFMS.

49. The effect of changing concentration in an equation using FMS rather than RFMS as an independent variable would almost certainly differ from both the direct effect reported by Marion, *et al.*, and from the total effect we have discussed.

50. In commenting on this paper, Professor Cotterill suggests that our focus should be on the change in total industry profits both from changes in concentration and from changes in individual firm market shares, not just the effect of the change in concentration. (See Cotterill (1992)) The problem with such an approach is that it fails to differentiate between any increase in efficiency rents and any increase in monopoly profits, even though the two have vastly different implications for economic efficiency. While we should seek to eliminate monopoly profits, we should seek to encourage the economic efficiency that generates efficiency rents.

51. Of course, this does not imply that it is inappropriate to bring antitrust actions to stop grocery-retailing mergers in specific cases. When the antitrust agencies undertake an investigation of a specific grocery retailing market, they will develop information on much more than just concentration and will therefore have much more information on which to base a decision.

52. Scherer (1979).

References

Anderson, K. B. 1990. *A Review of Structure-Performance Studies in Grocery Retailing*. Bureau of Economics, Federal Trade Commission.

Buijink, W., and M. Jegers. 1989. Accounting Rates of Return: Comment. *American Economic Review* 79(1):287-9.

Clarke, R., S. Davies and M. Waterson. 1984. The Profitability-Concentration Relation: Market Power or Efficiency? *The Journal of Industrial Economics* 32(4):435-50.

Cotterill, R. W. 1983. *The Food Retailing Industry in Arkansas: A Study of Price and*

Service Levels. Unpublished report submitted to the Honorable Steve Clark, Attorney General, State of Arkansas.

_____. 1984. *Modern Markets and Market Power: Evidence from the Vermont Retail Food Industry.* N.C. Project 117 Working Paper No. 84, Department of Agricultural Economics, University of Connecticut, Storrs.

_____. 1986. Market Power in the Retail Food Industry: Evidence from Vermont. *Review of Economics and Statistics* 68(3):379-86.

_____. 1992. A Response to the Federal Trade Commission/Anderson Critique of Structure-Performance Studies in Grocery Retailing. In *Competitive Strategy Analysis in the Food System,* ed. R. W. Cotterill, Boulder: Westview.

Cowling, K., and M. Waterson. 1976. Price-Cost Margins and Market Structure. *Economica* 43(171):267-74.

Demsetz, H. 1973. Industry Structure, Market Rivalry, and Public Policy. *Journal of Law and Economics* 16(1):1-9.

_____. 1974. Two Systems of Belief About Monopoly. *Industrial Concentration: The New Learning,* Harvey J. Goldschmid, H. Michael Mann, and J. Fred Weston, editors, 164-184.

Fisher, F. M. 1987. On the Misuse of the Profit-Sales Ratio to Infer Monopoly Power. *Rand Journal of Economics* 18(2):384-96.

Fisher, F. M., and J. J. McGowan. 1983. On the Misuse of Accounting Rates of Return to Infer Monopoly Profits. *American Economic Review* 73(1):82-97.

Geithman, F. E., H. P. Marvel, and L. W. Weiss. 1981. Concentration, Price, and Critical Concentration Ratios. *Review of Economics and Statistics* 63(3):346-53.

Grinnell, G. E., T. L. Crawford, and G. Feaster. 1976. Analysis of the Impact of Market Concentration on City Food Prices. Unpublished paper presented at the American Agricultural Economics Association's annual meetings, August 15-18, 1976. Reprinted in U.S. Congress, Joint Economic Committee. *Prices and Profits of Leading Retail Food Chains, 1970-1974: Hearings before the Joint Economic Committee, Congress of the United States,* 228-238.

Hall, L., A. Schmitz and J. Cothern. 1979. Beef Wholesale-Retail Marketing Margins and Concentration. *Economica* 46(183):295-300.

Kaufman, P. R., and C. R. Handy. 1988. *Determinants of Supermarket Prices and Price Differences: Results From a National Survey.* Unpublished paper presented at the American Agricultural Economics Association annual meetings, July 31—August 3, 1988.

_____. 1989. *Supermarket Prices and Price Differences: City, Firm, and Store-Level Determinants.* Economic Research Service, U.S. Department of Agriculture, Technical Bulletin No. 1776.

Kwoka, J. E., Jr., and D. J. Ravenscraft. 1986. Cooperation v. Rivalry: Price-Cost Margins by Line of Business. *Economica* 53(211):351-63.

Lamm, R. M. 1981. Prices and Concentration in the Food Retailing Industry. *Journal of Industrial Economics* 30(1):67-78.

_____. 1982. Unionism and Concentration in the Food Retailing Industry. *Journal of Labor Research* 3:69-79.

Liebowitz, S. J. 1987. The Measurement and Mismeasurement of Monopoly Power. *International Review of Law and Economics* 7:89-99.

Long, W. F., and D. J. Ravenscraft. 1984. The Misuse of Accounting Rates of Return: Comment. *American Economic Review* 74(2):494-500.

Marion, B. W., W. F. Mueller, R. W. Cotterill, F. Geithman and J. Schmelzer. 1979a. *The Food Retailing Industry*. New York: Praeger.

_____. 1979b. The Price and Profit Performance of Leading Food Chains. *American Journal of Agricultural Economics* 61:420-433.

Martin, S. 1984. The Misuse of Accounting Rates of Return: Comment. *American Economic Review* 74(2):501-6.

Newmark, C. M. 1989. *A New Bottle for the Profits-Concentration Wine: A Look at Prices and Concentration in Grocery Retailing*. Unpublished paper.

Parker, R. C. 1986. *Concentration, Integration and Diversification in the Grocery Retailing Industry*. Bureau of Economics, Federal Trade Commission.

Pautler, P. A. 1983. A Review of the Economic Basis for Broad-Based Horizontal-Merger Policy. *The Antitrust Bulletin* 28(3):571-651.

Peltzman, S. 1977. The Gains and Losses from Industrial Concentration. *Journal of Law and Economics* 20(2):229-63.

Ravenscraft, D. J. 1983. Structure-Profit Relationships at the Line of Business and Industry Level. *Review of Economics and Statistics* 65(1):22-31.

Salamon, G. L. 1985. Accounting Rates of Return. *American Economic Review* 75(2):495-504.

_____. 1989. Accounting Rates of Return: Reply. *American Economic Review* 79(1):290-293.

Scherer, F. M. 1979. The Causes and Consequences of Rising Industrial Concentration. *Journal of Law and Economics* 22(1):191-211.

_____. 1980. *Industrial Market Structure and Economic Performance*. 2d ed. Boston: Houghton Mifflin.

_____. 1986. On the Current State of Knowledge in Industrial Organization. In *Mainstreams in Industrial Organization—Book II*, ed. H.W. deJong and W.G. Shepherd, 5-22. Boston:Kluwer Academic Publishers.

Weiss, L. W. 1987. Concentration and Price—A Progress Report. In *Issues After a Century of Federal Competition Policy*, ed. by R. L. Wills, J. A. Caswell, and J. D. Culbertson, 317-332. Lexington, MA:Lexington.

11

A Response to the Federal Trade Commission/Anderson Critique of Structure-Performance Studies in Grocery Retailing

Ronald W. Cotterill

Introduction

Responding to Dr. Keith Anderson's critique of structure performance studies in grocery retailing, requires historical perspective. In May 1988 I testified at the invitation of then chairman of the House Judiciary Committee, Peter Rodino, before his Subcommittee on Monopolies and Commercial Law (Cotterill, 1988, 1989). Congressman Rodino was most concerned about the status of merger enforcement in the general economy and the food sector in particular. At that time the merger and LBO wave was at its high water mark and the Federal Trade Commission was on the sidelines. Based upon research of structure performance relationships in the food retailing industry (Marion *et al.*, 1977, 1979, Lamm 1981, Hall, Schmitz and Cothern 1979, Cotterill, 1983, 1984, 1985, 1986, de Maintenon 1984, Cotterill and Haller 1987, Marion 1987), I similarly criticized the F.T.C. for not enforcing the antitrust laws as rigorously as warranted to constrain the unprecedented wave of mergers and leveraged buyout in the food retailing industry (U.S. Congress, 1988 p. 81-132). Others, including state antitrust enforcement officials, also questioned the lack of FTC action.

The Subcommittee requested that the FTC respond to this testimony[1]. Subsequently the Bureau of Economics reveiwed the structure performance research on the food retailing industry and explained why this research should not be used as the basis for antitrust enforcement. Keith Anderson, staff economist, produced a paper in the Bureau of Economics, Economic Issues series titled "A Review of Structure Performance Studies in Grocery Retailing".

It appeared in June 1990. The paper he presents here is a shorter version of that report.

Since 1988, the legal staff of the enforcement agencies have in fact moved towards a more moderate and vigorous approach to antitrust enforcement. With regard to the analysis of entry barriers, for example, in 1985 the Reagan FTC adopted the Chicago School definition of entry barriers as additional long run costs that must be incurred by an entrant relative to the long run costs faced by incumbent firms. (Echlin Manufacturing Co., 105, FTC 410, 485). Under this definition, any sunk costs, diseconomies of small scale, product differentiation or other strategic conduct by incumbents are not entry barriers. By 1990, however, Kevin Arquit, the Director of the Bureau of Competition and Judy Whalley, Deputy Assistant Attorney General in the Justice Antitrust Division explained that the enforcement staff had moved away from Professor Stigler's definition. Arquit explained this shift as follows:

> The test in the Bureau is not whether based on some theoretical model, entry could occur, but whether in fact sufficient entry will likely take place in response to an anticompetitive increase, and on a timely basis so as to deter or prevent supracompetitive pricing (Arquit, 1989, pg. 4).

The authors of an article in the American Bar Association magazine, *Antitrust* titled "Justice, FTC Signal Tougher Merger Enforcement Standards" conclude:

> the message . . . is clear. Reliance on the Stigler formulation or an intuitive approach based on the nature of the industry in question . . . will not be sufficient at the agency enforcement level (Bell and Herfort, 1990, p. 7).

With regard to analysis of the ability to exercise market power in more concentrated markets with barriers to entry, enforcement staff similarly have retreated from an exclusive focus upon collusive pricing. James F. Rill, Assistant Attorney General for Antitrust emphatically makes this point stating:

> it is important to consider both coordinated and noncoordinated views of competitive effects when analyzing a merger of firms in a highly concentrated market where entry is not likely. The term "noncoordinated" refers to firms' independent decisions about price and output-decisions that do not rely on the concurrence of rivals or on coordinated responses by rivals. In contrast, the term "coordinated" refers to such conduct as either tacit or overt collusion, price leadership, and concerted strategic retaliation-conduct that requires the concurrence of rivals to work out profitably. The Department considers both noncoordinated and coordinated effects, but often the parties to a merger or their counsel are prepared only to discuss collusion or other coordinated effects (Rill, 1990, p. 51).

This retreat towards moderation suggests a renewed commitment to empirical analysis of industries, however, theoretic analyses of "rational" behavior that is essentially devoid of empirical content (hypothesis testing) still seems to dominate.[2]

However, research during the 1980's produced a comprehensive critique of the empirical work, especially cross-section studies of the concentration profit relationship for the entire manufacturing sector of economy. Scherer and Ross in their 1990 text summarize the current state of knowledge. Citing Ravenscraft (1983) and subsequent research, they conclude that market share is positively related, and concentration is negatively or not related to profits.[3] For the entire manufacturing sector of the US economy, the relationship between industry profits and concentration now seems to be spurious and due to aggregation bias (Sherer and Ross, 1990, P. 430).

Scherer, Ravenscraft, Shepherd, and others have cautioned against moving from this result to the conclusion that the profits of large market share firms are only due to the superior efficiency a la Demsetz of large share firms. Scherer and Ross conclude their analysis of possible sources of the share-profit relationship by stating:

> The positive profit-market share relationships observed in line of business studies represents a still-unknown mixture of temporary efficiency differences and more or less durable monopoly power. Disentangling the relative importance of the two effects . . . is the great challenge facing empirical industrial organization researchers (Scherer & Ross, 1990, p. 433).

Similarly Shepherd states:

> Market share is the unifying basis for evaluating market power, pricing behavior, and restrictive actions. Market structure is not closely determined by costs; substantial excess market share exists. Reducing the issue to 1) collusion versus 2) an efficient structure hypothesis is wrong and misleading (Shepherd, 1986 p. 53).

Several industrial organization economists however, have ignored this view, and Keith Anderson is among them. He writes:

> the authors of some recent studies have suggested that a positive coefficient on market share is evidence of market power in some way. How this occurs is not very well specified [See in particular, Shepherd (1986a), pp. 34-35 and Shepherd (1986b), pp. 1205-1206. Also see Ravenscraft (1983), Mueller (1986), and Borenstein (1988) (Anderson (1990) p. 177].

Anderson accepts the efficient structure-collusion position as developed by Cowling and Waterson, (1976) and Clark Davies, and Waterson (1984) wherein the profits due to high market share are assigned by theory entirely to the Demsetz efficiency explanation. More specifically, Anderson argues that empirical work that uses the relative market share, four firm concentration specification at the business unit level is not well grounded in economic theory. I disagree. Decomposing market share at the business unit level into two components, relative market share and concentration, produces a more general model that has the Cowling and Waterson specification nested in it. This model allows one to test to see if that specification is appropriate, and work on the entire PIMS data set and for the food manufacturing sector indicates that it is not. (Cotterill and Iton, 1991). This more general operationalization of oligopoly theory provides a theoretical basis for the relative share concentration specification used in much of the earlier structure performance research on food manufacturing and retailing (Connor *et al.* 1985, p.335, Marion *et al.* 1977, 1979, Cotterill, 1986).

The Demsetz critique of the industry level concentration profit studies generated another approach to the analysis of market power, the evaluation of structure-price relationships within particular industries. Demsetz maintained that the observed concentration profit relationship may be due to lower costs instead of higher prices (Demsetz, 1973, 1974). However, if one can directly analyze the concentration price relationship, and document that it is positive, then one has a direct test for market power. Recently Weiss has published a set of structure price studies that tend to confirm a positive concentration-price relationship (Weiss, 1989). The structure-price study in food retailing completed by Marion *et al.*, at the University of Wisconsin for the Joint Economic Committee of the Congress is the pioneering work that stimulated Weiss and others to examine other geographically dispersed industries to measure structure price relationships. (Marion *et al.*, 1977, 1979).

Anderson maintains that the price and profit studies for the food retailing industry are so poorly done that they do not provide reliable guidance for policy. Empirical work can always be improved as more detailed data and more refined methods of analysis become available; however, many of the studies that Anderson critiques have been relied upon by state attorney generals and private third party firms to challenge mergers that the FTC did not challenge or approved subject to cosmetic consent decrees during the 1980s. The most notable example is the California Attorney General's successful challenge (1990) of the American Stores-Lucky merger. The Federal Trade Commission approved this multi-billion dollar merger subject to the sale of approximately 35 supermarkets. Also, the Federal Trade Commission has relied upon these studies in its challenge or negotiation of consent decrees in some merger matters, most notably the National Tea Applebaum merger (1979), the Safeway sale of its El Paso division to Furrs (1987) and the acquisition of Grand Union by Miller Tabak and Hirsch the investment holding company that owns P&C Markets (1989).

In this paper, I will respond to each of Anderson's particular points in the order that he makes them to facilitate comparison. First, I will discuss the structure-price studies and then the structure profit study. Thereafter, I will discuss some issues not raised by Anderson and comment on how research in this area might proceed.

Structure-Price Studies: Controlling for Costs

Anderson's basic criticism of the structure-price studies in food retailing is two-fold. First, he maintains that the studies do not adequately control for differences among markets in the cost of retailing food. The cost of retail labor, for example may be different in different cities and thus retail prices might be different. Second, he maintains that within a particular market different firms could have different prices because they offer different levels of service, including quality, to consumers.

With regard to the first point, Anderson claims that the general food price level will vary among markets to reflect difference in retail costs, and that failure to control for these intermarket variations in costs may explain why more concentrated markets have higher prices. Specifically he argues that we do not know if more concentrated markets have higher costs. If they do, it could be the cause of the concentration-price relationship. His first candidate is labor costs; yet, as he notes, all of the studies do control for wage rate differences in different markets, and the concentration-price relationship persists.

His second cost control candidate is the difference in the cost of goods sold. Anderson notes that Lamm (1981) and Hall *et al.* (1979) specify the BLS price index for food, but I don't believe Anderson realizes that this is a national index value and as such is constant across local markets in a point in time. It is of no use in a cross section study. In footnote 9 of his paper, Anderson correctly explains that in a cross section study such as the JEC, Vermont or Arkansas study, the branded and private label processed food products and the nonfood products included in the grocery basket of items are produced nationally and therefore there is relatively little variation in their price to integrated retailers except for transportation.

Changes in procurement practices and public policy since these price studies were conducted (1974, 1981, 1982) probably makes the constant procurement price but for transportation less tenable today. Yet even if procurement prices do vary, it is very unlikely that the chains analyzed in these studies now pay higher prices for food products in local market areas that have high retail concentration. Economic theory, and the decline in enforcement of the Robinson-Patman act proscription against secondary line price discrimination jointly predict that large local buyers would pay lower not higher prices.

Industry analysts from Goldman Sachs describe the current state of affairs as follows:

> Supermarkets just have to recognize that the marketplace is becoming a free-for-all, that Robinson-Patman is breaking down, that diverting is here to stay, that deals are here to say, that nobody really knows what their competitors are paying, and that buyers just have to be sharp and use their leverage to their best advantage. For a supermarket chain with a leading share position in a major market, that leverage is considerable. Most manufacturers go to market regionally, and thus a 30% or greater market share for a retailer represents powerful control over a limited commodity—shelf space. The rise of slotting allowances and display allowances is simple economic proof that retailers can charge increasing amounts of money for their "real estate" (shelf space) (Mandel and Heinbockel, 1989, p. 22).

Thus, it is not likely that procurement prices generate a spurious positive concentration-price relationship. Anderson provides no evidence that prices paid for products by integrated chains are higher in more concentrated local markets. In the 1990s it is very doubtful that anyone would find such evidence.

For the Arkansas/Vermont studies there is additional reason to conclude that procurement prices are not positively correlated to local retail market structure variables. Since the Arkansas study focuses upon 32 local markets in that state or near its border in six surrounding states, all procurement is located essentially at the same spot. This is even more true for the Vermont study. Twenty six of the 35 observations come from two leading chains and each chain had only one warehouse supplying the Vermont area. Moreover the concentration price relationship holds for observations from each chain as well as the full sample. (Cotterill, 1986, p. 383).

Anderson's alternative theory of variation in the cost of goods sold proffered in footnote 9 of his paper is not coherent. He suggests that supermarkets could and would purchase higher priced locally produced goods if they can charge higher prices for other goods. His argument could also apply to higher priced goods from any location worldwide. In addition to the fact that this is not profit maximizing conduct, this reasoning violates cross market price comparison methodology. The reason for conducting a structure-price study in an industry with local geographic markets is to compare the price of the same product across several firms and markets to see if firm and/or market structure influence the price. It makes no sense to compare the price, for example, of a locally produced food in one store to a leading national brand in another store.[4]

Anderson's next cost difference candidate is differences in the prices of real estate and utilities, and local taxes among local market areas. He recognizes that Lamm partially controls for variation in these with his binary variables for region of the country, and Marion *et al.* may do an even better job by specifying city size in their model. In my opinion, since these costs represent a very small fraction of the retail price of food, and since it is very unlikely that their prices are positively correlated with market concentration or a firms' market share, they do not offer an alternative explanation for the concentration-price or relative market share price relationship reported in the studies reviewed by Anderson. Leading firms in concentrated markets may, if anything, be more likely to receive price discounts on real estate because of their desirability as an anchor tenant in a shopping mall.

Next, Anderson does a flip flop and argues that more concentrated markets have lower prices. He conjectures that markets that are served by identical large supermarkets that are units in firms with identical and large market shares will have lower costs due to real economies of scale and that competition would force them to pass these on to the consumer as lower food prices. This is the contestible markets hypothesis. In addition to the structure price and profit studies reviewed by Anderson, work on entry by Cotterill and Haller (1987, 1992) and Marion (1987) document that retail food markets are not contestible.

Anderson provides no evidence that large stores enjoy economies of scale. To my knowledge the most recent study is in Marion *et al.* and it found no scale economies for traditional format supermarkets from one chain that ranged in size from 13,000 to 31,000 square feet. (Marion *et al.* 1979 p.136). Also this reasoning suggests that one would observe a trend toward very few firms with large equal market shares and uniformly large identical stores. Anderson provides no evidence on this point and a cursory review of recent new store formats indicates that it is not correct. Food Lion, for example, is doing very well building 25,000 square foot traditional supermarkets. (Poole, 1991). Albertsons has prospered over the past decade with the combination food-drug store format. Others have advanced with the warehouse format, and the superstore format is the most common new unit. Economies of scale at the store or local market level are not the primary drivers of the strategic plans of large supermarket chains.

Anderson's final point concerning intermarket costs differences is to return to his argument that high concentrated markets may have higher costs. He correctly notes that smaller markets tend to have higher concentration . Then he hypothesizes that firms would not be able to achieve economies of scale in these small markets and thus would have higher costs that they would need to pass on to consumers. There are four answers to this hypothesis. First, Anderson cites no evidence of economies of scale that are so large in this industry that small or medium SMA's would force firms to operate below minimum efficient scale. Second if such economies exist and are important then we would expect to observe a more rapid trend towards uniformly large stores operated by as fewer firms possibly even one firm in these smaller markets. Third, average unit costs for retailing are a function not only of possible economies of scale but also the price of the inputs. The fact that the price level for local inputs such as labor, real estate, and possibly utilities tends to be lower in small or medium cities probably more than offsets any diseconomies related to small sales volume. Fourth, is the issue of causality. If more concentrated markets do have higher retailing costs, these higher costs may be due to x-inefficiency. Retailers may share the benefits of market power with input suppliers including labor and real estate owners.

Structure-Price Studies:
Controlling for Variation in Services

Andersons thoughts about differences in the price-service mix and its impact upon structure-price relationships suffer from the fact that the price determination model implicit in his analysis is too restrictive. He assumes in equilibrium all firms in a local market will charge the same price and that any dispersion in the equilibrium price charged by a firm is due to differences in the costs of the services including quality that they provide. Yet studies by Devine and Marion (1979) and others have demonstrated that consumers have imperfect information on food prices. This suggests that different firms could charge different prices for the same price-service mix in a market. Also it is entirely possible, even with perfect information that one supermarket chain is able to differentiate its enterprise from others and charge a higher price then competing firms for a set of groceries and services.

These points indicate that firms within a market may have higher prices, not only because they need to cover the costs of more services, but also because they are able to exercise market power due to imperfect information and/or superior enterprise differentiation. Anderson, for example, would attribute any positive relationship between a firm's market share and its price level to cost differences related to the "superior price service mix" that large share firms provide. Again, this is not a fact. It is a hypothesis that requires testing. Moreover if it is true, then one would not observe as we do a strong relationship between a firm's relative position in a market and its profitability. Anderson cannot have it both ways at the same time, i.e., leading firms cannot have higher prices due to higher cost "price-service mixes" and have higher profits due to superior efficiency.

Anderson maintains that the only study that controls for differences in services is the Kaufman-Handy study (1990). They count up the services offered by a firm in a local market giving double weight to some to produce an index. Their index has a significant positive impact on price. However, Anderson needs more than this result to conclude that the positive price relationship structure reported by other studies are spurious because services are not specified in the model. If removing the service index from the Kaufman Handy model produces significant positive structure price relationships then we would have evidence that the structure price relationship of the other studies may be due at least in part to the costs of higher services in noncompetitively structured markets. However, given the major flaws that Geithman and Marion (Geithman and Marion, 1992) have uncovered in the retail price survey and index computation methods of the Kaufman-Handy study, it is most unlikely that any respecifications of their model will provide a reliable insight on the vigor of competition.

Moreover, even if higher costs exist in noncompetitively structured markets, if profits are also related to structure then enterprise differentiation is operative. Firms with larger relative market positions in markets with higher concentration have higher prices, part of which covers the costs of the differentiating services and part of which generates higher profits.

Looking more carefully at some of the other studies also suggests that their failure to explicitly introduce a services variable is not crucial. The JEC price study by Marion *et al.* was for the local market operations of only three chains in 1974, well before the explosion in store format and service options. Even today the essence of the chain store operation is shared common systems. Independent supermarket operators, rather than the chains are known for their flexibility and ability to adapt to particular local situations.

In the Vermont study there was a binary variable to identify independents and they did have higher prices. Also the fact that the structure price relationship held in the Grand Union and the P&C subsamples indicates that there is no bias due to pooling observations. Having visited several chain stores in the State, interviewed the staff of the Vermont Retail Grocers association, reviewed operations and pricing records of these chains as part of a court case, and having participated in a lengthy court trial, I observed that the issue of different store formats or service levels only arose in one fashion. Grand Union and P&C stores in Vermont in 1981 were old, provided relatively few services, and rarely did any sort of merchandising because there was no competitive stimulus to do otherwise.

The Arkansas study controlled for store format (traditional, superstore, warehouse) and like the Vermont study specified store size to capture the price-service mixes and cost conditions related thereto. The store size results are interesting because a significant quadratic relationship exists in both studies. The quadratic relationship also was reasonably robust for the individual chain regressions in Vermont. Smaller supermarkets have higher prices, moderate sized units have lower and the largest units which are most likely the newest with the broadest product and service assortment have higher prices. In Vermont with its older, smaller supermarkets, the least cost size was 16,000 square feet. (Cotterill, 1986 p. 384). In Arkansas and surrounding states where superstores as large as 65,000 square feet were in operation, the minimum was very near the 30,000 square feet cut off between the traditional and superstore formats. It was 33,000 square feet. (Cotterill, 1983 p. 118). Clearly, store

formats and/or store size do affect price levels. In my opinion there is no doubt
that diseconomies of store size affect very small supermarkets (less than 20,000
square feet). Operators of these stores survive by differentiating themselves in
one fashion or another. The largest supermarkets, rather than suffer
diseconomies, are able to differentiate themselves and not only cover the higher
costs of doing so but also generate more profit due to higher prices. Superstores
are commonly acknowledged to be more profitable than other stores (Mandel
and Heinbockel, p. 10).

Concentration and Profitability

The Joint Economic Committee study completed by Marion *et al.* at the
University of Wisconsin contained in addition to a structure price study a
companion study of the relationship between structure and profits. Anderson's
critique has four major points: first he raises the Fisher McGowan critique that
accounting profits are not economic profits. Second, he questions the nonlinear
functional form used for concentration in some models; third he mentions that
market share, not relative market share is the correct specification; and, fourth
he maintains that the model overstates the impact of concentration on profits,
and in certain cases profits actually decrease when concentration increases. Also
in the Economic Issues paper, Anderson critiques our treatment of the A&P
company and questions the validity of the results because the industry was under
wage-price controls during part of the five-year 1970-1974 period analyzed.[5]

Before launching into specifics, one general observation may be helpful.
This structure profit study is not a typical cross-section study of four digit
census industry profitability or firm profitability or line of business profitability.
It avoids many of the criticisms of these studies. Specifically this study is for
one industry. It moreover examines intrafirm profitability in different local
markets to assess how local market structure influences profitability. For
example, we can, and did estimate the structure profit model for 28 observations
from the A&P company. Rather than decry this lack of generality as Anderson
does, one should welcome the ability to test hypotheses for specific firms. To
my knowledge no other cross section concentration-profit study has been able
to hone in upon one management team and assess its strategic conduct in
markets that are essentially identical except for strategic factors such as market
growth, market share and concentration levels. As expected, A&P's profits were
significantly related to market structure prior to their W.E.O. campaign, and not
related to it during or after that massive price cutting exercise.[6]

Our larger samples include local market operations for 6 and 12 of the top
17 chains of 1972, and there are multiple observations from individual chains.
In fact chain identity as measured by each chains internal total company growth

rate, which we interpret as a proxy for a chain's managerial acumen, is the most powerful determinant of a firm's profitability. The idea that well-managed companies, as measured by their ability to expand by internal growth, are more profitable is clearly supported by this study. In the early 1970s and today, Albertsons, for example, is commonly acknowledged to be an excellent managed company; and, then as now it is significantly more profitable than a laggard such as A&P.

Yet, as the study demonstrates, this firm level effect does not detract from the fact that the profitability of companies in a local market is significantly influenced by its relative market position, seller concentration, growth in market demand, and other features of market structure. Also these models, once corrected for heteroskedasticity related to the proxy for firm managerial acumen, routinely explain over 80 percent and in certain instances over 90 percent of the variation in the five year 1970-1974 average profit sales ratio. When one has data that are as disaggregate as these data are, one can obtain a very complete evaluation of the impact of market structure upon firm and industry profitability.[7]

Now turning to particulars, Anderson raises the Fisher-McGowan critique. The methodological overview section of the intoductory chapter of this book explains and rebuts Fisher and McGowen. The JEC profit study uses the pretax profit-sales ratio for firm operations at the local market level. With regard to the Fisher-McGowan critique, Anderson states:

> the ultimate question is whether the errors (in using accounting profitability to measure economic profitability) that arise with the data for real firms are significant and whether they are correlated with variables included in the regression models being estimated (Anderson, 1991, p. 8).

In this study the structure-profit model was estimated for samples of local market operations from six and twelve leading chains and for 28 A&P observations. Accounting conventions within a particular firm and within a particular industry such as food retailing most likely do not vary from local market to local market. Also the nature of the investment and the shape of the cash flow streams from those investments is similar across these firms, suggesting that there is little interfirm bias of accounting profit sales ratios. Within a firm, if these profit rates solely reflect different accounting conventions, different cash flow patterns or the current expensing of long term assets such as advertising, and thus say nothing about economic profits, then how does a firm evaluate the performance of operations in different local markets? In fact they do use these local market profit rates as measures of economic returns, and we will too.

Anderson's point that market power is capitalized into the purchase price of a firm and thus return on assets and return on equity profit rates in the subsequent firm do not measure market power is correct. Given the ownership changes that occurred in the 1980s, this is a major problem today; however, the 12 large firms included in this study of profitability in the 1970-1974 period were not recently acquired and due to the stringent merger policy of that era had not made major acquisitions.

The criticism that the profit sales ratio represents price minus average cost, not price minus marginal cost (the Lerner index) and therefore is not an accurate measure of market power is over emphasized. All profit rate measures, including the Lerner index, have advantages and disadvantages as a measure of market power. The distinction between average and marginal profit sales ratios is of minor consequence in this industry for the following reasons. First this is primarily a study of long run, five year average profitability. Thus the relevant cost curves are long run constructs. A firm increases its market share in a local market in the long run by adding stores. This would influence long run average and marginal costs in all stores only if there are not constant returns to scale. No one maintains that there are long run real diseconomies of scale in local food markets. Although there is no research on long run real economies of scale at the local market level, I don't believe anyone maintains that they are significant. To the extent that there are real economies of scale, the price average cost margin would understate the Lerner index, but even then these profit measures most likely remain highly correlated and close in value.

Moving onto another issue, Anderson's comments about the need to adjust the net profit sales ratio by subtracting the competitive rate of return to produce a measure of economic profits are not as clear as they could be. In theory the cost of capital is included in costs, i.e., the net profit sales ratio is net of that cost as well as others. In practice the net profit sales ratio is the return to equity holders in the firm; and, thus we do not expect it to be zero in competitive equilibrium. Anderson confuses the cost of capital with the capital-sales ratio, and actually would like us to adjust for both. Yet the oligopoly theory articles that he so heavily relies upon, specifically Cowling and Waterson, and Clark, Davies and Waterson do not make these adjustments in their related empirical work. The problem is more serious for these studies than our study because we are analyzing variation in profitability for one firm and small sets of firms within one industry. The competitive rate of return required by investors and the capital sales ratio are not likely to vary much across our sample. Again, for it to explain our structure profit results, firms with large relative market shares and those in concentrated markets have to have higher required rates of return and higher capital sales ratio. On this latter point, there is no readily observable preference by firms for more capital intensive store formats in more concentrated market, or where they have leading shares.

Moving from measurement problems to model specification, Anderson's major contention is that using relative firm market share instead of market share has no basis in oligopoly theory and thus constitutes a serious misspecification.

Yet as demonstrated by Cotterill and Iton, one can generalize oligopoly theory by decomposing market share into two components, relative market share defined as market share divided by four firm concentration, and market concentration. This enables one to avoid the efficient structure-collusion dichotomy of the Cowling and Waterson model with its complete assignation of a positive market share-profit relationship to the Demsetz hypothesis. As Scherer and Ross, Shepherd, and many others have argued, that model begs the question.

If the industry is Cournot in our more general model, one obtains a linear specification of relative market share and concentration as explanatory factors for the profit sales ratio. Relative market share measures differential efficiency. The concentration ratio measures the size of leading firms relative to the size of the market, and as such indicates how close leading firm demand curves are to the market demand curve, and their ability to increase profits by exercising market power. In a more general model that allows firms to charge different prices due to enterprise differentiation, relative market share could also be related to profits because leading firms who have differentiated themselves are able to charge higher prices. In fact, the JEC study and the Vermont price study do find that firms with high relative share have higher prices. (Marion, *et al.*, 1979, Cotterill, 1986).

If the industry is not Cournot and one models, as prior theorists cited in this paper have, the conjectural variation parameter as a function of market structure, a nonlinear relationship between concentration and profitability can exist. (Cotterill and Iton, 1992). Anderson correctly maintains that the nonlinear concentration profit relationship found in Marion *et al.* does not identify a critical concentration ratio. We never said that it did. We looked for a critical point and did not find one. The observed relationship, however is not inconsistent with the generalized Cournot oligopoly model that I have discussed here. Profits are higher in more concentrated markets due to the exercise of market power.[8]

I agree with Anderson that concentration is not as important as are some other variables in the determination of firm profitability; however, it is a significant determinant of profits. To my knowledge, no industrial organization economist, has ever maintained that concentration is the most important determinant of firm performance. Based upon the research reviewed in this paper, a firm seems primarily to secure market power and related profits in local retail food markets through enterprise differentiation associated with a large relative market share position; but, high concentration is also beneficial.

Moving to Anderson's second criticism of relative market share, he argues that researchers who use it misunderstand competitive equilibrium and the role of the marginal firm or least efficient firm. According to Anderson, if they did understand this role, they would use market share not relative firm market share to measure differential efficiency. With due respect, it is Anderson who is confused. Using relative market share instead of market share does not negate or mismeasure the marginal or least efficient firm's role in determining efficiency rents.

Lets look at Andersons example as reproduced in Table 11.1. His marginal firm has a one percent market share and the next smallest firm has 10 percent. He argues that the difference in share position between firm 1 and firm 2 should not change when one shifts from example 1 to example 2. Citing the fact that RFMS declines from 25 to 12.5 percent, he maintains that the difference in share position does change and that this change distorts the test for differential profitability between these two firms. Note, however, that the relative share of the marginal firm also is halved from 2.5 to 1.25 percent, and that the number 2 firm remains 10 times larger than the fringe firm. In a single market, relative market share does no better or worse than market share in measuring the relative size distribution of firms. However, when one moves to cross section analysis of several different sized markets, relative market share is superior to market share. For example, if one examines a second market and it is twice as big as the market in Table 11.1 then all market shares will be one half of those reported in Table 11.1 but relative market share will remain at their reported values. Therefore in a cross section sample of several markets, relative share is the appropriate measure for the differential efficiency related to firm size.

One can also illustrate the superiority of relative market share, concentration specification as follows. If four firm concentration is 40 percent and all firms in the industry have 25 percent relative market shares (10 firms at 10 percent market share) and if market concentration increases to 80 percent with all firms retaining the same 25 percent relative share position (5 firms at 20 percent market share) then an observed increase in profits, if any, is due to market power, not increased efficiency due to higher market shares relative to fringe firms. There are no fringe firms. In this equal share case the four-firm

TABLE 11.1 Alternative Share Distributions in a Market

| | Example 1 | | Example 2 | |
| | Share (%) | Relative Share (%) | Share (%) | Relative Share (%) |
Firms				
1	1	2.5	1	1.25
2	10	25	10	12.5
3	10	25	20	25
4	10	25	20	25
5	10	25	20	25
6	-		20	25
CR_4	40		80	

concentration ratio is equivalent to the more familiar measure of concentration in the Cournot model, the number of firms in the market.

Relative market share is also preferable to market share on statistical grounds because by definition market share and four firm concentration are correlated and relative market share and concentration are not.[9]   When correlation between two explanatory variables is a sample problem gathering more data can mitigate multicollinearity. However, when the correlation arises from a theory that requires specification of two variables that by definition are correlated, one has to question the usefulness of the theory, and attempt to provide a more tractable theory.

Finally, as demonstrated by Cotterill and Iton (1992), aggregation of the market share, concentration specification from the firm to the industry level produces a very unattractive model with both concentration the four firm ratio and the Herfindahl index as determinants of industry profitability. This specification clearly illustrates the multicollinearity problems with the share, concentration specification at the firm level. The relative share concentration specification is more attractive. It aggregates to two variables that are a decomposition of the Herfindahl index: the four firm concentration ratio and the Herfindahl index divided by the four firm concentration ratio.

Anderson's final critique of the structure profit study, may very well be the most misleading. He maintains that the concentration profits relationship is overstated because relative market share is used instead of firm market. I would argue the exact opposite. The concentration—profit relationship is understated when market share is specified instead of relative market share (Cotterill and Iton, 1992). When we evaluate the impact of concentration on profits, we hold relative share constant, and when evaluating the impact of relative market share we hold concentration constant. In real markets both of these variables can and often do increase jointly. Anderson would increase concentration holding a firms market share constant and compute the change in the profit rate for that particular firm. But if concentration increases and one firms share doesn't increase then one or more other firms must have be increasing market shares and their relative market shares go up. Anderson's conclusion that profit rates fall in more concentrated markets focuses upon the firm that is left behind as the market becomes more concentrated. I agree that the laggard firm's profitability declines as it loses relative position to other firms. The real question, however, is what happens to total industry profits as the market becomes more concentrated.

Table 11.2 reproduces and expands Andersons' average and high market share cases to answer this question for market that has 1.0 billion in total sales. It uses the same estimated equation that Anderson uses, with a statistically significant nonlinear relationship between concentration and profits. Note in the high market share case in Table 11.2 with four firm concentration at 40 percent, a leading firm with a 25 percent market share has profits totaling 9.8 million

TABLE 11.2 Change in Industry Profits Resulting from Change in Relative Market Share and Curvilinear Concentration (CCR4)[c]

Firm	Market Share (%)	Relative Market Share (%)	Firm Sales (Million $)	Relative Firm Share Profits (Mil$)	CCR4 Component Profits (Mil$)	Total Profits (Mil$)
High Market Share Case						
Before Market Concentration						
1	25	62.5	250	9.844		
2	5	12.5	50	0.394		
3	5	12.5	50	0.394		
4	5	12.5	50	0.394		
Other[a]	1	2.5	600	0.945		
Profits (Mil$)				11.97	28.242	40.212
After Market Concentration						
1	25	31.25	250	4.922		
2	20	25	200	3.15		
3	20	25	200	3.15		
4	15	18.75	150	1.772		
Other[b]	1	1.25	200	0.158		
Profits (Mil$)				13.151	36.61	49.761
Change in Industry Profits (Mil$)				1.181	8.368	9.549

Low Market Share Case

Before Market Concentration

1	10	25	100	1.575		
2	10	25	100	1.575		
3	10	25	100	1.575		
4	10	25	100	1.575		
Other[a]	1	2.5	600	0.945		
Profits (Mil$)				7.245	28.242	35.487

After Market Concentration

1	20	25	200	3.15		
2	20	25	200	3.15		
3	20	25	200	3.15		
4	20	25	200	3.15		
Other[b]	1	1.25	200	0.158		
Profits (Mil$)				12.758	36.61	49.367
Change in Industry Profits (Mil$)				5.513	8.368	13.881

[a] 60 firms with 1 percent market share each.

[b] 20 firms with 1 percent market share each.

[c] Based upon equation 1b Table 3.5, Marion, *et al.* (1979) p. 82.

dollars. The other firms are much smaller and their profits due to relative position are much lower. Total industry profits consist of the profits due to relative position (11.97 million) and those due to the level of concentration (28.24 million). The total is 40.2 million dollars.

Actually, Anderson and I are not interested in the level of profits which depends upon the values of several other variables in the regression. The question is how do these numbers change when concentration increases to 80 percent. After this change in market concentration the relative share of the leader declines one half, and its profits due to relative position also declines one half to 4.922 million dollars. Is this an unreasonable result? I think not. Note that the relative shares of the second to fourth firm increase and their profits due to relative position increase. The leader has lost its commanding position, and consequently it is sharing industry profits with firms that are now more its coequals. Scherer and Ross show that a market with a Stakkelberg leader and *n* followers will have a lower price and higher output than a market where all firms hold Cournot conjectural variations (Scherer and Ross, 1990, p. 225).

One can find examples of this type of change in grocery retailing. Steve Weinstein, the editor of Progressive Grocer writes:

> Safeway has been the king of Seattle for many years. but while the chain still reigns there, with a market share of more than 30%, there are ample signs that some of its subjects are getting unruly . . . The chain at one time had a market share approaching 50%, although it would never acknowledge the figure was that high, according to one observer . . . 'Four or five years ago this was a me-too market', says one wholesaler official, but not anymore. Competitors no longer are content to follow them on pricing . . . (Weinstein, 1987, pg. 21).

Returning to Table 11.2, industry profits also increase because concentration increases. The total increase in industry profits is 9.599 million dollars. Since total market sales remains constant in this example at 1.0 billion dollars this amounts to an increase in the industry profit rate of .95 percent of sales. Looking at the low market share case in Table 11.2 the increase in industry profits when moving from 40 to 80 percent concentration is much greater. It is 13.88 million dollars or 1.388 percent of sales.

Appendix Table 11.A.1 contains an analysis of Anderson's high and low market share cases using a model that specifies concentration linearly. The changes in industry profitability when concentration increases from 40 to 80 percent are larger. In the high share case, the industry profit sales ratio increases by 1.16 percentage points. In low share case it increases by 1.59 percentage points. For comparison the average pretax profit sales ratio for these companies during the 1970-1974 period was 1.45 percent. (Marion *et al.*, 1979, p. 62). Therefore the changes in profitability due to changes in concentration significantly enhance the profitability of the industry, and in both low share cases the increase approximately equals the average profit rate for these firms.

The implications of this result for merger policy needs to be clearly drawn. Prior analysis of rivalry in more concentrated markets, most notably Kwoka and Ravenscraft (1986), have missed the distinction between changes in profits for a leading firm, and changes in total industry profits. If one defines market rivalry as a decrease in industry profits, then allowing smaller firms to merge to challenge a leaders reduces the leaders profitability, but at least in this industry it does not increase market rivalry. Industry profits go up. Rivalry is not inducing firms to pass profits on to consumers in the form of lower prices.

Additional Comments and Research Possibilities

In his conclusions to his Economic Issues paper, Anderson states that the Kaufman and Handy study is the only study that is an appropriate basis for policy in this industry, because it is the only study to control adequately for costs and services. As explained earlier, several of the other studies do control for costs and services by either the explicit introduction of control variables or sample selection. I would be more receptive to Anderson on this point; if deleting the cost and service variables in the Kaufman Handy regression produced positive and significant coefficients for relative share and concentration. I doubt that it does; moreover the Kaufman Handy study has unique flaws in its price survey and aggregation procedures that compromise its reliability (Geithman and Marion, 1992) Anderson is unaware of these problems.

Another comment in Anderson's Economics Issues paper suggests a serious lack of familiarity with this industry. The Joint Economic Committee study found that prices seem to rise more rapidly than profits as one shifts to more concentrated markets. Marion, *et al.* concluded that these results in the 1970-1974 period may indicate the presence of x-inefficiency in the operations of these large chains in noncompetitively structured markets. Anderson rejects this conclusion stating:

> The (high) level of forgone profits suggests that this is unlikely to be the explanation. It seems likely that such performance would quickly mark the firm as an attractive takeover target for someone with an eye to improving operations and profitability (Anderson, 1990, p. 76).

Anderson is unaware of the merger, takeover, and leveraged buyout wave that completely restructured this industry during the 1980's. Table 11.A.2 in the appendix documents the extent of those changes on the top 20 supermarket chains. Between 1979 and 1989 mergers leveraged buyouts, or leveraged recapitalizations affected 81.6 percent of the top 20 chain sales.[10] Investment analysts at Goldman-Sachs speaking before the Food Marketing Institute Financial Executive Conference, May 1989, document the extent of the change in another fashion. They write:

> the aggregate amount of debt assumed by supermarket chains as a result of leveraged buyouts or recapitalizations over the 1986-1989 period alone exceeds $20 billion, which is greater than the aggregate market value of all publicly traded supermarkets today (Mandel and Heinbockel, 1989, p.1).

The fundamental question concerning the performance of the food retailing industry today is where is the increased cash flow necessary to cover the massive debt load of the industry coming from? Permit me to quote the Goldman-Sachs analysts speech to supermarket finance executives again at length. The underlining for emphasis is in the original text.

> LBO buyers have been lucky thus far, not only because of a favorable overall economic and stock market environment, but also because <u>many of these buyers did not, in our opinion, foresee the positive structural economic changes that are occurring in the supermarket business: the power of large store formats, increasing leverage versus suppliers, and market concentration, all of which have led to more rapid operating margin improvement than we had forecast</u> . . . it is certainly important from your point of view to understand the implications of this LBO phenomenon on the economic structure of the supermarket business. We see three primary implications: 1) increasing market concentration in major metropolitan markets, 2) increasing pressure on vendors, and 3) opportunities for aggressive, well-capitalized operators to pick up abnormal market share. The LBO phenomenon has accelerated the process of market consolidation . . . weak markets are sold off. Instead of Safeway deluding itself into thinking that one day it would become number one in southern California, management sold to Vons and chose to be a stockholder (30 percent ownership), hopefully benefitting from the improved economics of the combined company . . . Kroger sold its northern California Fry's stores to Savemart, and so on . . . The market share changes that have occurred in the country's two largest markets—New York and Los Angeles—over the last five years illustrate the impact of increasing concentration. Five years ago, five chains split 55% of the Los Angeles market. Now, three chains—Ralph's, Vons, and Lucky control 65%. Not surprisingly, the current returns of Ralph's, Vons, and Lucky are far superior to their returns of five years ago. <u>The Los Angeles and New York markets have had a reputation for being two of the most ruthlessly competitive markets in the country, but the reality has been record operating margins for most of the chains in both markets (e.g., Ralph's EBITD (earnings before interest, taxes and depreciation) margin is 7%, and A&P's profitability is now close to that in the Metro New York region)</u> (Mandel and Heinbockel, 1989 p. 4-7).

Commenting in the wake of the 1988 Kroger leveraged recapitalization to avoid hostile takeover. Edward Comeau, an analyst with Oppenheimer and Company similarly explained:

> Kroger has a history of being the tough guy in most of the markets it's in, dictating pricing and promotions in those markets. But as a highly leveraged company, Kroger is likely to become less competitive and less aggressive than it's been (Zwiebach, 1988, p. 8).

Investment bankers, or at least their analysts, seem quite at home with the benefits of increased concentration and the exercise of market power against input suppliers as well as consumers to increase cash flow. Industry executives also recognize this reality. Erivan Haub, who owns a controlling interest in A&P via the West German firm, Tengelmann, explained in a 1988 interview in *Forbes* magazine how firms such as A&P benefit from LBOs in the supermarket industry when they are competitors of the affected firms: "Through leveraged buyout's and takeovers, A&P's competitors are becoming loaded with debt . . . They will pass along the cost of serving this debt by raising prices" (Fuhrman, 1988).

The question appears to be not whether leveraged firms will raise prices, but how fast will their shares erode? Lets look at some crude but suggestive evidence on this point. Table 11.3 shows how Safeway's (earnings before interest and taxes, EBIT) has increased since its LBO in 1986. It also shows EBIT for two chains that compete directly with Safeway for all of their sales (Quality Foods in Seattle, and Giant in Washington and Baltimore). The comparison, of course is far from perfect because Safeway operates in several

TABLE 11.3 Earnings Before Interest, and Taxes 1985-1990 for Safeway, Giant and Quality Food Centers

Year	Safeway*	Giant Food	Quality Food Centers
1985	2.18	4.73	3.26
1986	2.03	4.05	3.50
1987	2.28	5.24	4.62
1988	2.39	5.70	4.98
1989	3.23	5.75	6.48
1990E	3.65	5.92	6.55

E-Estimated.

* Include results of Canadian and Australian retail food centers as well as all U.S. operations.

Source: Mandel and Heinbockel, 1989, Corporate 10-K's.

other markets. Note that the earnings of all three chains increase dramatically from 1985 through 1990. Safeway goes from 2.18 and estimated 3.65 percent. Giants moves from 4.73 to an estimated 5.92 percent, and Quality Foods moves from 3.26 to an estimated 6.55 percent.

If expansion by the competitive fringe in these markets or entry by firms from outside the market was timely, and sufficient to restrain the exercise of market power one should see declines in the market shares of all of these firms, not just Safeway, and more competitive conditions might ultimately prevail. This has not occurred.[11]

In Seattle there was no entry during the 1985-1990 period. Quality Foods' share of grocery sales in Seattle actually increased from 6.1 in 1985 to 9.8 percent in 1991. Safeway's share in Seattle increased slightly from 24.8 to 25.4 percent between 1985 and 1990.

In Washington, D.C., Shoppers Food warehouse entered and captured an 8.5 percent market share by 1990. This move into a strategic group where Giant and Safeway do not have operations did not affect their market shares. Safeway's grocery market share in Washington remained roughly constant. It was 24.6 percent in 1985 and 23.1 percent in 1990. Giant's share, however, exploded, increasing from 33.2 percent in 1985 to 43.4 percent in 1990.

Safeway's leading competitors appear to have reinvested their high profits in new stores and have consequently achieved significant share gains at the expense of fringe firms rather than Safeway. Thus, there is little evidence that competitive pressures from the fringe or new entrants are eroding the positions of the high profit firms listed in Table 11.3. One of the primary goals of Safeway's restructuring program—one that they have achieved,—is to maintain a number one or two position in every local market in which it operates (Morgenson, 1988).

Clearly there is need for more research on the organization and performance of the food retailing industry. Anyone who maintains the industry is currently performing in accordance with the competitive market norm or the contestable market norm is misinformed. The industry is split between highly leveraged and unleveraged firms with very different short run requirements for survival. The likelihood for noncompetitive pricing in concentrated markets seems higher now than ever before. Moreover, fringe firm expansion or entry may not be timely and of sufficient scope to discipline firms that exercise power over price.

Anderson believes that a definitive new study of the structure-price relationship "would entail a major research effort, which would probably require several work years of effort to collect and analyze all of the necessary data" (Anderson, 1990, p. 48). Again, I disagree. Price product movement and merchandising information is now readily available in electronic form due to the use of scanners in retail outlets. Retailers and manufacturers have developed complex strategic price and merchandising models that use this data to improve their profitability. At the University of Connecticut Food Marketing Policy Center, we have purchased a comprehensive data base that provides price, market share, and 14 other merchandising variables for individual branded grocery products and private label counterparts in 51 local markets on a quarterly basis. This data spans most dry grocery product categories, bakery, dairy, and drink categories; but, it does not include fresh produce or fresh meat. Our initial work with the data is on cottage cheese. We find that a relatively simple model with variables such as the brand's local market share, and retailer concentration can explain over 50 percent of the variation in branded cottage cheese prices. Preliminary results indicate that both of these variables are positive determinants of a brands price in a local market, however, other variables are more significant (Haller, 1992).

In conclusion, I would like to thank Keith Anderson for the substantial effort that he devoted to this project. Although I do not agree with most of his points, they help define the research agenda. The strategic and public policy issues addressed by the studies critiqued by Anderson are important. There will be more structure price studies in the food sector. Moreover, the new data may enable us to unravel the role of product differentiation as a source of market power in a more definitive fashion than has heretofore been possible. No single study or new theoretical approach will ever answer the central questions of industrial organization that we have discussed here. In the meantime, we need to proceed as best we can using theory judiciously and devising tests of hypothesis that recognize and if possible take advantage of new data as they become available. As one economist put it when faced with the Fisher McGowan critique:

> having wandered into the jungle, spied some fresh elephant tracks and smelt an elephant, one must be prepared to conclude that an elephant has recently wandered by (Horowitz, 1984, p. 493).

Notes

1. See Cotterill, (1989), Appendix A or U.S. Congress (1988) for letters to Chairman Rodino from Daniel Oliver, chairman of the FTC and Commissioners Azcuenaga and Strenio.

2. See for example the new 1992 merger guidelines and the discussion of them by several Justice and FTC staff in *Antitrust*, Summer, 1992.

3. Ravenscraft, however, found that in some sectors, including food manufacturing, share and concentration both positively related to profits (Connor *et al.*, 1985,p.335).

4. This, in fact, is a serious flaw in the Kaufman-Handy study. See Geithman and Marion (1992) for an extensive discussion of this and related sampling problems in the Kaufman-Handy study.

5. See the appendix for a discussion of the A&P issue and see Marion *et al.*, 1979, p. 63 for a discussion of the price control issue. My only comment on this latter point is that if anything it would have reduced the liklihood of finding structure profit relationships because it limited upward movements and not downward movements, in gross margins for a few quarters of the 5 year period.

6. See the appendix for a discussion of this result.

7. This point is equally significant when comparing this study's methods and results to price theoretic studies that estimate conjectural variations or Lerner indexes. Most of those studies were aggregate industry level data. If and when they are able to use firm level or intrafirm level data, as this study does, then they may more accurately identify oligopolistic interdependence.

8. See Figure 11.A.1 in the appendix for a graph of the nonlinear relationship.

9. Please see the appendix of Cotterill and Iton (1991) for a proof of this proposition.

10. For more information see Ronald W. Cotterill, "Food Retailing: Mergers, Leveraged Buyouts and Performance", in Lawrence Duetsch ed. *Industry Studies*, Englewood Cliffs: Prentice Hall, (forthcoming 1992).

11. Data on the Seattle and Washington markets are from the 1986 and 1991 issue of Metro Market Studies, Inc. *Grocery Distribution Analysis and Guide.* The market shares are grocery market rather than supermarket shares, as such they are understated, but this does not affect analysis of changes in shares much.

References

Anderson, K. 1990. *A Review of Structure-Performance Studies in Grocery Retailing*, Federal Trade Commission, Bureau of Economics, Government Printing Office.

_____. 1991. Structure-Performance Studies of Grocery Retailing: A Review and Suggestions for Future Studies. Paper presented at the Competitive Strategy Analysis in the Food System Conference, 3-5 June, in Alexandria, Virginia.

Arquit, K. J. 1989. Remarks to the Cleveland Chapter of the Federal Bar Association, Federal Trade Commission.

Bell, R. B. and J. A. Herfort. 1990. Justice, FTC Signal Tougher Merger Enforcement Standards. *Antitrust* 4(3):5-8.

Borenstein, S. 1988. Hubs and High Fares: Airport Dominance and Market Power in the U.S. Airline Industry.

Clarke, R., S. Davies, and M. Waterson. 1984. The Profitability-Concentration Relation: Market Power or Efficiency? *Journal of Industrial Economics* 32 (June):435-450.

Connor, J. M., R. T. Rogers, B. W. Marion, and W. F. Mueller. 1985. *The Food Manufacturing Industries: Structure, Strategies, Performance, and Policies.* Lexington: Lexington Books.

Cotterill, R. W. 1983. The Food Retailing Industry in Arkansas: A Study of Price and Service Levels. A commissioned report submitted to the Honorable Steve Clark, Attorney General, State of Arkansas.

_____. 1984. Modern Markets and Market Power: Evidence from the Vermont Retail Food Industry. Working Paper No. 84, N.C. 117.

_____. 1985. *Effects of Electronic Information Technology on Employment and Economic Performance in the Food Manufacturing and Food Distribution Industries*,

Research Monograph in fulfillment of Office of Technology Assessment Contract No. 533-0635 U.S. Congress.

_____.Market Power in the Retail Food Industry: Evidence from Vermont. *Review of Economics and Statistics* (August 1986): 379-386.

_____. 1988. Mergers and Concentration in Food Retailing: Implications for Performance and Merger Policy. Testimony before the Subcommittee on Monopolies and Commercial Law, Committee on the Judiciary, U.S. House of Representatives.

_____. 1989. *Merger and Concentration in Food Retailing: Implications for Performance and Merger Policy*. Research Report No. 2, Food Marketing Policy Center, University of Connecticut.

_____. 1992. Food Retailing: Mergers, Leverage Buyouts and Performance. In *Industry Studies*, ed. L. Duetsch. Englewood Cliffs: Prentice Hall.

Cotterill, R. W. and L. E. Haller. 1987. Entry Patterns and Strategic Interaction in Food Retailing. In *Issues After a Century of Competitive Policy*, ed. R. Wills, J. Caswell, and J. Culbertson, 203-222. Lexington: Lexington Books.

_____. 1992. Barriers and Queue Effects: A Study of Leading U.S. Supermarket Chain Entry Patterns. *Journal of Industrial Economics* forthcoming, December.

Cotterill, R. W., and C. Iton. 1991. A PIMS Analysis of the Structure-Profit Relationships in Food Manufacturing. Paper presented at the Competitive Strategy Analysis in the Food System Conference, 3-5 June, in Alexandria, Virginia.

_____. 1992. A Pims Analysis of the Structure Profit Relationship in Food Manufacturing. In *Competitive Strategy Analysis in the Food System*, ed. R.W. Cotterill. Boulder:Westview.

Cowling, K., and M.Waterson. 1976. Market Structure and Price-Cost Margins. *Economica* 43:267-274.

de Maintenon, D. 1984. An Analysis of the Price-Service Mix in Supermarkets. Master's Thesis, Department of Agricultural Economics and Rural Sociology, University of Connecticut, Storrs.

Demsetz, H. 1973. Industry Structure, Market Rivalry, and Public Policy. *Journal of Law and Economics* 16 (April): 1-9.

_____. 1974. Two Systems of Belief About Monopoly. In *Industrial Concentration : The New Learning*, ed. H.J. Goldschmid, H.M.Mann, and J. F. Weston, 164-184. Little, Brown and Company.

Devine, D. G., and B. W. Marion. 1979. The Influence of Consumer Price Information on Retail Pricing and Consumer Behavior. *American Journal of Agricultural Economics* 61(2): 228-237.

Echlin Manufacturing Co., 105, FTC 410, 485, 1985.

Furhman, P. 1988. The Merchant of Mülheim. *Forbes*, Oct., 52.

Geithman, F., and B.W. Marion. 1992. Testing for Market Power in Supermarket Prices. In *Competitive Strategy Analysis in the Food System*, ed. R. W. Cotterill, Boulder:Westview.

Hall, L., A. Schmitz, and J. Cothern. 1979. Beef Wholesale-Retail Marketing Margins and Concentration. *Economica* (August): 295-300.

Haller, L. E. 1992. Branded Product Marketing Strategies in the Cottage Cheese Market: Cooperative versus Proprietary Firms. In *Competitive Strategy Analysis in the Food System*, ed. R. W. Cotterill, Boulder:Westview.

Horowitz, I. 1984. The Misuse of Accounting Rates of Return: Comment. *The American Economic Review* 74(3): 492-493.

Kaufman, P. R. and C. Handy.1989. *Supermarket Prices and Price Differences: City, Firm, and Store-Level Determinants*. Economic Research Service, U.S. Department of Agriculture, Technical Bulletin No. 1776. Government Printing Office.

Kwoka, J. E. Jr., and D. J.Ravenscraft. 1986, Cooperation V. Rivalry: Price-Cost Margins by Line of Business. *Economica* 53 (August): 351-363.

Lamm, R. M. 1981. Prices and Concentration in the Food Retailing Industry. *Journal of Industrial Economics* 30 (September): 69-79.

Mandel, S. F. Jr., and J. E. Heinbockel. 1989. *Food Retailing Industry*. Goldman Sachs, Presented at the FMI Financial Executives Conference, 22 May, in Charleston, South Carolina.

Marion, B. W. 1987. Entry Barriers: Theory, Empirical Evidence, and the Food Industries. In *Issues After a Century of Federal Competition Policy*, ed. R. Wills, J. Caswell, and J. Culbertson. Lexington: Lexington Books.

Marion, B. W., W. Mueller, R. W. Cotterill, F. Geithman, and J. Schmelzer. 1977. *The Profit and Price Performance of Leading Food Chains, 1970-1974*. Study prepared for the Joint Economic Committee, 95th Cong., 1st sess.

_____. 1979. *The Food Retailing Industry: Market Structure, Profits, and Prices*. New York: Preager.

Morgenson, G. 1990. The Buyout that Saved Safeway. *Forbes*, Nov., 88-92.Mueller, D.C. 1986. United States' Antitrust: At the Crossroads. In *Mainstreams in Industrial Organization*, ed. H. W. de Jong and W. G. Shepherd, 215-241. Boston: Kluwer Academic Pub.

Poole, C. 1991. Stalking Bigger Game. *Forbes*, April, 73-74.

Rill, J. J. 1990. Sixty Minutes with the Honorable James F. Rill, Assistant Attorney General, Antitrust Division, U.S. Department of Justice. *59 Antitrust L.J.*, 45.

Scherer, F. M. and D. Ross. 1990. *Industrial Market Structure and Economic Performance*. Boston: Houghton Mifflin Co.

Shepherd, W. G. 1986. On the Core Concepts of Industrial Economics. In *Mainstreams in Industrial Organization*, ed. H.W. deJong and W.G. Shepherd, 23-68. Boston: Kluwer Academic Pub.

_____. 1986. Tobin's q and the Structure-Performance Relationship: Comment. *American Economic Review* 76 (December): 1205-1210.

Ravenscraft, D. 1983. Structure-Profit Relationships at the Line of Business and Industry Level. *Review of Economics and Statistics* 65 (February): 22-31.

U.S. Congress. House. Committee on the Judiciary. Subcommittee on Monopolies and Commercial Law. *Mergers and Concentration: The Food Industries*. 100th Congress, 2d Sess. 1988.

Weinstein, S. 1987. Seattle: The Serfs are Rising. *Progressive Grocer*, Nov., 32-33.

Weiss, L. W. 1989. *Concentration and Price*. Cambridge: MIT Press.

Zwiebach, E. 1988. Kroger Restructuring to Fend off Takeovers. *Supermarket News*, Sept. 7, p8.

Appendix: Supporting Tables

Analysis of A&P Performance

A&P launched a massive price cutting campaign in all its stores as a last ditch effort to save its strategy of extensive vertical integration to produce a wide array of private label products for sale in small, older, supermarkets. This program was named W.E.O. (where economy originates). (Marion *et al.* 1979, Chpt. 3). It did drive at least one competitor into insolvency (Penn Fruit) and affected others profitability for a few quarters, however A&P lost millions and closed over a thousand stores in a retrenchment that ultimately recognized that the dominant retailing strategy is to merchandise a broader array of nonfoods as well as foods in larger stores, not a cost focused, private label dominated, supermarket with very few nonfood items. Consequently, we did specify a variable that identified whether a chain competed against A&P (A&P impact variable) and a variable that identified an A&P observation. Both are significant, however adding or deleting them from the models does not affect the estimation result (Marion *et al.* 1979, Chpt. 3).

Anderson argues that one should not accord one chain special treatment, and more specifically that a general theory of noncompetitive conduct is not validated if one controls for chiseling by one or more firms. This is a somewhat curious criticism because the most pervasive critique of structure-profit studies has been the excessive reliance upon broad cross section data sets that have ignored specific firm or industry characteristics. Here Anderson is critical because the model is too firm specific. I would have expected him to demand the opposite, i.e., that the models be estimated for individual firms and that the models incorporated more, not less dynamics of price competition.

Part of our special treatment of A&P was to estimate an abbreviated version of the model for the 28 observations from A&P. (Marion *et al.*, 1979, p.201). In my opinion, the results strongly support the structural explanation of profitability. In the years before W.E.O. the model performs pretty much as hypothesized with positive and significant coefficients for relative market share and concentration. However during the W.E.O. periods and its chaotic aftermath the model completely falls apart. Nothing explains A&P's performance during that period which probably is as it should be.

TABLE 11.A.1 Change in Industry Profits Resulting from Change in Relative Market Share and Linear Market Concentration[f]

Firm	Market Share (%)	Relative Market Share (%)	Firm Sales (Million $)	Relative Market Share Profits (Mil$)	Market Concentration Profits (Mil$)	Total Profits (Mil$)
High Market Share Case						
Before Market Concentration						
1	25	62.5	250	9.844		
2	5	12.5	50	0.394		
3	5	12.5	50	0.394		
4	5	12.5	50	0.394		
Other[a]	1	2.5	600	0.945		
Profits (Mil$)				11.97	10.4	22.37
After Market Concentration						
1	25	31.25	250	4.922		
2	20	25	200	3.15		
3	20	25	200	3.15		
4	15	18.75	150	1.772		
Other[b]	1	1.25	200	0.158		
Profits (Mil$)				13.151	20.8	33.951
Change in Industry Profits (Mil$)				1.181	10.4	11.581

Low Market Share Case

Before Market Concentration

1	10	25	100	1.575		
2	10	25	100	1.575		
3	10	25	100	1.575		
4	10	25	100	1.575		
Other[a]	1	2.5	600	0.945		
Profits (Mil$)				7.245	10.4	17.645

After Market Concentration

1	20	25	200	3.15		
2	20	25	200	3.15		
3	20	25	200	3.15		
4	20	25	200	3.15		
Other[b]	1	1.25	200	0.158		
Profits (Mil$)				12.758	20.8	33.558

Change in Industry Profits (Mil$)	5.513	10.4	15.913

[a] 60 firms with 1 percent market share each.

[b] 20 firms with 1 percent market share each.

[c] based upon equation 1b Table 3.5, Marion, et al. (1979) p. 82.

TABLE 11.A.2 Top Twenty Retail Chains of 1972, 1979, and 1989, and Ownership/Finance Changes Between 1979 and 1989

Rank	Name Sales ($ million)/ Share (%) 1972[b]	Name Sales ($ million)/ Share (%) 1979[a]	Changes 1979-1989	Name Sales ($ million)/ Share (%) 1989[a]
1	A&P (6,369) 7.21	Safeway (13,718) 7.52	(LBO-KKR 1986)	American (22,004) 6.27
2	Safeway (6,057) 6.86	Kroger (9,029) 4.95	(RECAP-G.Sachs 1988)*	Kroger (18,832) 5.37
3	Kroger (3,791) 4.29	A&P (6,684) 3.66	(acquired by Tengelmann 1979)	Safeway (14,325) 4.08
4	ACME (American) (2,025) 2.29	American (6,121) 3.36	(acquired by Skaggs 1979)	A&P (11,100) 3.16
5	Jewel (2,009) 2.28	Lucky Stores (5,816) 3.19	(acquired by American 1988)	Winn-Dixie (9,151) 2.61
6	Lucky (1,988) 2.25	Winn-Dixie (4,931) 2.70		Albertson's (7,420) 2.11
7	Food Fair (1,980) 2.24	Grand Union (3,138) 1.72	(LBO-Mgmt, 1988, acquired by Miller, Tabak, Hirsch 1989)	SGC (6,299) 1.79
8	Winn-Dixie (1,834) 2.08	Jewel Cos. (2,818) 1.54	(acquired by American 1984)	Publix (5,386) 1.53
9	Grand Union (1,380) 1.56	Albertson's (2,674) 1.47		Vons (5,200) 1.48
10	Supermarkets GC (SGC) (1,194) 1.35	SGC (2,370) 1.30	(LBO-Mgmt, 1987)	Food Lion (4,717) 1.34

Rank				
11	National Tea (1,090) 1.23	Stop & Shop (1,879) 1.03	(LBO-KKR, 1988)	Stop & Shop (4,636) 1.32
12	First National (849) .96	Publix (1,800) .99		AHOLD[d] (3,630) 1.03
13	Stop & Shop (774) .88	Dillon (1,792) .98	(acquired by Kroger, 1983)	Giant Food (3,250) .93
14	Albertson's (682) .77	Von's (1,500) .82	(LBO-Mgmt, 1985 from Household Int.)	Grand Union (2,717) .77
15	Publix (676) .77	Food Fair (1,492) .82	(bankrupt, exited 1986)	H.E. Butt (2,586) .74
16	Fisher Foods (650) .74	First National (1,365) .75	(LBO, acquired by AHOLD 1985)	Ralphs (2,556) .73
17	Giant Food (496) .56	Fisher Foods (1,336) .73	(merged with Riser Foods, 1988, divested main division Dominick's)	Fred Meyer (2,285) .65
18	Dillon (406) .46	Giant Food (1,243) .68		Bruno's (2,134) .61
19	Waldbaum (394) .45	Waldbaum (1,103) .60	(acquired by A&P, 1986)	Dominick's (2,000) .57
20	Fred Meyer (349) .40	Fred Meyer (1,060) .58		Hy-Vee (1,800) .51
Top Twenty Sales	34,993 37.49%	71,869 38.38%		132,028 37.61%
Total Grocery Sales	93,328	187,242[c]		351,000

[a] 1979, and 1989 sales reported by Progressive Grocer Marketing Guidebook 1981, 1991; Bureau of Census, Statistical Abstract, 1981.

[b] Cotterill and Haller, 1987; Bureau of Census, Statistical Abstract, 1977.

[c] After 1977, Census reports establishments with payroll, the 1979 figure is adjusted upward based on the ratio of total sales to payroll sales for the 1977 census.

[d] Includes Giant Food Stores, Carlisle, PA., Bi-Lo, and First National.

[e] In response to hostile takeover by Haft family, Kroger with Goldman-Sachs did a leveraged recapitalization. Operationally it is equivalent to a LBO.

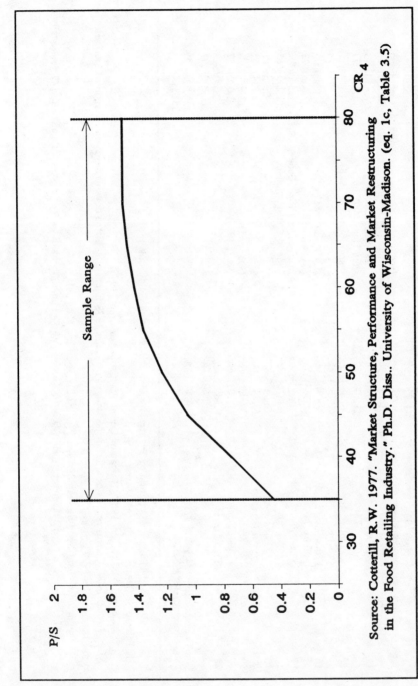

Source: Cotterill, R.W. 1977. "Market Structure, Performance and Market Restructuring in the Food Retailing Industry." Ph.D. Diss.. University of Wisconsin-Madison. (eq. 1c, Table 3.5)

FIGURE 11.A.1 Graph of the Concentration-Profit Relationships in Food Retailing

12

Testing for Market Power in Supermarket Prices: A Review of the Kaufman-Handy/ERS Study

Frederick E. Geithman and Bruce W. Marion

Introduction

Supermarkets dominate grocery retailing. Because supermarkets capture nearly 4 out of 5 dollars spent in retail food stores, it is important to understand what fans or dampens the fires of competition among supermarkets. Economic theory tells us that market structure usually has some influence on competition.

Supermarkets as sellers compete in local markets. Thus, the structure of Metropolitan Statistical Areas (MSAs) (or smaller areas) has generally been used to study the linkage between structure and performance for supermarkets. Concentration of supermarket sales in MSAs is high and increasing. In the explosive decade of 1977-1987, supermarket four-firm concentration (SCR4) in 164 comparable MSAs jumped from 71 to 77 (Cotterill 1991). Whereas in 1977, 27 percent of MSAs had a SCR4 of 80 or more, by 1987 46 percent had that level of SCR4. By any criteria these are very high levels of concentration.

Does market concentration affect supermarket performance? In addition to the academic interest in this question, important policy issues also depend on the answer. The courts and antitrust agencies (state and federal) must decide what to do about mergers between supermarket chains and alleged instances of predation, monopolization, and restraints of trade.

If there is no linkage between supermarket concentration and supermarket performance, the antitrust agencies can largely ignore this industry. If supermarket concentration has no influence on prices, entry barriers are likely to be nil. And if entry barriers into supermarketing are nil, even a single firm monopoly in an MSA would have limited market power. Predation or conspiracies might yield some control over prices, but the control would be temporary. Market power without entry barriers is a transient power. On the other hand, if supermarket concentration substantively affects supermarket performance, a very different stance is called for by the antitrust agencies. Horizontal mergers should be carefully scrutinized. Predatory behavior and restraints of trade are plausible behaviors and may yield enduring market power. Thus, the evidence on this matter has major policy implications.

Several studies over the last two decades have tested the hypothesis that the structure of local retail grocery store markets is a major determinant of retail prices (Marion *et al.* 1979; Hall, Schmitz and Cothern 1979; Lamm 1981; Meyer 1983; Cotterill 1986; Marion, Heimforth and Bailey 1991). The Kaufman-Handy (K-H)/ERS study (1989) is a recent addition to this literature. Because its results are in conflict with the vast majority of the literature, both in grocery retailing and other industries (Weiss 1989), because it was a costly and ambitious effort, and because it has important policy implications, the K-H/ERS study deserves a careful critique.[1]

In evaluating empirical research, at least two issues are important to consider:

1. Does the research represent good science? Is the theory sound, the data accurate, and the analytical techniques appropriate?
2. Are the results generalizable? If the research is carefully done, how broadly do the results apply? Is the instance studied typical or atypical?

In this critique, we focus mainly on the first issue. Was the ERS study good science? The generalizability of a study's findings is only an issue if one first concludes that the research is solid. This article focuses particular attention on the data used by K-H. Although K-H's theory and analytics can be criticized, these weaknesses can be overcome if the underlying price data are sound.

Because of the complexity of the ERS effort to measure prices, part of our objective is simply to explain their methodologies. In the process, we will identify some of the unusual characteristics and critical weaknesses of the ERS study. Hopefully, as a result of our discussion, the pros and cons of alternative research designs and approaches will come into focus.

Before examining the details of the ERS study, we provide an overview of its most critical characteristics.

Overview

Designing Research To Compare Supermarket Prices: Avoiding the Apple v. Orange Problem

There is enormous variation in the products and competitive emphasis of supermarkets. Warehouse stores emphasize low prices and spartan store facilities while up-scale super stores emphasize great variety, extensive departmentalization and high quality products. Clearly, any study that attempts to examine the presence of market power in grocery retailing must either avoid comparing prices of different store types or effectively control for differences in store types.

In a similar way, there are enormous variations in the number and quality of products carried by supermarkets. Do we learn anything about retail market power by comparing the price of Michelob beer in one store with the price of Busch beer in another? Or by comparing the price of Kellogg's Fruit Loops in one store with the price of Post Alpha Bits in another? Clearly not. One of the essential characteristics of a valid study of supermarket prices is to eliminate as much spurious variation as possible through careful research design. *Ideally*, the prices of identical products would be compared across stores and across markets. And, the prices in similar types of stores would be compared across markets.

To illustrate the problem of comparing non-comparable products, assume a survey of three firms in a city reveals the following prices for beer:

Brand	Firm A	Firm B	Firm C
W	$1.00	$1.00	$1.00
X	.80	.80	NA
Y	.60	NA	NA
Z	NA	NA	$1.00
Ave. Price	$0.80	$0.90	$1.00
Mkt. Share	50%	35%	15%

If we consider only the average price in each store, it appears that market share is inversely related to beer prices. However, further information reveals that products W and Z are super premium beers, product X is a premium beer, and product Y is a popular beer. This tells us that the average price involves comparing "apples and oranges." The missing price data result in the average beer price for each store being an inaccurate indicator of beer prices—since quality (or product differentiation) are not held constant.

One of our fundamental points of disagreement with Kaufman and Handy is the methodology thought to be necessary to accurately test for retail market power. K-H believe the impact of different store types on prices can be accounted for by variables in the econometric model. While we do not reject this possibility out-of-hand, we believe that this source of variation can best be controlled for by careful research design and/or by analyzing subsamples of similar stores.

K-H also apparently believe that the *average* price in the above example is a valid measure of beer prices in the three stores. On this point, we strongly disagree. The higher price for Firm C in this example is solely because the two brands priced in that store were super premium beer.   I.e., the mix of quality/product differentiation in the three stores is not held constant.

ERS went to great lengths to improve certain weaknesses of previous studies. Unfortunately, they failed to avoid the apple-orange problem.[2] This was a serious mistake that, in our opinion, jeopardizes the usefulness of the entire study.

The ERS Research Design and Methodology

The ERS study was originally designed in the late 70's by Charles Handy and Gerald Grinnell.   One of their objectives was to correct some of the perceived weaknesses of the "JEC study" (Marion *et al.*, 1979). In particular, they elected to employ random sampling procedures in several aspects of the study.   They also decided to include all store types and all products and departments in the stores. Had Handy and Grinnell chosen instead to ignore meat and produce products because of enormous problems of comparability and to concentrate on similar stores across markets, the study could have been greatly simplified—and improved.

Handy and Grinnell selected the product categories/subcategories to be priced using probability sampling. That is, if white bread accounted for twice the sales of diet soft drinks in the average U.S. supermarket, bread was twice as likely to be selected. This is an acceptable approach. However, once a detailed product category was selected, Handy and Grinnell decided to price *all* brands that were carried by a store—i.e., to do a census. This was a pivotal decision. Handy and Grinnell could have used two other procedures to select brands that would have avoided doing a census and would have reduced error by pricing the same brand across stores:

1. They could have continued the principle of probability sampling to select the brands to price.   For example, if processed cheese was selected as a product category to price, and if we assume Kraft had a 50% share of processed cheese while Borden had a 10% share, the Kraft brand would be 5 times as likely to be selected as the Borden brand.[3] But—once a brand was selected, the same brand would be priced in all stores.
2. They could have used judgement to select one or two of the leading brands in each product category—brands that were likely to be carried by most stores. Here also, the same brands would be priced across all stores.[4]

The decisions to price a sample of all products in the store and to do a census of all brands within those products sampled resulted in an enormously complex data collection task. The target number of items in the sample was set at 500 per store—in part to allow prices to be collected in 4 hours.[5] The survey form used by enumerators was 57 pages long. A great deal was also left to the enumerator's judgement. For example, they were supposed to price *all* pre-sweetened cereals in a certain package size, *all* brands of domestic beer sold in 6 packs, etc. Enumerators were supposed to measure the diameter of certain produce items such as Red Delicious apples and Temple oranges, and to weigh those produce items sold by the head. They were also asked to complete a questionnaire on store characteristics. All in all, it was an enormous task for the enumerators.[6]

Given the enormity of the data collection task, experienced and skilled enumerators should have been a high priority. However, instead of selecting a well-known marketing research firm, USDA chose the Pinkerton Agency, a firm whose primary expertise is in surveillance, background investigations and security work. Phone conversations with Pinkerton employees indicate that this company rarely gets involved in the ERS type data collection. It is outside the realm of their real expertise.

It is perhaps not surprising, given the above, that, on average, prices on 170 items were collected per store—roughly one-third of the target sample. And, equally important, the number of products priced and the brands of products priced varied greatly from one store to another.

Finally—in the aggregation of the price data, one other ERS decision proves critical: that is to treat all brands within a category as the same, to weight them equally, and to calculate the "price relative" from the simple average of branded product prices in a category. Thus, Coca Cola, Shasta Cola and Jolly Time Cola were considered as equal and their prices were given equal weight in determining a store's average price for branded cola. Similarly, Michelob, Budweiser and Old Milwaukee were considered equal and their prices weighted equally.[7]

The consequences of the above aggregation procedure would not be as egregious had the same census of brands been priced in all stores. E.g., if Michelob, Budweiser and Busch had been priced in all stores, there would still be some bias because the price for Bud (35% market share in 1980) and Busch (3% market share) are given equal weight. But, the problem would be far less than the one confronted in the ERS study in which the beer price in one store may be based upon Michelob, in another store on Busch, and in yet another store on Budweiser.

Alternatively, price relatives could have been calculated for each brand. That is, a store's price for Michelob could be indexed to the all-store average price for Michelob. To illustrate the difference between this approach and ERS's procedures, consider Table 12.1.

The results of the two approaches are sharply different. Whereas the ERS approach would show a price index of .90 for Store C because of the brands that happened to be priced, the more appropriate index—we believe—shows store C having the highest price of the three stores with an index of 1.10. Had ERS

TABLE 12.1 Illustration of Two Alternatives for Calculating Store Price Indices for Beer

		Store A	Store B	Store C	All Store Avg
	Michelob	$2.00	$2.50	NA	$2.25
	Budweiser	NA	1.50	1.75	1.625
	Busch	1.00	NA	1.25	1.125
A)	ERS Approach:				
	Store Avg. Price	1.50	2.00	1.50	1.667
	ERS Index =	.90	1.20	.90	
	(Store Avg./All Store Avg.)				
B)	Price Index for Individual Brands:				
	Michelob	.89	1.11	--	
	Bud	--	.92	1.08	
	Busch	.89	--	1.11	
	Store Avg. Index:	.89	1.015	1.095	

followed the B procedure of computing price relatives, some of their apple-orange problem would have been avoided.

An index such as that shown as (A) in Table 12.1 is of little value. It would be highly misleading to conclude that beer prices in Store B are 30 percent higher than in stores A or C. And, increasing the sample size will do nothing to overcome this problem. Product quality and differentiation are not being held constant.

Because enumerators were instructed to collect prices on all branded products in a category, the brands actually priced varied greatly. In one store, Stokely canned corn might have been priced, in another Del Monte, in another Libby, in a fourth Green Giant, and in a fifth all four. If all manufacturer brands were expected to carry equal prices, this situation would not present a problem. However, the expected prices of different brands are not equal. This is true not only for products like beer, where even the casual observer would contend that the prices for super premium (e.g., Michelob), premium (Budweiser) and popular (Busch) beer brands should not be compared. It is also the case for products in which competing brands are physically similar and not sharply different in product differentiation (e.g., Del Monte, Libby and Stokely brands of canned corn).

Wills and Mueller (1989) examined the relative prices of different brands of 133 food products (selected because the quality of competing brands was similar). Setting the price of the leading brand equal to 1.00, they found the average price of other brands to be as follows:

	1979	1980
Brand 1	1.00	1.00
Brand 2	.97	.96
Brand 3	.93	.92
Brand 4	.94	.94
Other brands	.91	.88
Private Labels	.83	.82

These results indicate it can make a big difference which brand is selected for pricing. If the price for Green Giant canned corn (1.00) in one store is compared to Libby canned corn in another (.93), the difference may simply reflect differences in product differentiation of the brands priced—not differences in store prices.

How pervasive were these problems? Are these "possible" problems that do not in fact materialize? In most cases, did enumerators price the leading brand in a category? To check this, we examined the brands priced in the 63 edible grocery categories in the 152 stores of the 10 largest chains (Table 12.2). I.e., we selected the group of stores with the largest number of prices. The brand that was most frequently priced in each category (e.g., Folger's drip coffee, 16 oz.) was determined. In only one category was the same brand priced in 90% or more of the 152 stores. In only 7 of the 63 categories was the same brand priced in three-fourths of these stores. In only 20 of the 63 categories was the same brand priced in half or more of the stores. The actual brands priced in different stores varied substantially.[8] The problems we speak of are real and pervasive.

An analogy on this point can be made from the Bureau of Labor Statistics (BLS) methodology for collecting prices for the Consumer Price Index (CPI). Whereas most studies of market power in food retailing are concerned about accurately measuring supermarket prices across markets (MSAs), BLS is concerned about accurately measuring prices over time. For the metropolitan areas selected by BLS, the outlets and specific items priced are held constant from month to month.[9] For example, if a 14 oz. box of Wheaties is selected to price in Store X, that is the item priced each month. Very heavy emphasis is placed on holding quality (or product differentiation) constant over time.

The ERS procedures are analogous to BLS allowing the brands priced to change from month to month. This month Wheaties might be priced, next month Cheerios and the following month Post Raisin Bran. For obvious reasons, BLS does not allow this to happen. Just as BLS places heavy emphasis on holding quality and store types constant over time to accurately measure price changes, studies of market power need to try to hold quality and store type constant across markets.

K-H respond to this criticism by contending that it is impossible to price a market basket of identical items across a large number of stores. We have two reactions to this. First, the evidence does not support their contention. Nielsen

TABLE 12.2 Sixty-Three Edible Grocery Categories in Wave 2 Sample with Most Frequently Priced National Brand in 152 Stores of the Top 10 Chains

Categ. No.	Description	Unit Measure	Most Frequently Priced Brand			TOTAL NR. NAT'L BRAND PRICES
			BRAND	STORES WITH BRAND	% OF STORES	
8	Tomatoes, stewed	GE 5 lbs.	Hunt's	0	0.0	0
23	Brown Sugar	1.2-3.9 lbs.	U and I	0	0.0	0
26	Rice, white, medium-grain	32-47 oz.	Not Listed	0	0.0	27
27	Rice, brown	33-48 oz.	Not Listed	0	0.0	0
40	Potato Chips, BBQ flavored	7-8 oz.	Not Listed	0	0.0	150
62	Frozen Cauliflower	GE 17 oz.	Frosty Acre	0	0.0	1
9	Olives, blk. pitted ripe	8-15.9 oz.	Early Calif.	1	0.7	1
15	Peanut butter, creamy	29-36 oz.	Peter Pan	1	0.7	19
31	Cereal, presweetened	16.1-18 oz.	K Fruit Loops	1	0.7	159
81	Jelly Roll	3-5 oz.	Dolly Madison	2	1.3	2
10	Pinto Beans	40-50 oz.	Randall	4	2.6	6
61	Frozen turnip greens	LE 10 oz.	Bird's Eye	6	3.9	8
5	Peaches, cling, heavy syrup	GE 5 lbs.	Stokely	8	5.3	10
21	Mayonnaise	128 oz.	Best Foods	10	6.6	25
38	Bread sticks, plain	5.5-8 oz.	Stella D Oro	12	7.9	12
35	Crackers, bacon flavored	8.1-11 oz.	Keebler	13	8.6	16
36	Cookies, oatmeal	GE 15 oz.	Archway	15	9.9	48
64	Frt flav, ice on stick, single	20-24 count	Popsicle	15	9.9	39
49	Beer, domestic	12 oz. ret. bottles, 6 pk.	Budweiser	23	15.1	112
32	Cereal, Raisin Brand	10.1-12 oz.	K Raisin Brand	24	15.8	25
60	Frozen mixed veg.	10.1-16 oz.	Bird's Eye	25	16.4	75
68	Frozen Spaghetti entree	11-15 oz.	Stouffer	31	20.4	35
22	Sugar, white granulated	2.1-5 lbs.	Domino	35	23.0	138
51	Malt liquor, domestic	16 oz. cans, 6 pk.	Schlitz Malt	39	25.7	79
63	Frozen green beans, Fr. cut	LE 10 oz.	Bird's Eye	39	25.7	58
17	Cocoa, instant, indiv. env.	4.1-8 oz.	Alba	41	27.0	48
20	Salmon, pink	14-16 oz.	Bumble Bee	43	28.3	171
16	Milk, instant (powdered)	14.1-20 oz.	Carnation	50	32.9	51
66	Frzn. grape juice conc.	LT 12 oz.	Welches	53	34.9	57

42	Candy, caramel chews	12.1-14 oz.	Kraft Caramels	59	38.8	59
50	Beer, regular, domestic	12 oz. cans, 12 pk.	Budweiser	61	40.1	397
45	Cat Food, beef flav. canned	LE 10 oz.	Kal Kan	62	40.8	172
46	Dog Food, moist, burgers w/ch	36 oz.	Ken L Ration	63	41.4	104
24	Cake decorator icing, yellow	4.25-4.5 oz.	Cake Mate	65	42.8	112
12	Spaghetti sauce (meat flav.)	13-16 oz.	Ragu X-Thick	66	43.4	282
7	Tomatoes, whole peeled	14-17 oz.	Hunt's	67	44.1	171
44	Roll candy, mint flavor	LE 1 oz.	Lifesaver Wint.	69	45.4	148
69	Frozen bagels, egg flavor	LE 12 oz.	Lenders	69	45.4	77
2	Coffee, flaked (non instant)	16-31 oz.	Folgers	71	46.7	80
41	Fabricated potato chips	5.1-10 oz.	Pringles	72	47.4	75
67	Frozen fried chicken	25-32 oz.	Banquet	74	48.7	127
80	White bread, slices	20-24 oz.	Wonder Bread	74	48.7	285
47	Cola-flavored soft drink	NR bottle LE 1/2 liter	Coca Cola	75	49.3	200
33	Cereal, 40% bran	14.1-16 oz.	K 40% Bran	76	50.0	151
43	Candy, choc. covered peanut	8-9 oz.	Reese's Mini	80	52.6	82
34	Crackers, cheese flavored	8.1-11oz.	S Cheez-its	82	53.9	219
48	Beer, domestic, light	12 oz. NR bottles, 6 pk.	Miller Lite	82	53.9	335
6	Apple juice	25-32 oz.	Treetop	86	56.6	224
13	Salad/cooking oil	GE 64 oz.	Wesson	95	62.5	289
29	Dehydrated potatoes, instant	9-16 oz.	Pills. Hungry	96	63.2	252
4	Pear Halves, heavy syrup	15-17 oz.	Del Monte	103	67.8	125
11	Worchestershire Sauce	5.1-10 oz.	Lea & Perrin	106	69.7	240
30	Cereal, presweetened	10.1-12 oz.	K Fruit Loops	106	69.7	667
39	Tortilla chips, cheese flav.	9-13 oz.	Doritos Nachos	106	69.7	161
3	Coffee, ground, perk	16 oz.	Folgers	109	71.7	303
14	Peanut butter, crunchy/chnk	13-18 oz.	JIF	111	73.0	303
19	Chicken veg. soup	10-15 oz.	Campbells	114	75.0	114
1	Coffee, ground, drip	16 oz.	Folgers	116	76.3	251
37	Cookies,(choc. w/cream fill.)	8.1-15 oz.	N Oreo	127	83.6	263
25	White flour, all-purpose	5 lbs.	Gold Medal	128	84.2	306
28	Macaroni & cheese dinner	6.1-10 oz.	Kraft	131	86.2	219
18	Cream of Mushroom soup	10-15 oz.	Campbells	135	88.8	136
65	Frzn. orange juice conc.	7-12 oz.	Minute Maid	139	91.4	314
total				3566		8615

NEIS data (to which ERS had access for 1979 and 1980) shows the percentage of stores nationally that carry a particular item. For a large number of products, the leading brand and package size was carried by over 75% of the stores. For example, of the edible grocery categories selected for Wave 2, NEIS data were found for 42. In 2/3rds of these, the most popular brand and package size was available in over 75 percent of the stores. Even in products with strong regional brands, it is often possible to select 2 or 3 brands (in the most popular package size) that are carried by most stores. E.g., in coffee, Folgers has historically been strong in the West, while Maxwell House has dominated the East. Both Folgers and Maxwell House could be included in the sample, but their price indices calculated separately (method B in Table 12.1).

From our own experience in the JEC study, we also know it is possible to price the same brand and package size across chains and MSAs. In the JEC study, the chains' own price check data allowed us to construct a market basket of 94 comparable grocery products. It can be done, particularly if one concentrates on chain stores.

Second, if we accept K-H's argument that it is impossible to accurately measure prices, holding quality/product differentiation constant, then why do it? If it can't be done right, why expend several hundred thousand dollars and 10 years of staff time to come up with questionable results?

If the market basket price of one firm is based upon Michelob Beer, Shasta Cola, Kellogg's Fruit Loops, Yoplait Yogurt and Heinz Catsup, it makes no sense to compare it to the market basket price of another firm that is based upon Busch Beer, Coca Cola, Post Alpha-Bits, Dean Yogurt and Hunt Catsup. Yet this is precisely the dilemma one confronts with the ERS data set.

Finally, one other characteristic of the ERS study merits note. The firms to be price checked were notified in advance by ERS researchers. Supermarket chains and their trade association were well aware of the pending study. While a chain was not told the specific week of the data collection, they were told which MSAs would be checked. And, after the first data collection in March 1982, firms knew which stores would be price checked in Waves 2 and 3. We have no way of knowing whether supermarket chains adjusted their prices in certain MSAs to remove the evidence of market power. However, the supermarket industry has a long history of tampering with the academic marketplace. For example, documents obtained in the JEC study indicated that the National Association of Food Chains "had allocated either $50,000 or $100,000 to sponsor research by selected academicians to disprove the 'Mueller hypothesis' that market power and high concentration in food retailing results in higher margins and profits" (JEC Hearings, p. 150-151).

Thus, there are at least two issues concerning the ERS study. First, is it good science? Does it employ sound theory, data development and statistical analysis? Does it measure what it purports to measure? Second, are the results (if sound) generalizable? Or was this study atypical because firms knew in advance that their prices would be checked and had an opportunity to change them? In the remainder of this critique, we focus on the first issue.

Unusual Characteristics of the ERS Study

In their report, K-H show the results of 6 regression models. Both firm market share and market concentration have a consistent negative relationship to firm prices, the dependent variable. In most cases, these variables are not statistically significant. However, concentration is significant in 2 of the 6 models.

Juxtaposed against the broad literature of industrial organization, these are rare findings. Weiss (1989) reviewed a large number of price-concentration empirical studies—most of which have been completed since Demsetz challenged the interpretation of concentration-profit studies in the early 1970s. Of some 70 studies reviewed by Weiss, 73% found a significant positive relationship between concentration and price. An additional 19% found a non-significant positive relationship. Only 9% found a negative relationship—with 3% finding a significant negative relationship. Weiss concludes, "We have found what I feel is very strong evidence that concentration raises price" (p. 270).

Schmalensee (1989) also reviewed the literature on concentration and price. He concludes, "This work generally provides strong support for" the proposition that "seller concentration is positively related to the level of price" (p. 987-88).

In addition to being inconsistent with the theory and the bulk of the empirical evidence in the field of industrial organization, the ERS study is also an outlier when compared to other concentration-price studies in food retailing. Of the eight studies done using relatively modern econometric techniques, six found a significant positive relationship between concentration and price (Marion *et al.* 1979; Hall, Schmitz and Cothern 1979; Lamm 1981; Cotterill 1983; Cotterill 1986; Marion, Heimforth and Bailey 1991).[10] Only the ERS study and a methodologically questionable study by Newmark (1990) found no significant linkage between market concentration and prices (see Werden 1991). When a study's findings are so inconsistent with the bulk of the literature, we believe the authors bear a heavy burden to be sure they have employed sound scientific procedures, and to clearly explain their methodologies.

Perhaps even more remarkable than their concentration-price results is the ERS finding that entry is positively (and significantly) related to price. This is contradictory to virtually all theories of entry. We know of no other empirical study that has found entry to have a significant positive effect on firm prices.

The ERS study is also unusual for the lack of explanatory power of its econometric models. The ERS report includes the results of six regression equations. All but one equation, where New York City is dropped, are run with the full sample of 321 firms. All six equations use the log-log functional form.

We re-analyzed the ERS data to test the sensitivity of the results to different functional forms and various subsamples. The results are summarized in Table 12.3.

Equation 1 is the one reported by K-H in the body of their report. Six of the 12 independent variables are statistically significant. The R^2 is .33; the F-value for the equation is 13.96. However, these results are almost totally dependent on the single store firms in the sample.[11] Equation 2 is the same as equation 1, but is run with only the 65 observations of the top 10 chains from the ERS data set.[12] Only one variable is marginally significant. R^2 drops to .02; the F-value drops to 0.86.

Equations 3 and 4 use a linear functional form instead of log-log. Equation 3 (321 firms) is somewhat less robust overall than equation 1. Once again the results are decimated when only chains are included in the data set. Not a single variable is significant in equation 4.

These results provide a good indication of whether variations in store type were controlled for in K-H's econometric model. Ostensibly, the variables included in the K-H model to control for store types and levels of service were sales * size, firm integration and store services. If these successfully capture the price effects of different store types, we would expect the remaining variables to show little change from the full sample to the more homogeneous chain store sample. This is not the case. Occupancy costs, market growth, market rivalry, market entry and the warehouse store variables all drop to insignificance with the chain store sample.

The ERS data were also run using models similar to the JEC study (Marion *et al.*, 1979). Whereas the JEC results generally explained over 60% of the variation in prices for the 36 firm-in-market observations, equations 5 and 6 using the ERS data explain 11 and 10 percent of the variation. Once again no variables are significant when the 65 observations of the top 10 chains are used.

Finally, the range in prices found by ERS is unusual for food retailing. Using an index of 100 for the firm with average prices, ERS found firms with price indices that ranged from 73 to 121. A 48% range in prices across "supermarkets" is far greater than any we have encountered. We suspect the extreme observations either represent data errors or are for very unique discount and gourmet stores.[13] We also expect that the extreme observations of some single store firms have a powerful influence on the K-H regression results.

The ERS Research Design

The ERS study sought to improve upon prior studies by drawing a random sample of grocery product prices. Probability sampling, when properly employed, allows the researcher to make statistical statements or inferences about parameters being estimated and the reliability or precision of those estimates (Churchill, 1983). Nonprobability sampling, such as the "market basket price index" used in many prior retail studies (e.g., Marion *et al.*, 1979), cannot rely on the rigor of statistical theory to validate its parameter estimates. Rather, the reader or researcher must judge the sample based on his own experience and/or education.

K-H state that random sampling procedures were employed to select each sample used in their analysis (p. 4).[14] A sample of supermarket products was priced from a sample of supermarket stores, selected from a sample of supermarket firms, from a sample of markets. The report, however, includes no rigorous discussion of the sampling techniques or the statistical level of confidence that can be placed on estimates of store price, firm price, market price, or national price. Nor will the reader find a description of the products sampled, how large each sample was, or any justification for the sample size.

Sampling Frame

The ERS analysts constructed a national sampling frame which consisted of "about 95 percent of supermarket products" (p. 4). From this sampling frame, three independent samples were drawn for Waves 1, 2, and 3.

For sampling and aggregation purposes a supermarket was stratified into five departments. Each department (e.g., produce) was further divided into categories and subcategories. Table 12.4 provides an example of the actual survey form used.

The term "category" is not well defined in this study. The produce and meat department categories often have *multiple* subcategories (e.g., iceberg, bib and romaine are subcategories within the category "lettuce and escarole"). In the dairy, edible grocery, and non-edible grocery departments, each category had only one subcategory. National sales shares for each category and/or subcategory were derived from a variety of sources. Since we focus primarily on the edible grocery department in our review we will generally simply use the term category to avoid reader confusion.

An "item" is a detailed category further identified by a particular brand. Each item surveyed was assigned a seven digit item code (See Table 12.4). Except for prelisted brands (e.g., Kellogg's Fruit Loops in Table 12.4), item codes *were not strictly defined* across stores, firms, or markets. That is, two different brands of the same product (or category) could have the same item code in different stores. Thus, except for the prelisted brands, there is no way of knowing which brands of a category were priced in a particular store.

Survey Sample

The probability for selection of each category was based upon its national sales share.[15] The researchers "randomly selected individual subcategories, until the desired sample size was achieved" (KH, p. 4). The desired sample size was 500 *items*.[16]

TABLE 12.3 Regression Results That Test the Sensitivity of Kaufman-Handy Results

		K-H Models				JEC Model		
Equation		1	2	3	4		5	6
Sample		321 Firms	Top 10 Chains	321 Firms	Top 10 Chains		321 Firms	Top 10 Chains
Function Form		(Log)	(Log)	(Lin)	(Lin)		(Lin)	(Lin)
Intercept	Intercept	4.59 (45.05)[b]	4.85 (23.08)[b]	98.92 (40.08)[b]	97.29 (17.38)[b]		103.32 (53.22)[b]	102.03 (25.03)[b]
Mkt. Share	Rel. Mkt. Share	-.003 (-1.172)	+.000 (.042)	-.028 (-.584)	-.025 (0.368)		-.026 (-.971)	-.015 (-.407)
HERF4	CR4	-.008 (-1.574)	-.006 (-.622)	-3.091 (-.579)	-3.942 (-.427)		-2.150 (-.860)	-1.713 (-.447)
Sales * Size	Avg. Store Size	-.015 (-5.184)[b]	-.013 (-2.245)[b]	-.000 (-3.828)[b]	-.000 (-.953)		-.000 (-4.269)[b]	-.000 (-.458)
Firm Integ.	Mkt. Size	-.001 (-.081)	NA	-1.391 (-1.982)[a]	NA		.000 (2.409)[b]	.000 (.119)
Occup.		.055 (3.233)[b]	.004 (.138)	.053 (3.243)[b]	.001 (-.022)			
Store Service		.045 (3.874)[b]	.044 (1.477)[c]	.012 (.112)	.282 (1.241)			

	Labor Cost			
Labor Cost	-.002 (-.184)	-.007 (-.303)	-.094 (-.862)	-.011 (-.053)
Wrhse. Stores	-.059 (-4.695)	.016 (.742)	-7.347 (-6.026)ᵇ	2.383 (1.092)
Market Growth	.382 (3.716)ᵇ	.044 (.287)	32.692 (3.142)ᵇ	.308 (.021)
Market Rivalry	-.016 (-2.130)ᵃ	-.003 (-.212)	-13.305 (-2.233)ᵃ	-4.330 (-.369)
Mkt. Turb.	-.005 (-.682)	-.005 (-.430)	-.594 (-.755)	-.878 (-.688)
Entry	.002 (2.731)ᵇ	.001 (.862)	.201 (2.600)ᵇ	.104 (.775)
NOBS	321	65	321	65
Adj. R^2	.327	-.024	.246	-.097
F-Value	13.965ᵇ	.865	9.717ᵇ	.486

	Labor Cost	
Labor Cost	-.071 (-.617)	-.102 (-.532)
Market Growth	31.542 (3.766)ᵇ	5.402 (.527)
Market Rivalry	1.676 (.282)	4.359 (.460)
NOBS	321	65
Adj. R^2	.114	-.100
F-Value	6.899ᵇ	.174

t statistics in parentheses.
[a] 1 percent level of significance.
[b] 5 percent level of significance.
[c] 10 percent level of significance.

TABLE 12.4 A Sample of the Kaufman-Handy Survey Form and Categories Priced

Dept. Nr.
  Category Nr.
    Subcategory Nr.
      Item Nr.

OTHER FOODS

ITEM CODE	ITEM NAME (DESCRIPTION)	UNIT OF MEAS.	CONTAINER SIZE	METRIC	CONTAINERS SOLD TOGETHER	
					NUMBER	$ PRICE ¢
Pre-Sweetened Cereal, Ready-to-Eat; 14.1-15 oz.						
7-30-01-11	x Kelloggs Fruit Loops	oz.				.
7-30-01-12	x Kelloggs Sugar Frosted Flakes	oz.				.
7-30-01-13	x Kelloggs Apple Jacks	oz.				.
7-30-01-14	x Post Alpha-Bits	oz.				.
7-30-01-15	x Kelloggs Sugar Corn Pops	oz.				.
7-30-01-18	x (Items 16-17 omitted)	oz.				.
7-30-01-51	P/L	oz.				.
7-30-01-52	P/L	oz.				.
7-30-01-71	Gen.	oz.				.
Beer, Domestic, Refrigerated; 12 oz. Cans Sold in Six-Packs. (Exclude "Light" or Low-Calorie Beer, and All Imported Beers)						

7-32-01-11	x Budweiser	oz.	12	6	.
7-32-01-12	x Busch	oz.	12	6	.
7-32-01-13	x Coors	oz.	12	6	.
7-32-01-14	x Miller	oz.	12	6	.
7-32-01-15	x Michelob	oz.	12	6	.
7-32-01-16	x Pabst Blue Ribbon	oz.	12	6	.
7-32-01-17	x Schlitz	oz.	12	6	.
7-32-01-18	x Old Milwaukee	oz.	12	6	.
7-32-01-23	x (Items 19-22 omitted)	oz.	12	6	.
7-32-01-51	P/L	oz.	12	6	.
7-32-01-71	Gen	oz.	12	6	.

Beer, Domestic, Non-Refrigerated; 12 oz. Returnable Bottles Sold in 24-Bottle Case. (Exclude Bottle Deposit from Price)

7-33-01-11	x Busch	oz.	12	24	.
7-33-01-12	x Budweiser	oz.	12	24	.
7-33-01-13	x Schlitz	oz.	12	24	.
7-33-01-16	x (Items 14-15 omitted)	oz.	12	24	.
7-33-01-51	P/L	oz.	12	24	.
7-33-01-71	Gen	oz.	12	24	.

15

Table 12.5 shows the five departments, the target item sample size, the number of categories and subcategories selected for pricing in the three waves, and the number of computer draws to determine the department sample. Department sales weights (not shown) derived from secondary sources were used to aggregate department price indices (or price relatives) and to determine the department item sample size. Department sample sizes were calculated by multiplying the sales share for each department times the target sample of 500 items. The 500 item sample size was arbitrarily determined but conditioned on the ability of an enumerator to collect prices within 4 hours per store.

Category Selection

ERS researchers listed out all supermarket categories for which they had information. Under each category, subcategories were identified. Shares of total sales were calculated for each subcategory. A random number computer program was then used to pick a random number (with replacement). The number selected identified a subcategory listed according to accumulated sales share. After a subcategory was determined, the expected number of items to be priced in a supermarket was estimated. As subcategories were selected, the expected number of items was accumulated until the target sample size of 500 items was achieved.[17] Thus the number of categories or subcategories sampled was a function of the number of items (brands) expected to be priced in a supermarket. As a result, the number of edible grocery categories in the samples ranged from 49 in Wave 3 to 64 in Wave 1 (Table 12.5).

Category Prices[18]

Product categories were divided into the following brand classes:

- national brand or advertised brand items (e.g., Arnold 100% whole wheat)
- private label or store brands (e.g., Safeway white bread)
- generic items
- other (unbranded) items[19]

Items falling in the same brand class were treated as being identical in terms of quality and sales share. For example, Kellogg's Fruit Loops was treated as identical to Post Alpha-Bits. Similarly, Safeway peanut butter was treated as identical to Kroger peanut butter. The prices of each item *within* a brand class were given equal weight. For example, Coca Cola prices were weighted the same as RC Cola and C&C Cola prices. In calculating the all-brand category price index, the prices for each brand class were weighted according to the share of sales (e.g., national brands might account for 80 percent of category sales with private label brands accounting for the remaining 20 percent).

TABLE 12.5 Number of Categories, Subcategories and Computer Draws Included in the Three Price Collection Waves, by Department

Department	Target Number of Items in Sample	Wave 1			Wave 2			Wave 3		
		Cat	Number of Subcat	Draws	Cat	Number of Subcat	Draws	Cat	Number of Subcat	Draws
Produce[b]	42	21	106	39	19	76	42	23	94	61
Dairy	54	20	20	26	22	22	26	20	20	28
Meat	101	30	179	45	23	171	30	26	153	67
Edible Grocery	222	64	64	65	62	62	63	49	49	51
Non Foods	80	15	15	17	15	15	19	16	16	18
total	500[a]	150	384	192	141	345	180	134	332	225

[a] Sum is off due to rounding.
[b] The produce department included fractional weights (e.g., 0.5).

Price Relatives

Item prices were averaged within brand classes for each category within a given store. Where more than one store was priced for a given firm within an MSA, the category brand price was simply averaged across stores (unless one of the stores was a warehouse store).[20] The average category prices for the various firms in an MSA were weighted by the market shares of the firms in the sample to determine market (or MSA) level average category prices—by brand class. Market level category prices were then simply averaged to form U.S. or national average brand class prices for each category.

The national average *category* price for each brand class was used to index each store's average category price according to brand. That is, a store's relative national brand price for a given category was its average national brand price divided by the U.S. average national brand price for that category. Price relatives for private label and generic brands were similarly computed. The resulting category brand class index values were then combined using national brand, private label, and generic sales shares (see ERS Report, App. D, IIc). The resulting weighted category index value was termed the "all-brand price relative."

Department and Store Prices

Department all-brand price relatives were determined by averaging the all-brand price relatives for all categories priced (ERS Report, Appendix IIe). The weight for each category was determined by the number of times the computer randomly selected that category, divided by the sum of computer draws for that department (see Table 12.5). For example, in Wave 2 each category in the edible grocery department received a weight of 1/63 with the exception of one category which received a weight of 2/63 since it was selected twice.[21]

The department level price relatives were combined using each department's share of total supermarket sales to calculate store level all-brand price relatives. In Wave 2, there were all brand price relatives or index values for 503 stores. Given that more than one store was sampled from some firms, these 503 store index values were combined (simple averages) to form 321 firm-in-local-market price relatives.

One should note that the number of categories *actually priced* was important in calculating the firm level price relatives. If only 30 categories were priced in a given store, then each category was weighted by 1/30th, not 1/63.

Reliability of Data

Kaufman and Handy have described the ERS study as employing an "innovative sampling design." However, there is no statistical analysis of the actual drawn sample in the study. For example, for each store and firm in each market, and for each wave, there exists a standard error. Had the standard errors and their computations been included in the report, readers could more readily make a judgement concerning the reliability of the price data.

The authors have provided no statistical substantiation for their claim of superior data. The only piece of information made available in their technical bulletin was that over 300,000 item prices were collected over the three waves for an average of about *170 item prices per store per wave* (p. 7). These numbers, while impressive at first blush, pale when compared to the pre-survey *target of 500 items per store* (i.e., the survey results were about 330 items short per store).[22]

We believe that sampling errors are large and that nonsampling errors are so pervasive that the sample can only be viewed as a judgement sample. That is, the quality of the data can only be subjectively evaluated by individual readers. Nonsampling errors dominate our view of this study's data problems. However, even if there were no nonsampling type errors, we believe sampling errors at the store (firm) level may be so large that comparisons of mean firm level prices may be meaningless.[23] Our basis for this statement is that the *number* of items and categories actually sampled in each store varied so greatly in the single wave that we had an opportunity to examine, that the store level variances almost certainly are very large (e.g., see Table 12.7). We could not make estimates of the variances because we were not provided all of the detailed category sales shares used in drawing the probability samples.[24]

ERS Post Survey Sample Size Evaluation

An ERS statistician estimated confidence intervals for the price data for the 503 stores surveyed in Wave 2.[25] Here, the statistician is asking, given the estimated variance for the category price relatives within a store, how many categories (or price relatives) need to be priced in order for the sample price relative to be within plus or minus two percent of the true store mean price at the 95 percent confidence level. This calculation assumes that the measured category price relatives are equal to the true category price relatives, an assumption we believe is highly unlikely because of the mixing of non-comparable products.

For individual stores we were provided the sample coefficient of variation, sample relative precision (r), the actual sample size (number of categories), and the sample size needed to achieve a relative precision of 2 percent, 3 percent, and 5 percent at a 95 percent level of confidence. Table 12.6 summarizes a portion of the ERS statistical analysis.

For all 503 stores the average number of categories priced was 83. The minimum and maximum categories priced were 16 and 115. The second column shows the number of stores where the sample results indicated a relative precision of 0-2%, 2-3%, 3-4%, and greater than 4%. The median level of precision (confidence interval) is about $\pm$ 3.6 percent. Only 3 stores had estimated relative precisions (r) within 2 percent, the level we believe necessary

TABLE 12.6 Wave 2 Store Sample Relative Precision at a 95 Percent Confidence Level

Estimated Store Relative Precision	Number of Stores	Actual Average Catagory Sample Size[a]	Average Estimated Sample Size to Achieve 2% Relative Precision[a]
0 < r ≤ 2%	3	87	71
2 < r ≤ 3%	133	91	165
3 < r ≤ 4%	198	85	254
4 < r	169	73	-
Total	503	83	-

[a] Rounded to the nearest whole number.

TABLE 12.7 Summary Statistics of Missing Edible Grocery Categories for 10 Large Multimarket Supermarket Chains[a]

Company	Number of Missing Edible Grocery Categories per MSA				Mean as a Percent of Total[c]
	MSAs	Min	Max	Mean[b]	
Safeway	8	13	18	15	23.8
Kroger	11	14	38	22	34.9
A & P	8	15	27	21	33.3
American	9	19	41	23	36.5
Lucky	6	16	29	22	34.9
Winn Dixie	5	20	47	27	42.9
Grand Union	6	14	27	20	31.7
Jewel	2	21	46	34	54.0
Albertson's	7	15	32	23	36.5
Supermarket General	4	15	28	22	34.9

[a] In the Edible Grocery (Dept. 7) group there was only 1 subcategory per category. This group includes Baked Goods, Other Foods, and Frozen Foods.

[b] Rounded to nearest whole number

[c] There were a total of 63 categories/subcategories.

for studying monopoly power in the supermarket industry.[26] The average actual sample size (Column 3) did not vary greatly. However, the estimated sample size needed to achieve an r of 2% varied substantially (Column 4).

There were 169 stores where the calculated precision (r) of store price index was greater than plus or minus 4 percent. For some of these stores, the estimated sample sizes were so great that the simple average of the estimated store sample sizes would not provide meaningful information (e.g., the needed sample size for two stores to achieve a relative precision of 2 percent was greater than 10,000 categories). The table demonstrates that the ERS sample size was inadequate for measuring store level price relatives according to their own analysis. We believe, however, that the calculation is still incorrect and that a sufficient sample size remains an open question.[27]

In a very real sense, the *entire exercise* reflected in Table 12.6 is a charade. The ERS sample size calculation implicitly assumes that the price relatives for each store are based on comparable products.[28] That is, however, not the case. The beer category included in Table 12.4 illustrates our point. Because of the large number of missing prices, the price relative for beer might be based upon Busch Beer in one store (relative to the national average price for all brands of beer), on Michelob Beer in a second store and on the simple average of Budweiser and Michelob in a third store. The price relatives in this example are meaningless. The second store may have a price relative 10 percent above store one simply because different brands were used to measure beer prices. And, if the price relatives in this example are meaningless, so also are the estimated variances of the price relatives in each store.

Our point is simple, yet central. If the price indices for different stores are based upon sharply different products (that are all indexed relative to a common national average price), the indices may tell us nothing about store price levels. And, the problem of non-comparable products is so pervasive in the ERS data set that all of the price relatives are suspect. No sample size is sufficient to provide an accurate comparison of prices if non-comparable products are being priced. Similarly, if the products priced are noncomparable across stores, statistically meaningful calculations of sample size cannot be made.[29]

Nonsampling Errors

Nonsampling errors are widespread and serious in this study. Churchill (1983) states, "Nonsampling errors can occur because of errors in conception, logic, misinterpretation of replies, statistics, arithmetic, errors in tabulation or coding, or in reporting the results."[30] Each of the topics cited by Churchill apply to the ERS study. We have, however, four major points we would like to make with respect to nonsampling errors:

1. Their missing data imputations result in inflated standard errors of estimated mean firm prices, and likely result in biased estimates.
2. The different brands (items) priced resulted in noncomparable firm price relatives.
3. Anecdotal evidence suggests that enumerator and data handling errors may have been large.
4. Item weighting in the aggregation to firm level relative price indices led to biased firm indices.

Missing Data[31]

The price relative index value for a given category indicates how much higher or lower a given store's average category price is relative to the all-store simple average price for that category. Missing data values are implicitly replaced with the *store* all brand price relative for the products that are priced. The authors fail to recognize that the standard error of the market basket price is inflated by this procedure. That is, as the categories in the sample are reduced, the confidence interval around the market basket price estimate becomes wider. In the ERS study, the number of categories actually priced in the edible grocery department varied across stores from 63 to 13. Under these circumstances, meaningful inferences from the estimates are more hazardous.

Table 12.7 provides a summary of the number of missing edible grocery *categories*, per MSA, for the ten largest supermarket firms in 1982. The edible grocery department accounted for 44.4 percent of supermarket sales.[32] Safeway, the nations largest chain in 1982, had the best survey response rate. On average, Safeway stores were "only" missing prices for 24 percent of the edible grocery categories. A quick review of this table demonstrates that two of the nation's largest chains had MSA prices based on as few as 16 (Winn Dixie) and 17 categories (Jewel). Moreover, analyses of the data revealed that firms with the largest stores or the largest market shares typically had a larger number of prices in the edible grocery department.

Table 12.2 shows some of the commonly missing categories (right-most column) in the edible grocery departments of the ten largest chains for Wave 2. The Wave 2 sample included many products in less frequently purchased container sizes. The container sizes probably explain a large part of this missing data problem. However, some of the more popular products missing in all stores (e.g., yogurt and nonfilter cigarettes) strongly suggest an enumerator or data entry problem.[33]

Other survey results also support the assertion of an enumerator problem. For the largest 10 chains we calculated the number of prices recorded for each of the prelisted brands in the edible grocery department. There were 43 categories (out of 63) where the most frequently priced prelisted brand was priced in less than half the stores (Table 12.2). Priced in less than half the stores were Kraft caramel chews, 12.1-14 oz. (59 stores); Welch's frozen grape juice concentrate, less than 12 oz. (53 stores); and Domino sugar, white granulated, 2.1-5 lbs. (35 stores), to name a few.

Pinkerton enumerators were supposed to survey each store in four hours. As indicated in the overview, Pinkerton is not generally in the market research business. The 57 page survey used in Wave 2 would test the ability of any enumerator to complete, let alone one with little or no expertise in this line of research. We believe there is sufficient evidence to support our belief that "enumerator and data handling errors" also played a large part in the "missing data" problem.

Missing Data Imputations

Are the firm index values biased as a result of the missing data imputations? We believe they are. Missing data imputations varied according to whether or not all items were missing at the category level.[34] Where one or more items were priced within a brand class (e.g., national brands), brand class index values were based on only the items that were priced. The *all-brand* indices were calculated at the category level and were based on the three (or fewer) *brand class* index values. When all brand classes for a category were missing, the *imputed* missing value was equal to the (remaining) all-brand index value for all categories priced in a store.

Table 12.8 depicts hypothetical price relatives or index values for firms X, Y, and Z for some category. Where a price index exists, one or more items were priced within that brand class. In this example it is assumed that national brand's sales share is 75 percent with private labels accounting for the remaining 25 percent sales share. These weights are used in calculating the all-brand index values.

What is immediately apparent from this table is the potential bias resulting from the ERS method of handling missing data. Firm Z has the highest all-brand price index—only because its all-brand index is based upon its private label price relative. Firm X and Y have equal national brand price relatives but unequal all-brand indexes because in one firm, private labels were priced and in the other firm they were not. Thus, the aggregation procedure used by ERS implicitly imputes prices for missing observations.

Are the biased price relatives systematically related to market share? We can't say for sure. We are reasonably confident that the biases resulting from the missing data imputations add considerable noise to the data, inhibiting the ability of regression analysis to detect mean level price differences among firms and markets. Moreover, the noise is greater for those firms with more missing data; hence the error variances are likely nonconstant. We found that the number of observed categories and items per store were related to store size and firm market share. A regression of a firm's market share or average square feet on the *number of* national brand categories priced was positive and statistically significant. That is, low market share firms or firms operating smaller stores had fewer observed prices.

The effect of the missing data was that for each store and/or firm the price relative was estimated from a different sample size. This is masked by the aggregation procedures. The actual number of edible grocery categories for which national brand average prices were calculated for each store varied dramatically.

TABLE 12.8 Illustration of Procedures for Calculating the All-Brand Price Relative Index Values for a Category

Firm	National Brand (NB)	Private Label (PL)	All-Brand (AB)
Firm X	104.00	110.00	105.50
Firm Y	104.00	Missing	104.00
Firm Z	Missing	110.00	110.00

TABLE 12.9 A Summary of the Number of Categories Where Prices Were Actually Surveyed for the Edible Grocery Department, Wave 2

Number of Categories[a] (S)	NB No. of STORES	PL No. of STORES	Generic No. of STORES
0 < S < 10	6	233	220
10 ≤ S < 20	35	222	0
20 ≤ S < 30	131	31	0
30 ≤ S < 40	288	0	0
40 ≤ S < 50	43	0	0
50 ≤ S < 60	0	0	0
60 ≤ S	0	0	0
	503	486	220

[a] There were 63 categories surveyed in Wave 2.

We've summarized the number of national brand (NB), private label (PL), and generic edible grocery categories actually priced by frequency classes in Table 12.9. Only 43 of the 503 stores had national brand prices in more than 40 categories. For roughly one third of the 503 stores, national brand prices were collected on less than 30 categories—that is, less than half of the 63 categories selected for pricing. Here we're talking about *whole categories* in which not a single national brand price was collected. In the Wave 2 survey, prices were supposed to be collected in 53 categories for private label items and in 35 categories for generic brand items.[35] However, very few categories per store were actually priced for PL and generic brands. For most categories, the all-brand price relative was effectively the national brand price relative.

Although we have focused our examples only on the edible grocery department, the problems are likely at least as serious for the other departments.

TABLE 12.10 Wave 2 Edible Grocery Category 30, Presweetened RTE Cereal, 10.1-12 Ounces; Summary Statistics and Number of Priced Items by Item Code

Item	Item Name	Unit Prices			NOBS
		Min	Max	Mean	
11	Kellogg's Fruit Loops	0.11727	0.19909	0.14198	330
12	Kellogg's Sugar Frosted Flakes	0.10818	0.13833	0.12678	8
13	Post Alpha Bits	0.09500	0.14909	0.12512	5
14	Kellogg's Sugar Corn Pops	0.0	0.0	0.0	0
15	(Write-ins for other	0.09416	0.21727	0.13421	357
16	manufacturer brands)	0.09000	0.17181	0.13463	324
17	-	0.07750	0.19100	0.13447	288
18	-	0.08750	0.17181	0.13480	244
19	-	0.08916	0.18090	0.13513	211
36	-	0.12083	0.12083	0.12083	1
51	Private Label	0.13416	0.14833	0.14133	3
71	Generic	0.13181	0.13666	0.13423	2
					2540

A Case Study of Presweetened Cereals

Since category 30 (presweetened cereal, 10-12 oz.) had the largest number of observed prices of the 63 edible grocery categories, we decided to use this category to illustrate the USDA price index methodology. Prices for at least 26 different national brand items were recorded by enumerators. Tables 12.10 and 12.11 contain some of the results.

In Table 12.10, the four prelisted brands on the survey form are shown as items 11-14. We have also included summary statistics for national brand items 15-19. Of the prelisted brands, only Kellogg's Fruit Loops had a sizeable number of observations—330 for the 503 supermarkets. The other three prelisted brands were apparently seldom stocked, at least in that package size.

Enumerators wrote in 22 national brands (items 15-36) for this category in at least one store. National brand item 15 recorded the largest number of prices. However, there was no attempt to record the same brand under the same non-prelisted item number. Thus, the number of items and the enormous range in price of item 15 (9.4-21.7) probably reflects the fact that several different brands were recorded at this item number.

Item 51 is private label and item 71 is generic presweetened cereal. The prices here are infrequent but surprisingly high relative to the national brand

TABLE 12.11 Private Label and Generic Survey Results for Category 30, Wave 2. Presweetened Ready-to-Eat Cereal 10.1-12.0 Ounces

				Portland, ME				
	Firm = 30					Firm = 34		
			Unit					Unit
Obs	Store	Item	Price		Obs	Store	Item	Price
1	1	15	0.12416		1	1	11	0.15363
2	1	16	0.12416					
3	1	17	0.12416		1	2	11	0.11727
4	1	18	0.10416		2	2	15	0.12390
5	1	19	0.12416		3	2	16	0.12636
6	1	20	0.12416		4	2	17	0.14272
7	1	21	0.13181		5	2	18	0.14272
					6	2	19	0.14272
Mft. brand average			0.12240		Mft. brand average			0.132615
8	1	71	0.13181		7	2	51	0.14833

prices. Because of the dearth of PL and generic observations, the category all brand index values are primarily national brand average prices.[36]

A frequency distribution of the number of category 30 national brand items priced per city/firm (not shown) indicated the following: thirty-five firms (including six large chains) had no price observations for pre-sweetened cereal in the 10.1-12.0 oz. package size;[37] nearly half of the firms (151 of 321) had 5 or fewer brands.

Table 12.11 shows the survey results in Portland, Maine for firms carrying private label or generic brands of presweetened cereals. There were only five private label or generic price observations for this category in all 503 stores.[38] Data for two firms in Portland, Maine are shown. Firm 30 had 8 price observations—all in the same store. Items 15-17 and 19-20 had identical unit prices of $0.12416. Item 71 was the generic brand and was equal in price to the highest national brand. No prices were found for the prelisted items. Firm 34 had prices for two stores. Store 1 was priced for only Kellogg's Fruit Loops (item 11). Store 2 also carried Item 11, but at a 24% lower price.[39] Store 2, in addition, carried five other national brands and a private label (item 51). The private label item was the highest priced cereal in Store 2.

The prices shown in Table 12.11 strike us as suspect. However, that is not our main purpose for including this table. How was the average national brand price for category 30 determined by ERS in these cases? For *firm* 34, the *firm* average national brand price would be a simple average of the two average store prices. In store 1, the average price would be based upon one observation (0.15363); in store 2, six national brand observations would be totalled and divided by six (= 0.132615). The firm national brand average price is 0.14312.[40] Note that although there was a 24 percent price difference on the one identical item priced in the two stores, by the time five other brands are averaged into the price for store 2, the price difference shrinks to 13 percent. This reduction in the price differential is based on a comparison of average prices from two different product mixes.

A comparison of firm 30 and firm 34 would indicate that firm 34 is more expensive. Firm 30 has a significantly larger market share than firm 34. Firm 30 had two stores surveyed with an average selling square footage of 24,528.[41] Firm 34 also had two stores surveyed, with average selling area of 8,100 square feet. Each firm offered a different product mix in terms of store size and available products. Are we comparing apples and oranges? Could the ERS econometric model sufficiently control for the level of product differentiation that exists between these two firms if an analysis of all their categories demonstrated similar results? We doubt it.

Given the numerous and serious limitations in their price data discussed above, K-H have no basis for claiming superior quality data. Relative to most of the studies that have examined concentration-price relationships, we believe the ERS data are decidedly inferior. For example, the JEC study (Marion *et al.*, 1979) had a dependent variable based upon the prices of 94 identical grocery products—products that the chains themselves considered important enough to price check. By comparison, ERS's price data are based upon an average of 83 products that are not identical across stores or markets and that vary greatly in their importance.

K-H contend that we have failed to prove or document our charge that their price data have massive deficiencies. On this they are correct. The only way we could document the level of errors in the ERS data would be to compare their indices to one that is developed so as to avoid the problems noted in this critique. Such an index cannot be developed from the ERS data.

A Brief Comment on the ERS Econometric Results

In light of the severe price data problems we will limit our discussion to pointing out some of the more significant limitations of the ERS econometric model and regression analyses.[42]

- The ERS market concentration and market share data were calculated from 1982 USDA Food and Nutritional Services (FNS) data. Comparison of Census 1982 four-firm concentration ratio's indicated substantial differences between the two data sources.

- The market entry variable was calculated using 1982 market shares for firms that entered the market within the 5 preceding years, and had achieved a market share of 5 percent or more in 1982. Both de novo entry and entry by merger were included as entry. Entry by merger undermines the basic logic that entry will create additional capacity, reduce market concentration and enhance price competition. We believe that the two forms of entry (de novo and merger) should be modelled separately and the length of the entry period shortened.

- The method of controlling for the effect of warehouse stores is inadequate. For some reason, warehouse store prices were co-mingled with non-warehouse store prices where firms operated more than one format in an MSA. Thus, if a firm's warehouse store accounted for 25% of its sales in a market, and several super stores accounted for 75%, a weighted average price was calculated. However, the independent variable used was a simple zero-one binary that identified firms that operated one or more warehouse stores. What the authors failed to recognize was that the set of prices identified by the binary variable was not restricted to warehouse store prices. Furthermore, we believe that a "warehouse store in market" variable should have been included in the model to measure the impact of warehouse stores on overall market prices. Marion, Heimforth and Bailey (this volume) found that markets more heavily populated by warehouse stores tend to have lower overall market prices.

- The authors only present the results of regressions where the data are expressed in logarithms. The logarithmic specification suggests significant multicollinearity problems. Both linear and polynomial regressions should have been presented, especially linear.

- An analysis of the residuals based on the data and regression model presented in the report indicates that the regression error terms are heteroskedastic. There was no attempt to correct for this.

Although there are significant weaknesses in the econometric model employed by K-H, most of these are correctable. Clearly, the central problem in the K-H/ERS study is the dependent variable. Unfortunately, the weaknesses in the measurement of prices appear to be uncorrectable.

Notes

1. The K-H results are inconsistent with the bulk of the literature in that they found nonsignificant negative relationships between supermarket prices and both market concentration and firm market shares and a significant positive relationship between prices and market entry. This study is also unusual in the amount of self-praise—both in the report itself and in ERS press releases about the report. Claims that "This study will be the 'benchmark' against which other food pricing studies will be evaluated..." argue for a careful review of their research methods and data.

2. The apple-orange problem was brought to Handy and Grinnell's attention in September 18, 1980 letter by Bruce Marion: "If I am correct, your procedure will not provide you with price data for a comparable market basket across firms and markets . . . You'd be comparing apples and oranges" (personal correspondence).

3. BLS follows this procedure to select specific items to price at the store level. That is, for each category selected, BLS uses within store sales to probability sample specific items.

4. Karmel and Jain compared conventional random sampling with model-based judgement sampling on data from over 12,000 Australian businesses. They found model based sampling was by far the most efficient. I.e., judgement sampling can provide equal or superior results at much lower costs.

5. Kaufman-Handy indicated a target sample size of 500 in written correspondence. They now contend this was not the target sample size but have repeatedly refused to tell us what the sample size was or how the number of categories selected was determined. MacDonald and Nelson (1991), relying on ERS data, report a market basket of 600 items.

6. BLS notes that following a highly stratified sampling procedure allows them to reduce the sample size (gain efficiency) and reduce enumerator errors. Churchill discusses the increase in non-sampling errors when attempting a census. Indeed, sampling is used to verify a census.

7. See K-H/ERS Report, Appendix D, Step Ic(1). Note that while brands were averaged within brand classes to calculate the U.S. average price for each subcategory the all-brand price relatives were aggregated at the subcategory level (Appendix D, Step IIc).

8. The authors do not recognize product mix differentiation as undermining their ability to test the relationship between market structure and price. Kaufman wrote to us: "Your concern about the wide range of values in the data is unfounded, given the nature of the study and its objectives. To the contrary, we expected considerable variability in the data, having designed a methodology to accommodate this feature."

9. Roughly every 5 years, the outlet sample and the item sample is adjusted by BLS. When this occurs, BLS uses a bridging technique to change from the old to new sample.

10. Probably the most widely cited of these is the JEC study (Marion *et al.*, 1979). That study was reviewed by 27 independent economists. Doug Greer, author of the leading industrial organization textbook, described the study as "undoubtedly the best piece of research ever done on the topic" (JEC Hearings, p. 198).

11. For Wave 2 there were 209 single store firms and 112 multi-store firms.

12. A single observation was deleted because the store format resulted in extreme values. This exclusion did not materially affect the results.

13. In order to obtain the identity of the ERS markets and firms, we agreed to confidential treatment of specific information about a firm in a market. As noted by K-H in their response, the extraordinarily low price indexes were generally box store observations. The box stores and warehouse stores were dealt with in their analysis by a single dummy variable. Interestingly, one of the highest price indices (121.2) was for a warehouse store observation. The average index value for the 20 firms operating warehouse stores was 93.0.

14. The authors indicate that previous judgement or convenience type samples were biased due to item selection procedures and handling of missing data (see pp. 2-9).

15. The fact that national sales shares were used likely explains the inclusion of institutional size containers and other "oddball" size containers often not found in supermarkets.

16. Phil Kaufman recently wrote to Bruce Marion (8/14/91): "The 500-item sample size was only used as a baseline for study planning purposes. As such it provides an upper-bound on the number of items potentially enumerated in a given store." While this is inconsistent with prior communication, it still begs the question of what was the sample size for each wave and what were the statistical qualities of each wave's sample results.

17. Much of this information came from an unpublished paper, "Methodology and Procedures for a National Study of Supermarket Prices," USDA/ERS 10/83, pp. 19-23. This was sent to us by one of the authors. Apparently, after a subcategory was selected it was redefined as a category.

18. Prices actually recorded were per unit prices. Thus if a 12 oz. can of cola was priced at $.50, its per unit price would be $0.04167/oz. This allowed for some variation in container sizes.

19. The "other" classification was used only in the produce and meats departments.

20. Warehouse store and possibly box store prices were not kept separate from the prices of conventional supermarkets and super stores. Instead, prices were combined with "other format" stores using firm store format sales weights. For example, if warehouse stores accounted for 10 percent of a firm's sales in a given MSA, then warehouse stores received a weight of .10 and the other stores were weighted by .90. This made it impossible to eliminate the effect of warehouse stores in order to hold store format constant.

21. There were, in fact, 63 categories surveyed with one category selected twice. We used 1/63rd since K-H dropped category 8 in their weight data set.

22. ERS's actual language: "When a subcategory was selected, the number of items expected to be found in an average supermarket was estimated and sample selection proceeded until the target number of items was reached." (See note 14.)

23. The missing data imputations compound the error problems; however, that is likely best viewed as a nonsampling error. The biases inherent in the missing data imputations will be discussed in detail later.

24. Detailed subcategories were often not developed until after a broader subcategory was drawn (e.g., Other Milk & Milk Beverages). All of these detailed breakdowns were not recorded on the list of subcategory shares sent to us. See Cochran, pp. 308-381 for a discussion on one possibility for estimating the standard errors. We point out, however, that the task of proving or disproving the quality of the ERS sample should not be ours. It is, in our view, the task of the authors given their claims of data superiority and analytical results.

25. At the June 3-5, 1991 NE165 Conference held in Alexandria, Virginia, Kaufman-Handy announced the results of a statistical analysis that supported the accuracy of the estimated store level index values and actual sample sizes. Our review and discussions with the statistician who completed the analysis disclosed an error in the sample size formula. The correct formula indicated the results shown in Table 12.6.

26. We requested that the authors estimate the sample size necessary to achieve a mean relative price within two percent of the true mean (precision) at the 95 percent confidence level. We requested the relative precision to be 2 percent to correspond with supermarket industry profit levels. During the period of analysis profit to sales ratios generally ranged between 0 and 2 percent (German and Hawkes). The ability to raise and maintain prices 1-2 percent above competitive prices suggests a significant profit difference on the order of 100-200 percent. The authors, based on correspondence with us, have failed to recognize the distinction between precision and confidence levels. Thus they would have us measure the mean relative prices within ± 5 percent, based on typical 95 percent confidence intervals for hypothesis testing. Being satisfied with relative prices that are only within ± 5 percent of the true mean price would eviscerate the ability to identify or test for market power in the supermarket industry. (See Churchill (Ch. 9-11) for a lucid discussion of these points).

27. U.S. Department of Labor, 1988.

28. The U.S. average category price used in the index calculation is based on average prices for a variety of brands, stores, and markets.

29. We believe that the ERS procedure resulted in substantially greater variance in price relatives than would have been the case if identical brands had been compared across stores, firms and markets. Thus, if identical brands had been priced, the sample size necessary to achieve the desired level of accuracy would likely have been significantly smaller than those estimated using the ERS data. BLS recognizes this feature in their sampling procedures for the CPI.

30. Churchill, p. 400. The sampling design undermined the authors' statistical model (i.e., OLS) with the decision to sample the six leading firms with certainty. While this is a fairly common sampling procedure, the procedure undermines the ability to use OLS regression analysis. (This was pointed out to us by Robert Miller, Chairman, Department of Statistics, UW-Madison.)

31. Items could be missing for four reasons: (1) store did not sell the item; (2) item was out of stock; (3) enumerator could not find the item; and (4) the observation was lost due to computer or computer entry error. It is impossible to calculate the number of missing data. Our best guess would be to take the difference between the number of prices found and the number expected. Apparently the number expected for the store was 500 and the number found, on average, was 170. Hence, there were approximately 330 items per store missing.

32. We did not review the produce and meat departments because we believe that these departments are so heterogeneous that firm-market comparisons are inherently hazardous. A second problem with these two departments is the city location impact on price. For example, wholesale meat prices are higher at the coastal markets, reflecting the transportation cost from the midwest. K-H attempted to control for this in their regression analysis (p. 24).

33. The cited examples are from the Dairy and Non-Foods departments. Non-filter cigarettes were to be priced by the carton and yogurt, fruit-filled in packages of 6 ounces or less. Other categories missing were Cheddar Cheese, sharp, 6 oz, or less; Cheddar Cheese, mild, random weight; Jarlsberg Cheese, random weight; Butter, salted in tubs of 8 oz. or less; and Sour Cream, plain, 8 ounces or less.

34. See K-H/ERS Report, Appendix D, IIc or IId. Keep in mind that for dairy, edible, and nonedible grocery departments, categories and subcategories are synonymous.

35. This was determined by the number of subcategories where PL and generic were listed as having sales shares.

36. PL and generic each had brand class share weights of only 1 percent.

37. We use city/firm to indicate that firms which operate in more than one of the sampled markets are treated as separate firms in each MSA.

38. Hence, the national average price for PL products in this category was based on 3 observations.

39. A reviewer of our critique assumed that this was a promotion price. Hence a comparison of Fruit Loop prices between stores would have resulted in a biased price relative. The reviewer felt that since the ERS method reduced the price differential this reduced the biased price relative. Strangely, the reviewer did not question why two supermarkets operated by the same firm carried such different product mixes. Nor did he wonder why PL and Generics were generally more expensive.

40. All-brand average unit prices were not calculated by ERS. National brand, private label, and generic prices were only combined in the price relative calculations. The U.S. average unit prices for category 30 national brand, private label, and generic brands were respectively $0.136, $0.141, and $0.134. Private label and generic brands each had sales share weights of 1 percent. The firm 30 all brand category index value would have been as follows: Firm 30, Store 1 = .9008 or 90.08 percent. ([((.1224/.136) x .98) + ((.13181/.134) x .01)]/.99) = [(.9 x .98) + (.98 x .01)]/.99. Similarly, firm 34, Store 1 = 112.96 percent and Store 2 = 97.59 percent. Store price relatives were only combined at the department level.

41. One store apparently did not carry RTE presweetened cereal, at least not in the size containers priced.

42. We are nearing completion of a working paper which will contain a more complete discussion of the ERS econometric model and analysis.

References

Churchill, G. A., Jr. 1983. *Marketing Research: Methodological Foundations*. Third ed. Hinsdale, IL:Dryden Press.

Cochran, W. G. 1977. *Sampling Techniques*. Third ed. New York:Wiley.

Cotterill, R. W. 1983. The Food Retailing Industry in Arkansas: A Study of Price and Service Levels. Unpublished report submitted to the Honorable Steve Clark, Attorney General, State of Arkansas, January 10.

_____. 1986. Market Power in the Retail Food Industry: Evidence from Vermont. *Review of Economics and Statistics* 68(August):379-386.

_____. 1991. Food Retailing: Mergers, Leveraged Buyouts, and Performance. Food Marketing Policy Center, Research Report 14. Univ. of Connecticut. Sept.

Cryer, J. D., and R. B. Miller, ed. 1991. *Statistics for Business: Data Analysis and Modelling*. Boston:PWS-Kent.

German, G. and G. Hawkes. 1983. Operating Results of Food Chains, 1982-83. Cornell University, College of Agriculture. Sept.

Hall, L., A. Schmitz and J. Cothern. 1979. Beef Wholesale-Retail Marketing Margins and Concentration. *Economica* 46(August):295-300.

Karmel, T. S. and M. Jain. 1987. Comparison of Purposive and Random Sampling Schemes for Estimating Capital Expenditure. *Journal of the American Statistical Assoc.* (March):52-57.

Kaufman, P. and C. Handy. 1989. *Supermarket Prices and Price Differences*. ERS Tech. Bull. 1776. U.S. Dept. of Agriculture. Washington, D.C., Dec.

Lamm, R. McFall. 1982. Unionism and Concentration in the Food Retailing Industry. *Journal of Labor Research* 3(Winter):69-79.

MacDonald, J. M. and P. E. Nelson. 1991. Do the Poor Still Pay More? Food Price Variations in Large Metropolitan Areas. *Journal of Urban Economics* 30:344-59.

Marion, B. W., W. F. Mueller, R. W. Cotterill, F. Geithman and J. Schmelzer. 1979. *The Food Retailing Industry*. New York: Praeger.

Marion, B. 1984. Strategic Groups, Entry Barriers and Competitive Behavior in Grocery Retailing. Working Paper 81, U of Wisconsin-Madison.

Marion, B. W., K. Heimforth and W. Bailey. 1992. Strategic Groups, Competition and Retail Food Prices. In *Competitive Strategy Analysis in the Food System*, ed. R. Cotterill. Boulder: Westview.

Meyer, P. J. 1983. Concentration and Performance in Local Retail Markets. In *Industrial Organization, Antitrust, and Public Policy*, ed. J. V. Craven. Boston:Kluwer-Nijhoff.

Newmark, C. M. 1990. A New Test of the Price-Concentration Relationship in Grocery Retailing. *Economics Letters* 33:369.

Porter, M. 1979. The Structure Within Industries and Companies' Performance. *Rev. of Economics and Statistics* 61(May):214-27.

Scherer, F. M. and D. Ross. 1990. *Industrial Market Structure and Economic Performance*. Boston:Houghton Mifflin Co.

Schmalensee, R. 1989. Inter-Industry Studies of Structure and Performance. In *Handbook of Industrial Organization*, ed. R. Schmalensee and R. Willig. Vol II. New York:North-Holland.

U.S. Congress. Joint Economic Committee Hearings. *Prices and Profits of Leading Retail Food Chains, 1970-74*. U.S. Gov't Printing Office. 1977.

U.S. Dept of Labor, Bureau of Labor Statistics, *BLS Handbook of Methods*, April 1988, Bulletin 2285.

Weiss, L. ed. 1989. *Concentration and Price*. Cambridge, MA:MIT Press.

Werden, G. J. 1991. A Review of the Empirical and Experimental Evidence on the Relationship Between Market Structure and Performance. U.S. Dept. of Justice, Economic Analysis Group Discussion Paper EAG 91-3, May.

Wills, R. and W. F. Mueller. 1989. Brand Pricing and Advertising. *Southern Economic Journal* 56(Oct):383-395.

ADDENDUM

ERS's Latest Precision Estimates

In the earlier versions of their response, Kaufman and Handy included ERS's calculation of the relative precision of *store level* price indices for Wave 2.[1] We critique their precision estimates in the section on sample size evaluation, p. 273. Now, in their latest response, Kaufman and Handy report *firm level* estimates of relative precision.[2] We will limit our addendum to a discussion of their section *Issue 3. Sampling Error Results in Unreliable or Biased Price Index Estimates.*

Once again, ERS's analysis of the level of precision of their price data is simply wrong. Based upon what we have learned about their estimation procedure, their new calculations (Table 13.2 and text discussion) have the following errors:

1. ERS aggregate their price data in a different way for their latest estimates of precision than in the data used in their regression analysis. The two methods of aggregation used by ERS can best be shown by example. Below we show the price indexes for two stores of the same firm. In store 1, five product categories are priced. The store average (simple mean of categories) is 103.4. In store 2, only two categories are priced. The store average index is 109. The price data used in K-H's regression was the average of the two store means; i.e., 106.2. That is, each store was given equal weight in calculating the firm price for the MSA.

Cat	Store 1	Store 2	Firm Sum	Firm Average
1	102			
2	100			
3	104			
4	108	108		
5	103			
6	-	110		
Sum	517	218	735	105.0
Store Means	103.4	109.0		106.2

In their latest calculations of precision, ERS pool the price indices for all categories in all stores and take the simple average. In the example, the seven price observations are totalled—the mean is 105.0 instead of 106.2. With this approach, each category price is given equal weight.

The price index K-H used in their regressions would be 106.2 in this example. However, they calculated their precision estimates using the 105.0 index and aggregation procedure.

2. ERS used inappropriate sample sizes in calculating the degree of precision. In the above example, 6 categories are priced for a particular firm. However, ERS simply add the total categories priced in all stores of the firm—in this example 7.

Had all six categories been priced in both stores, ERS would consider that a sample of 12. We believe it should be considered a sample of 6. The impact on the calculation of precision can be substantial.[3]

Firm Versus Store Results

ERS's latest calculations at the firm level would differ from their previous store analysis only for multi-store firms. There were 209 single store firms and 112 multi-store firms in Wave 2. ERS had previously found that only 3 of the 503 stores had confidence intervals of 2.0% or less (at the 95% level of confidence). All three were single store firms. At most, they could have only 115 *firms* with relative precisions of 2.0% or less. ERS's recent calculations of *firm* level data precision show that only 39 firms (12%) had relative precision of 2.0 percent or less for their price data.[4] In general, ERS's firm level precision estimates do not alter the results (at the store level) presented in our discussion of Table 12.6.

A Note on Confidence Intervals and Relative Precision

Kaufman and Handy have misunderstood the intent of the sample size calculation. Rather than compute individual confidence intervals for each store to demonstrate the level of precision (or lack thereof) of the Wave 2 sample results, we requested the sample size needed to achieve a given level of relative precision. Put another way, what sample size is necessary so that 95 times out of 100, the sample mean price will be within ±2% of the true mean. If the actual sample size was substantially smaller than the required sample to achieve the 2% relative precision, we would know that the confidence interval (CI) surrounding the store index value was greater than ±2%.

K-H contend that a 5% level of precision is sufficient. However, accepting this level of precision would largely eliminate the chances of detecting market power in the supermarket industry. For example, consider a monopolist with a price index of 108. If a precision level of 5% is accepted, the confidence interval for this firm would be 108 ± 5.4 (i.e., 102.6 to 113.4). Now—assume a second market that is competitive and that has price indices of 100. The 5% confidence interval in that market would be 95-105. A statistical test of the difference in means in these two markets would reveal no difference at the 95% confidence level.

The differences in the prices of a given supermarket chain across markets are not so large that prices that are within ±5% of the true price can be relied upon to detect the exercise of market power. K-H/ERS have never been able to provide us with information on the level of precision originally sought in the ERS research design.

Some Final Comments

We continue to believe that an appropriate calculation of confidence intervals that accounts for the variation in category, store, and firm level indexes would demonstrate that the ERS index values are inadequate to analyze the relationship between price and market structure in the supermarket industry.[5]

However, we also remind the reader of the point made in our critique: the entire exercise to calculate the precision of the price data is somewhat of a charade. If the prices in different stores are for non-comparable products, no sample size is sufficient to correct this problem. We do not believe that it is possible to control for product mix differentiation among stores and firms with the ERS model.

A final point. We strongly encourage researchers to examine the underlying data themselves. Only then will they know whom to believe. All three of the wave data sets (especially the "Other Foods or Edible Grocery Department") coupled with the surveyed items, summary statistics, weight data sets, and a detailed description of the aggregation procedures should be made available for review and replication by the USDA.[6]

Notes

1. At the June workshop at which these papers were presented, K-H presented their first attempt to estimate the precision of their price data. In that analysis, they concluded that nearly all of the 503 stores had adequate size samples. Indeed, in all but 36 stores, a sample size of *less* than 10 subcategories was found by ERS statisticians to be sufficient to achieve the desired level of pricing accuracy ($\pm$ 2%). Common sense and 30 years of studying this industry told us this was impossible. Further investigation revealed that the wrong formula had been used. The correct formula indicated that only 3 of 503 stores had the desired level of precision (our Table 12.6).

2. Although one of the authors initially refused to send us a copy of their manuscript, we obtained a copy from the editor. Subsequently, we were able to talk to the statistician who did the calculations and obtained from the authors a copy of the firm level calculations.

3. The formula for calculating the relative precision (r) is: $r\mu = t_{.05} * S/\sqrt{N}$, where S is the estimated standard deviation and N is the sample size. Solving for relative precision: $r = t_{.05}/\mu * S/\sqrt{N}$. Clearly as N increases, *ceteris paribus*, the calculated value for r declines (i.e., relative precision improves). Thus, following our example, the new ERS approach increased N from 5 to 7. In the ERS sample, N more than doubled for most of the 112 multi-store firms. The increase in N likely resulted in the new lower estimates of relative precision.

4. Our own analysis of the ERS precision estimates demonstrated that by aggregating category indexes for each firm rather than averaging the *store indexes* and *store relative precisions*, ERS was able to reduce their estimates of relative precision, on average, by 1.35 percentage points for 106 multi-store firms. For this calculation we excluded six multi-store firms that operated warehouse stores and supermarkets. ERS did not make any adjustments for these firms in their own analysis of firm relative precision. Note that simply averaging the store relative precisions allows us to give only an approximation of the effect of increasing the sample size by pooling the store category indexes.

5. For example, in the determination of the store and firm level estimates under both the original and the new procedure, item indexes were treated as parameters and not estimates. That is, it was assumed that each of these values were measured without error.

6. We must caution, however, that we received, piecemeal, various parts of the data set sometimes only after repeated requests. At one point we were informed we would have to pay for computer time to generate summary statistics from the original data. Ultimately, two data tapes were sent to us (Wave 2 and Wave 3 plus summary tables), both of which were incorrectly documented. We were unable to read portions of the latter tape but were able to recover the Wave 3 data. Often summary tables sent to us were poorly identified, if at all, and in some instances they were not summaries of the underlying data actually used in the final regressions. One request for a verification of facts in our review was simply ignored.

13

The Geithman-Marion Review of the ERS Supermarket Pricing Study: A Response

Phil R. Kaufman and Charles R. Handy[1]

Geithman and Marion raise many serious allegations in their critique: "Testing for Market Power in Supermarket Prices: A Review of the Kaufman-Handy Study" (Geithman and Marion, 1992). Their conclusion that the ERS study is "fatally flawed", largely due to price measurement, obviously requires greater scrutiny of the facts beyond what limited information and analysis is provided in their review. In this response, we address the questions raised by Geithman and Marion (G-M) concerning validity of the data used to conduct analysis of the concentration-price relationship. We show that the ERS study is not an outlier among the concentration-price relationship literature, as is claimed. We identify features of the research design used to preclude or minimize the potential for unlike item comparisons, and show how both on conceptual and empirical grounds, the G-M review arguments for price bias are insupportable. We respond to the G-M review allegations about sampling error through both store and firm-level examples, and refute their standards for relative precision of price indexes. We confirm the reliability of calculated price indexes, both within firms operating in multiple sample cities, and within a city, among multiple store firms. We also refute claims that nonsampling errors are greatly exaggerated and misrepresent the facts. Their assertions about biased price indexes, integrity of the data, and other numerous *ad hominem* arguments merely raise doubt without providing compelling conceptual, statistical or empirical support.

The Geithman-Marion review also criticizes the ERS report for omitting information they deemed important for evaluating the data and analysis. While we disagree, we have taken this opportunity to provide additional background of the methodological development and detail of procedures used in the study.

Methodological Development and
Empirical Considerations

In the process of developing our survey methodology, the ERS research team systematically reviewed the methodology used in previous supermarket price studies. We discussed the pros and cons of alternative methodologies with academic researchers such as Joe Uhl (Purdue) and Bruce Marion (Wisconsin). We met with representatives of the Food Marketing Institute, National Grocers Association, the Bureau of Labor Statistics, the Federal Trade Commission, (FTC) and private research firms such as Case and Company. ERS also held an all-day seminar attended by researchers from USDA, other government agencies, and universities to review and critique our proposed methodology.

We drew from the knowledge and insights gained from the above review plus the experience we gained in conducting three surveys of supermarket prices among competing supermarket firms in the Washington, D.C. area during 1979-80, in developing the methodology used in the ERS study of supermarket prices.

We explicitly considered and then rejected the methodology suggested in the G-M review as inappropriate for a nationwide survey of price differences among supermarkets and supermarket firms. After a six-year study, the Federal Trade Commission also concluded that the market basket, identical item, methodology similar to that suggested by G-M was unworkable for a multimarket, multi-firm study.

In 1973, the Federal Trade Commission initiated a project to develop a survey design capable of accurately measuring statistically significant differences in overall prices among retail food stores. A major goal of the project was to develop a practical survey methodology to help the Commission eliminate deceptive comparative pricing claims and, at the same time, to provide consumers with information they could use to make valid price comparisons among competing supermarkets.

In the (January 15, 1973) *Federal Register* the FTC published, for comment, a proposed protocol for the Federal Trade Commission Retail Food Price Survey. Because prior chain-to-chain comparative surveys had generally compared identical items at competing stores, the FTC staff initially assumed that this was the best procedure. The FTC conducted pilot surveys in Boston, Atlanta, and Washington, D.C. during 1973. An FTC Staff Report concluded that:

> However, Commission pilot surveys as well as industry submission amply demonstrated that competing stores do not, for the most part, sell the same items. The vast majority of the items surveyed during the pilot surveys were simply not available at all major chains in a city. . . . An F.T.C. pilot survey of food stores in Atlanta on June 21, 1973, revealed that an average of only 11 percent of all items at any given store were available at all major

competitors surveyed. . . . Therefore, it was decided that products, not items, should be surveyed.

It was also decided that all items of each product selected to be included in the survey would be surveyed. [FTC] staff felt that surveying all items within certain products would eliminate some of the problems inherent in a food price survey. One of these was the unavailability of items. In a survey in which certain specific items are preselected for comparison, the unavailability of an item in one or more stores would limit the usefulness of the survey results (FTC, pp. 14-15).

Furthermore, the top selling brands may vary widely from one region of the United States to another. For example, Jays potato chips may be the number one selling brand in the Midwest, but not sold at all in another region of the country. But even if the same advertised brand is available in all stores, what does this tell us about a store's overall price level or about retail market power? Again, quoting from the FTC Staff Report:

However, food chains tend to advertise different items than their competitors. Thus, interstore comparisons are difficult. Even if retailers advertised the same items at the same time and thus comparisons were possible, such advertising could easily present a misleading picture. By reducing prices on advertised products and raising prices on unadvertised items, a food store can give the appearance of having reduced prices when, in fact, its overall prices have not been reduced.

This is demonstrated by the results of the F.T.C.'s pilot surveys in June 1973. One surveyed food chain had apparently surveyed its competitors on national brands and lowered its prices on many of these items. It subsequently published a comparison price chart showing lower prices on most, if not all, national brand items in order to support the chain's claim of having the lowest prices in town. The F.T.C. surveys indicated that this chain did have the lowest prices on national brands. However, the market basket totals revealed that the chain was not the lowest priced chain overall. Instead, it was identified at the highest priced in one survey, and second highest priced in another (FTC, pp 4 & 5).

Geithman and Marion fail to acknowledge the practical problems of an identical-item price comparison survey. They cite an A.C. Nielsen sales directory to support their claim that most leading brands and package sizes are sold by 75 percent or more of supermarkets. Our study consisted of both leading and non-leading advertised brands, as well as private/store label and generic products. We consider a price survey of leading brands (as is implied) to be inadequate both for empirical and conceptual reasons that will be outlined later in this response. The A.C. Nielsen sales data cited by Geithman and Marion were derived from a geographically dispersed sample of 150 supermarkets. Sales volume and other product information were obtained from warehouse withdrawal shipment data during a 2-month period in 1979 and 1980. Our review of product categories contained in the Wave 2 item price survey *did not* confirm sales by 75 percent or more of stores. To the contrary, even among more common categories such as bottled apple juice, presweetened ready-to-eat cereal, and ground coffee, the share of stores in the A.C. Nielson sample carrying identical items of leading brands averaged much less than 50 percent and varied considerably between both advertised/national brands and product categories.

Geithman and Marion also chose to ignore the sampling error problem that arises when a set of specific items are selected for price comparison under a subjective regime, as they advocate. Selection of one or more leading brands (on a national sales basis) of a given product category may appear to minimize differences in the mix of brands across stores, and control for potential quality differences. What they overlook, however, is the degree to which one or more leading branded items (or any other subjective sample) adequately reflects price levels of *all* items within a given product category or subcategory. Without such information, the judgement sample advocated by Geithman and Marion contains potentially large sampling error and bias, both within and across stores. A critical premise of scientific sampling is that repeated samples from the same universe would result in similar sample estimates. The type of judgement sample proposed by the reviewers systematically excludes certain items and cannot be relied upon to produce consistent, repeatable, and unbiased price indexes. As a result, the ability to generalize the calculated market basket index to other possible market baskets similarly selected is severely constrained. An assessment of the statistical accuracy, reliability, and bias of the price indexes is not possible. Thus, one is unable to estimate the true market basket index with any degree of confidence or relative precision.

Issues Raised by Geithman and Marion

The Geithman-Marion review can be distilled into four central points of contention. These issues are: (1) The ERS study findings are inconsistent with similar studies in the literature, (2) The research design does not adequately control for differences among supermarkets, or for differences in products between stores, factors they deem critical to measuring market power, (3) Sampling errors are large, resulting in unreliable price index estimates, and (4) Nonsampling errors due to the number of "missing items" and the treatment of missing item data leads to biased price index estimates. We will respond to each of these in turn.

Issue 1. The ERS Study Findings are "Inconsistent With the Vast Majority of the Literature" Involving Similar Industry Studies

Contrary to Geithman and Marion, a review of the literature shows that the ERS supermarket pricing study is in fact, one of a number of studies that found little or no evidence of market power in grocery retailing. One of most ambitious undertakings was that of the National Commission on Food Marketing (NCFM), a study initiated by the U.S. Congress in the 1960's. The Commission was specifically formed to address the question of market power in the food industries. Their study of 6,000 supermarkets operated by nine food retailing chains found no systematic relationship between market share and prices. A study of 21 Midwestern cities and towns by Gorman and Mori (1966) found no evidence that market concentration was related to higher city average prices. Newmark (1990) analyzed supermarket prices in 14 cities across the U.S., and 13 cities in Florida using a market basket of 35 common grocery items. He found a negative, but nonsignificant relationship between the market basket index and metro area four-firm grocery store concentration. Reviews of the grocery pricing study literature have been conducted by Anderson (1990), Padberg (1991), and Weiss (1987, 1989). Among those studies finding a positive concentration or market share-price relationship, Anderson and Padberg questioned the quality of the price data and the generalizability of results. The Marion, *et al.* (1979) study for the U.S. Congress Joint Economic Committee was also reviewed separately by Padberg (1981). He found the price data "were sketchy and generally considered inadequate."[2] Price levels among stores of the same chain varied as much 15 percent, which in his opinion was "far too large to be accommodated within variation of the retail margin" (Padberg, 1991). Both the Lamm (1981) and Cotterill (1986) studies cited in the G-M review which found evidence of market power have serious limitations. Lamm used as his dependent variable price data collected by the Bureau of Labor Statistics (BLS) for its consumer price index series. The BLS has warned that their series is not designed for price comparisons between cities (Rothwell, 1965). Cotterill chose one SMSA and 17 other towns in Vermont for his study of market power in grocery retailing. Supermarket concentration averaged 96.1 percent, with 11 of the 18 market areas dominated by two large chains. We question the generalizability of the   Cotterill study to more geographically dispersed markets, or to a sample of cities with greater concentration variability. Weiss (1987) reviews his own and others' studies which found a positive concentration-price relationship, including industries such as cement, airlines, banking, gasoline and the Marion, *et al.* grocery retailing study (1979). He notes that "settings where concentration had no significant effect on price are fairly easy to find."

Issue 2. The Research Design does not Adequately Control for Sources of Variation—the "Apples and Oranges Problem".

At the outset, the Geithman-Marion review cites the "enormous variation in the products and competitive emphasis of supermarkets." In their view, only comparisons between identical items should be made across a homogeneous sample of stores and firms. Geithman and Marion would exclude comparisons between supermarket formats, between different brands within a product category, and among categories that contain products with considerable quality variation such as fresh meat, produce, and dairy items. By violating these principles, the G-M review states that the ERS study has "failed to avoid the apple-orange problem."

The "apples and oranges problem" is a potential source of error common to all pricing studies, however. After conducting an extensive review of methodologies used in prior analyses of this kind, we concluded that the approach advocated by Geithman and Marion contained serious problems and pitfalls. Rather than repeat these same mistakes, ERS developed a methodology which addressed many of the inherent limitations and weaknesses of the earlier studies. Our objective was to solve many of the practical problems inherent in a nationwide survey of supermarket prices, and to develop representative samples of products, stores, firms, and market areas. Through scientific random sampling procedures, the ERS study captures the wide range of retail outlets, products, and services available to the consumer in a given market area. Our aim was to represent the diversity of competitors and competitive settings in which supermarket retailing takes place. Failing to accomplish these objectives would severely weaken any findings and conclusions.

Other factors guiding the choice of methodology and procedures used in the ERS study were the study objectives themselves. Geithman and Marion advocate certain methods and procedures that focus only on the question of market concentration and its relationship to store and firm prices. In contrast, our goal was to examine all factors hypothesized to influence supermarket prices and price differences. These include market structure and firm market share, as well as characteristics and cost differences between stores, firms, and markets[3]. In this way, the net effect of market concentration on price could be measured without statistical bias.

Controls for Variation in the Data. Given the objectives of representativeness and statistical reliability, a number of controls were essential in the price measurement and analysis phases. Geithman and Marion cite the need to eliminate "as much spurious variation as possible" by severely restricting the range of item, store, and firm comparisons. We chose instead to incorporate the diversity which naturally occurs in a market. Our solution was to provide controls for as much of the sources of variation as possible, preferably as an integral part of the research design. In the ERS study, these features include:

a) Enumeration of all items contained in each homogeneous product category or subcategory selected[4]. A census of each detailed product category, such as Green Beans, canned, 15-17 oz., captures the range of brand-price combinations available to consumers. A census permits all products meeting the category or subcategory definition to be included in the sample. According to Geithman and Marion, the enumeration of only specified identical items in each category, such as Del Monte Green Beans, in 16 oz. cans is appropriate. Restricting the sample to a single brand in each category or subcategory would not accurately reflect prices charged for comparable brands, resulting in biased price indexes. Their approach would severely restrict the number of item-price comparisons between stores, however. Recent evidence shows that even in a single mid-size city, matching of identical brand items between stores poses a serious problem. In a 1987 price survey reported by Dahlgran, *et al.* (1991), 23 retail supermarkets operating in Tucson, Arizona were surveyed. The sample included 16 stores from five major chains, six independent stores, and one discount store. A 44-item market basket was used based on the market basket used by Uhl, Boynton, and Blake (1981). The study reported that data were not available from all stores for all 44 items. "Missing data were particularly problematic in constructing the standardized national-brand market basket, where complete data were available for only 22 items (Dahlgran, *et al.* 1991, p. 90)." Obviously, the approach advocated by Geithman and Marion of comparing only identical items would be impractical in any survey representative of U.S. cities.

b) Item prices recorded within each product category or subcategory were grouped according to brand-type (national/advertised brand, private/store label brand, and generic), and an average price per unit was subsequently calculated for each brand-type group. This procedure addresses several potential sources of variation in the price data. First, price variation due to quality differences are controlled by limiting item comparisons to those within brand types. Second, item price averaging minimizes the potential for bias due to promotions and specials. Thus, the category-brand type average produces a value that is representative of the range of item-price combinations available to consumers in a given store.

Claiming that biased prices will result, Geithman and Marion use a single category, Beer (no other size or type specifications given) to compare hypothetical prices between three stores (G-M, Table 12.1). Their example attempted to show how differences in product mix could lead to price bias and erroneous price relationships in the analysis. The item price differences between brands are not reflective of actual data collected in the survey, however. In developing the sample, all product categories and subcategories were scrutinized and tested for comparability of the items they contained within product categories, subcategories, and brand types. Subsequently, all item prices collected were systematically reviewed to ensure accuracy. Outlier checks were also made to detect both data errors and noncompatibility problems. Furthermore, the ERS study does not make price comparisons between stores on the basis of an individual product category (such as Beer). Rather, a sufficiently large number of product categories was used to calculate an overall index for stores and firms. A systematic pattern of quality-based price differences would be required to affect a given store's overall price index relative to other sample stores. The reviewers had the opportunity to test their hypothesis about product mix price bias using the Wave 2 data made available to them and chose not to do so. Their example amounts to an assertion lacking either conceptual or empirical support.

Geithman and Marion extend their objection to non-identical item pricing to include any product category or subcategory where the mix of brands compared varies from store to store. They cite nationwide average prices of leading brands as evidence of systematic price differences between brands, even when there are little quality differences (Wills and Mueller). Significant price differences across stores are likely, however. There are any number of regional and even local brands that often occupy leading positions in many market areas. More importantly, pricing of leading brands may vary considerably between retailers. Some retailers may price leading brands higher than competing brands, appealing to the consumers sense of quality. Other retailers may set very competitive prices on leading brands as a promotional effort to attract new customers.

The ERS method of price averaging within brand-types precludes any systematic pattern of store index bias due to the presence or absence of certain higher-priced or lower-priced branded items. The extent to which comparisons involving non-identical brands produces bias, as is claimed in the G-M review, and the magnitude of potential bias in price index measurement and analysis is an empirical question that could be tested. Geithman and Marion merely raise the issue but do not demonstrate any significance in the data. Our evaluation of the item price data did not suggest significant bias due to differences in the mix of brands enumerated across stores.

c) Random sampling procedures that allow for differences in the mix of brands, brand-types, product categories and subcategories, and departments. Because random sampling was used to select categories and subcategories within a department, non-matching price comparisons can be made across sample stores. Succeedingly higher levels of price index aggregation (product subcategory, category, department, and store, for example) are accomplished by averaging together indexes at each level. Geithman and Marion claim that the presence of non-matching price comparisons produces greater "noise" or error in the index estimates. In Table 12.8 of the G-M review, the ERS procedure used to calculate category All Brand price indexes for Firms Y and Z implicitly assumes that brand-type price indexes are in proportion to those brand types for which price indexes were observed. For a given store and category, the formula is:

$$P_k = \sum (p_{ij} w_{ij}) / \sum (w_{ij}),$$

where: P_k = price index, category k.
p_{ij} = price index of category i, brand-type j, and
w_{ij} = weight for category i, brand-type j[5].

Thus, unbiased price comparisons can be made across stores having non-matching samples, if we assume that relative pricing within a category is consistent for a given store. Had a judgement sample been used to determine store price indexes as Geithman and Marion advocate, non-matching price comparisons would not have been possible.

Issue 3. Sampling Error Results in Unreliable or Biased Price Index Estimates

As part of their review of the ERS study, Geithman and Marion requested additional information concerning the reliability of price index estimates and measures of sampling error for Wave 2 store indexes[6]. ERS conducted a separate post-survey analysis, consisting of measures of the coefficients of variation and the level of relative precision for each store[7]. Although the G-M review contains some of our findings, their selective reporting of the ERS analysis produces a distorted picture of the reliability of the price data.

In their presentation of the ERS post-survey sampling error analysis, Geithman and Marion chose to confine their comments to the question of adequacy of sample size given the coefficient of variation (c.v.) and level of relative precision of the store price index estimates. They use a 2-percent standard for relative precision to compare actual sample size with required sample size. Our objective in the post-survey analysis was to determine c.v.'s and levels of precision for store and firm price indexes. To our knowledge there is no reference in the literature to any absolute standard for relative precision[8]. To support their argument for a "within 2 percent" standard for relative precision of price indexes, Geithman and Marion cite in a footnote, the example of a firm that unilaterally raises prices 1-2 percent. A profit increase of 100-200 percent would result, they argue. Here we have yet another example of appeal to hypothetical fiction. We know of no instance where such a price increase could be made without affecting the firm's sales volume and costs, hence limiting profit gains. In the real world, competitors would also adjust prices up or down depending on their assessment of market conditions. Only a monopolist could approximate the magnitude of profit increases suggested by Geithman and Marion. In that case, the presence of market power would be self-evident[9].

Moreover, Geithman and Marion apparently misapply the concept of relative precision in their attempt to show that a 1-percent price increase would not be detected if the level of relative precision was +/-5 percent. Relative precision is used to indicate the range about the estimated price index mean that contains the true population mean. It is a function of the sample size and variance. Changing the relative precision level from +/-5 percent to +/-2 percent would not improve upon the ability to measure a 1-percent price change, however. The price change requires calculation of a new estimated mean and a new range about the estimate, based on the desired level of precision. This is so because the price increase consists of a different sample of item prices, having a different mean and variance. Thus, relative precision has no relevance for the ability to detect price changes or the ability to discern between two means. Geithman and Marion therefore have neither conceptual nor statistical support for the use of a 2-percent relative precision standard.

Estimates of Relative Precision. In the G-M review of the ERS post-survey analysis, some important omissions are apparent. First, use of Wave 2 store-level indexes are not representative of the actual level of price index aggregation used to test hypotheses about factors influencing supermarket prices and price differences. Regression analyses was conducted at the *firm* level of aggregation. Although approximately 140 supermarkets in each wave were single-store firms, the remaining firms' price indexes were calculated from two or three store indexes in each SMSA[10]. To adequately measure error variance in the price index estimates, the computational method should take all sources of variation into account. These sources include variation at the subcategory, category, store department, store, firm and survey Wave levels. All econometric analysis made use of a firm-level index representing the average of each firm-level index of the three survey waves conducted.

Second, in their summary of the Wave 2 store-level results (G-M, Table 12.6), Geithman and Marion group together all store indexes having relative precision levels of greater than 4 percent. Although judgement is clearly called for, we believe a more useful standard for precision is the +/-5 percent level, which is common to many statistical tests of validation[11]. Of the 503 store indexes in Wave 2, all but 57 had relative precision levels of five percent or less. Of those 57 stores, all but eight had relative precision levels of less than 10 percent (Table 13.1). Based on these results, we find the store item counts to be largely sufficient for conducting analyses of the price data.

The price index used as the dependent variable in the regression analysis was an aggregation of store level indexes. In Wave 2, there were 321 firms represented by 503 stores. One would expect, all else being equal, that for those firms represented by more than one store, the sampling error is reduced relative to that of its individual store index components[12]. This hypothesis was confirmed by results of the *firm* level coefficient of variation and their corresponding measure of relative precision of the price index estimates[13]. Of

TABLE 13.1 Wave 2 Store-Level Relative Precision[a]

Estimated Pecent Relative Precision (r)	Number of Stores	Percent of Stores
0 < r ≤ 5	438	87.1
5 < r ≤ 10	57	11.3
r > 10	8	1.6
total	503	100.0

[a] All estimates conducted at the 95 percent confidence level.

TABLE 13.2 Wave 2 Firm-Level Relative Precision[a]

Estimated Pecent Relative Precision (r)	Number of Firms	Percent of Firms
0 < r ≤ 5	289	90.0
5 < r ≤ 10	29	9.0
r > 10	3	1.0
total	321	100.0

[a] All estimates conducted at the 95 percent confidence level.

the 321 firms represented in Wave 2, all but 29 (less than 10 percent) had a relative precision level of greater than 5 percent (at the 95 percent confidence level, Table 13.2)[14].

Geithman and Marion consider only one of the many potential sources of variation addressed by the study design. To judge pricing accuracy based on store level variation is insufficient and out of context within the overall objectives of the ERS study. The sampling error of individual store indexes cited by Geithman and Marion are much less relevant when viewed in this larger context of firm-level analysis.

Issue 4. Nonsampling Errors are Large, Due to the Number of "Missing Items" and the Treatment of Missing Item Prices in the Data

The discussion of nonsampling error in the G-M review is cluttered with incomplete analyses, errors, distortions, and unsupported allegations concerning sources of bias in the price data.[15] Geithman and Marion cite the presence of nonsampling error from a variety of sources, including "missing data imputations," differences in the mix of items between stores and firms, enumerator and data handling errors, and bias in the item-price to store-level index aggregation procedure.

As a motivation for the discussion of sampling error, the G-M review cites the gap between a "target sample" of 500 items and the Wave 2 store average of 170 items as an indication of a "missing item problem." We have stated numerous times that we never expected to obtain prices for 500 items in any store. That number is based on our assumptions of the largest number of brands any store might carry in each product category or subcategory, and then reached only if a store carried every size, variety, and flavor of products selected in our sample. We did not estimate in advance the number of items expected in any given store, but it would have been much less than 500. We also did not expect all supermarkets to carry all product categories selected. The sampling and aggregation procedures were designed to permit accurate price index estimates knowing that stores handle a variety of brands, sizes, and flavors of products. An item would not be "missing" in the ERS survey unless an enumerator failed to report the price of an item that was present. Geithman and Marion are trying to raise concerns that are unfounded. For these reasons, it is erroneous and misleading to hinge the validity of the ERS study largely on the basis of actual versus potential sample size.

To support their claim that there were extensive "missing data problems," Geithman and Marion use a subsample of product categories in Wave 2 referred to as "edible grocery" categories. Their subsample is incomplete however, because it is limited to a) observations for the ten largest chain firms in the study, and b) consists of food categories corresponding to a most frequently priced leading brand requirement (see G-M, Table 12.2). As a result, a significant number of food categories such as fresh meat, produce, processed meat, and many dairy categories were excluded from the sample.

The G-M review summarized the extent of missing categories in the subsample for ten supermarket firms (G-M, Table 12.7). They neglected to explain that one firm operated only limited assortment stores which accounts for most of the missing items found in that firm. Geithman and Marion do not support the claim that missing items were evidence of an "enumerator or data entry problem," when, as we have shown, leading brands are not sold in all stores, nor are all product categories. Thus, the authors' claims remain unsupported.

Non-Matching Price Data. Geithman and Marion then ask whether the firm index values are biased due to missing data. They argue that for many categories, the price index is calculated from observations of a single brand-type such as a national brand or private label brand. As such, the reviewers fail to make a distinction between "missing" and non-matching price comparisons. As we have stated above, the aggregation procedure accommodates non-matching price comparisons without bias. A brand is "missing" only if the item was carried by the store but not enumerated.

In their analysis of presweetened cereals, Geithman and Marion cite the wide mix of items involved in comparisons of price indexes between stores and firms. They refer to the large differences in prelisted national/advertised brands across stores and firms as evidence of missing item bias. To the contrary, the observations by the authors concerning lack of sufficient matching of identical items between stores offer added evidence that the market basket approach they advocate is unworkable and constrained by serious practical and conceptual limitations. We refer to their own example of price comparisons for presweetened cereal between two stores of firm 34 in Portland, Maine (G-M, Table 12.11). They cite observations consisting of a single branded item in Store 1, while Store 2 had six branded items. Following the identical item market basket method advocated by Geithman and Marion, comparison of the two item prices yields a 24 percent difference between the two stores. However, following the ERS procedure, all six branded items in Store 2 are averaged together. The resulting price difference between stores is only 13 percent for branded presweetened cereal between the two stores. By including all six branded items in Store 2, the resulting index is more representative of the choices available to consumers. Reliance on the single branded item would not accurately reflect Store 2's price for presweetened cereal. Clearly, without detailed store-by store-information, reliance on a fixed set of specified items in each product category poses the potential for significant price comparison bias.

Data Validity and Integrity. The ultimate test of the quality and accuracy of the data collected and price indexes used in the analyses is the "reality check", in which we compare the ERS results with existing price data and reasoned expectations about price relationships both within and across firms. Our analysis of store and firm price indexes are internally consistent. Geithman and Marion assert in footnote 29 that, "We believe that the K-H procedure resulted in substantially greater variance in price relatives than would have been the case if identical brands had been compared across stores, firms, and markets." This "belief" is merely speculation and is not based on factual evidence. In fact, when comparing price differences across stores of multimarket firms, just the opposite is true. The Joint Economic Committee study (1977), which used only identical brands, had a greater variation in prices across stores of the same firm in different cities than did the ERS study.

In the ERS study, price indexes of individual stores of multimarket chains demonstrate a fairly consistent strategic pricing pattern. While a firm's prices may vary from store to store and from city to city, the variation typically occurs within a rather narrow range. For example, one large Western U.S. multimarket chain operated in six of our sample cities. This chain has a well-known reputation for having a relatively low price pricing strategy. A total of fifteen stores of the firm were price-checked in the six cities. For Wave 2, the 15 store-level price indexes ranged from a low of 91.4 to a high of 96.3 (100 = all 28-city average). This firm's average price index across all six cities was 93.7 with a standard deviation of 1.7. Another large Eastern U.S. firm followed a different pricing strategy in 1982. Twenty of the firm's stores in eight different cities were included in our sample. The price indexes for this firm in the eight cities ranged from 100.5 to 106.2, with an average index of 103.3 and a standard deviation of 1.9.

In contrast, the Joint Economic Committee study reported substantially greater variation in price indexes across stores even though the sample was limited to 94 items in the dry grocery department. The average cost of the market basket for firm K across seven cities ranged from $86.27 to $100.42, a variance of 16 percent. The average market basket cost for firm H across 17 cities ranged from $85.97 to $97.07, a variance of 13 percent (JEC, pp 58 & 59). This price variation between identical format stores of the same firm is much wider than can be accommodated within the variation of the retail margin and much wider than the ERS study found for any of the multimarket firms included in its sample. The unusually wide price variation in the item sample was limited to branded packaged grocery products. The wholesale cost for those items across cities should be nearly identical for any given firm.

In addition, the G-M review asserts that, "the range in prices found by Kaufman-Handy is unusual for food retailing." In fact, the ERS price indexes perform exceedingly well in their ability to discriminate among major store formats. The authors claim that the ERS price indices, ranging from 73 to 121 is a variation "far greater than any we have encountered." However, our range of store price indexes is what one would expect to find when the sample included the full range of types and sizes of supermarkets and firms in the sample. The sample consisted of firms operating limited assortment stores, full service supermarkets, independent (non-integrated) operators, small chainstore firms, and regional and multi-regional supermarket firms.

Three firms operated limited-assortment box stores in two of our sample cities. While these stores generally had the lowest number of items priced due to their limited assortment, the price indexes for these stores are exactly in the range one would expect ex ante. In a large Eastern city, the price indexes of the three box stores were 85.0, 74.9, and 78.0. In a large Midwestern city, the box store price indexes (2 different firms) were 74.2, 73.7, and 75.7.

Our price indexes were able to accurately discriminate between store formats operated by the same firm in the same city. For example, in a Southeastern city, one firm had three stores selected for our price survey; two were conventional supermarkets and one was a warehouse store. The price indexes for the conventional supermarkets were 100.8 and 100.4, while the warehouse store's price index was 89.4. The same situation occurred for another firm in a Northeastern city. This firm's price indexes for its

TABLE 13.3 Comparison of Survey Results in San Diego

Firm	Relative Price Rank (from lowest to highest)	
	ERS	CalPIRG
Fedmart	1	1
Food Basket (Lucky)	2	2
Von's	3	3
Alpha Beta (Amer.)	4	4
Big Bear	5	5
Safeway	6	6
Mayfair	7	7

conventional supermarkets were 101.4 and 101.3, while its warehouse store's price index was 90.2. These price indexes are what one would expect to find. Yet Geithman and Marion assert that our price indices have "very serious problems." What evidence do Geithman and Marion offer? They, in fact, have presented no evidence of bias in our dependent variable. They have only made assertions based on hypothetical situations and limited sample data.

Finally, there is direct evidence to corroborate our relative price rankings of firms in one of our sample cities. The California Public Interest Research Group (CalPIRG) conducted their regular semi-annual survey of supermarket prices during the same month as our Wave 2 survey (April 1982) in San Diego. The six leading chains plus an independent were included in both surveys allowing a direct comparison of their relative price rankings in San Diego. Their relative price rankings were identical for all seven firms as shown below (Table 13.3). This is a strong corroboration of the accuracy of the two surveys.

Notes

1. The views expressed herein are those of the authors and do not represent official policy or opinions of the U.S. Department of Agriculture.

2. The Joint Economic Committee subpoenaed each of 17 supermarket chains to supply any price comparison checks they had conducted during October 1974. The quantity and quality of the price data submitted was so uneven that price comparisons were limited to only 3 of the 17 supermarket chains. A "market basket" of 94 items was constructed consisting primarily of dry grocery categories. In total, the categories accounted for about 50 percent of store sales.

3. The objectives of the ERS supermarket pricing study are specifically stated in the report (Kaufman and Handy, p.2).

4. Product categories containing comparable items were randomly selected. If differences remained within products of a category, a dissaggregation into multiple subcategories was performed in order to isolate similar characteristics such as package size, flavor, and variety, for example. Item comparisons were limited to similar but not identical package sizes. For example, a #303 can, a commonly used container size, may contain from 15 to 17 ounces by weight, depending upon the manufacturer. All item prices were subsequently converted to a price-per-unit basis, thereby virtually eliminating bias due to size differences.

5. The brand-type weight represents the share of category sales accounted for by each brand-type (national/advertised brand, private label/store brand, and generic/unbranded). The combined brand-type weights account for all of category sales. If no price indexes were calculated for one or more brand-types, the category brand-type weight would sum to less that 1.0.

6. Estimates of sampling error were conducted for each store using the all department, all brand-type index.

7. All measures of sampling error were conducted at the 95 percent confidence level.

8. Furthermore, the reviewers ignore the well-known cost of obtaining each incremental improvement in the precision level of estimates through larger sample size, an factor that every survey must consider.

9. The broader issue concerns the validity of the hypotheses testing of the firm market share and four-firm concentration explanatory variables. ERS firm, all-wave price indexes are unbiased estimates of their respective true indexes, a fact the reviewers have not disproven. Given price indexes that are unbiased, they must then demonstrate empirically the extent to which the higher standard of relative precision would alter hypotheses tests of the explanatory variables.

10. The number of stores selected to represent each firm varied from one to three supermarkets, depending on the firm's total number of stores operated in an SMSA.

11. Clearly, use of a 2-percent standard for precision is too restrictive. To meet this level, some store indexes would need to have more than 10,000 subcategory prices, based on their sample c.v.'s. Such sample size would not be practical for surveys of this kind.

12. In their Addendum, Geithman and Marion claim that one of the ERS authors refused to send them a copy of a revised draft containing new firm-level coefficient of variation analysis (table 13.2), with the implication that the reviewers were being prevented access to these additional results. However, prior to the request by Geithman and Marion, the Editor had received the revised ERS draft with instructions to forward a copy to the review authors. We strongly object to the implications made in the Geithman-Marion Addendum concerning this matter.

13. Geithman and Marion observe that the estimation procedures used for store and firm coefficients of variation are not directly comparable to the actual aggregation procedures used to calculate price indexes. The authors argue that price indexes of identical categories between stores of a firm were not first combined (averaged) before computing c.v.'s. The authors maintain that the ERS firm-level c.v. estimates are reduced soley because we did not combine identical category prices. They ignore the

effect of larger sample size on c.v. estimates when combining store indexes into firm indexes, all else being equal. Furthermore, Geithman and Marion offer no evidence from the price data that would allow us to gauge the alleged bias in the estimates of precision. They offer only a hypothetical example to support their assertion that relative precision estimates are biased.

14. These are firm-level indexes for only one wave. Actual regression analysis combined firm-level indexes for all three waves, as stated above. Although measures of relative precision have not been calculated at the firm, all-wave level of aggregation, we expect that all price index observations would meet the +/-5 percent standard.

15. Geithman and Marion include among these the question of "tampering" by firms included in the store sample. Retailers did not know which of their stores were selected prior to the first survey wave, nor at what time the subsequent survey waves would be conducted. Given that 95 percent of all supermarket product categories had the potential for inclusion in the item-price sample, the ability of firms to anticipate and act on the information ERS supplied in advance in order to influence the outcome of the results is extremely unlikely.

References

Anderson, K. B. 1990. A Review of Structure-Performance Studies in Grocery Retailing. Bureau of Economics, Federal Trade Commission.

Cotterill, R. W. 1986. Market Power in the Retail Food Industry: Evidence from Vermont. *Review of Economics and Statistics* (August): 379-386.

Dahlgran, R. A., M. Longstreth, M. D. Faminow, and K. Acuna. 1991. Robustness of an Intermittent Program of Comparative Retail Food Price Information. *The Journal of Consumer Affairs* 25(11):84-97.

Federal Trade Commission, 1979. Staff Report on the Federal Trade Commission's Retail Food Price Survey. Washington, D.C.

Geithman, F. and B. Marion. 1992. Testing for Market Power in Supermarket Prices: A Review of the Kaufman-Handy Study. In *Competitive Strategy Analysis in the Food System*, ed. R. W. Cotterill. Boulder:Westview.

Gorman, W. and H. Mori. 1966. Economic Theory and Explanation of Differences in Price Levels Among Local Retail Markets. *Journal of Farm Economics* 48(5):194-1502.

Kaufman, P. R. and C. R. Handy. 1989. *Supermarket Prices and Price Differences: City, Firm, and Store-level Determinants.* ERS, USDA, TB-1776 (December).

Lamm, R. M. 1981. Prices and Concentration in the Food Retailing Industry. *Journal of Industrial Economics* 30(5):67-77.

Marion, B. W., W. F. Mueller, R. W. Cotterill, F. E. Geithman, and J. R. Schmelzer. 1979. *The Food Retailing Industry: Market Structure, Profits, and Prices.* New York: Praeger.

National Commission on Food Marketing. 1966. *Organization and Competition in Food Retailing.* Technical Study No. 7. Washington, DC.: U.S. Govt. Printing Office (June).

Newark, C. M. 1990. A New Test of the Price-Concentration Relationship in Grocery Retailing. *Economic Letters* 33:369.

Padberg, D. I. 1981. Review of Marion, Mueller, Cotterill, Geithman, and Schmelzer. *Journal of Consumer Affairs* (Summer):180-186.

_____. 1991. Generalizability of Industrial Organization Studies. *Proceedings*. International Agribusiness Management Association, Boston, Mass. (March):102-109.

Rothwell, D. P. 1965. Calculation of Average Retail Food Prices. *Monthly Labor Review* 88(1):61-65.

Uhl, J. N., R. D. Boynton, and B. F. Blake. 1981. Effects of Comparative Foodstore Price Information on Price Structures and Consumer Behavior in Local Food Markets. Final Report to U.S. Dept. Agr. Purdue Univ. (February).

U.S. Congress, Joint Economic Committee (1977), *Prices and Profits of Leading Retail Food Chains, 1970-1974: Hearings before the Joint Economic Committee, Congress of the United States, Ninety-Fifth Congress, Fisrst Session*, U.S. Government Printing Office.

Weiss, L. 1987. Concentration and Price—A Progress Report. In *Issues After a Century of Federal Competition Policy*, ed. by R. L. Wills, J. A. Caswell, and J. D. Culbertson. Lexington, MA: D.C. Heath and Co.

PART SIX

Strategies in
International Markets

14

Non-Cooperative Game Theory: An Application to Branded Food Product Licensing

Ian M. Sheldon and Dennis R. Henderson[1]

Introduction

Traditionally, firms market their products overseas either through exports or direct investment in foreign production and marketing facilities. Empirical evidence for and analysis of these activities by food manufacturing firms is becoming fairly well documented (see for example, Handy and MacDonald, 1989, Henderson and Frank, 1990, and Handy and Henderson, 1990). However, a third form of international transaction, branded product licensing, has been identified in work by Sheldon and Henderson (1990), but has received scant attention in the agricultural economics and economics literature. For example, most of the work on licensing has focussed on process technology rather than branded products (see Tirole, 1989, for a survey). Potentially, product licensing represents a means of expanding international market opportunities for processed foods and their agricultural ingredients.

International food product licensing can be defined as an economic transaction whereby a food manufacturer, based in one country, with legal title to a brand name, licenses the rights for production and marketing of this brand to another firm operating overseas. As such, it is uniquely relevant to firms that have a clearly established right to market a product under either a recognized name or trademark. Observation suggests that international licensing of the production and marketing of branded food and related products is an increasingly important aspect of the globalization of food marketing. A survey of the world's leading food manufacturing firms indicates that at least half of those with international operations are engaged in product licensing. In terms of volume, licensed products sales may actually be greater than trade through imports and exports. This is likely to be beneficial to firms from both the perspective of licensing their brand(s) to a foreign firm and producing a foreign branded product under license.

The purposes of this paper are: first, to provide examples of international food product licensing, focussing on licenses from the U.S. to foreign firms (outbound), production and marketing by U.S firms under license from overseas firms (inbound), and third-country licensing; second, to analyze the strategic motivation for food manufacturing firms to license their branded products to overseas firms using non-cooperative game theory. Specifically, in order to focus on an equilibrium between a licensor and licensee, product licensing is modelled in the context of a simple entry game where licensing enters explicitly into the strategy space of both a potential licensor and licensee. Utilizing a non-cooperative, sequential game framework, a number of licensing equilibria are derived that describe different motivations for licensing.

Assuming a complete information, "one-shot" game with no pre-commitment by the incumbent firm, it is shown that, with certain restrictions on the cost function of the incumbent firm, licensing will emerge as an alternative strategy to direct market entry. With credible pre-commitment, a licensing equilibrium can also be derived whereby the licensor attempts to extract some share of monopoly rents from the licensee's market, the amount depending on the nature of the bargain struck over the terms of the license.

It is also shown that the previous results will hold in the case of repetition of the game with complete information. However, once incomplete information about the incumbent's payoffs is allowed for, licensing may emerge in a sequential equilibrium in the sense that, reputation-building behavior by the incumbent firm will deter entry such that, a failed entrant will offer a license in order to recoup losses from entry and reap future profits, and the incumbent will accept the offer in order to delay/deter future entry. This result is obtained by adapting the solution method suggested by Easley, Masson and Reynolds (1985).

Branded Food Products and Licensing

International product licensing is characterized by a food or beverage manufacturing firm with a well established brand name in one country, the licensor, licensing a firm in another country, the licensee, to manufacture and sell the branded product in the licensee's and/or third-country markets. As well as exclusive use of the brand name in the assigned markets, the licensor often provides some technical production assistance and quality control regime. In addition, depending on the specific circumstances, the licensor may provide the product formula or recipe, some critical ingredient(s) such as a flavoring syrup, and some financial assistance towards market development (sometimes in the form of foregone royalties). In turn, the licensee has production, marketing and distribution rights and obligations for the licensed product in the specified market(s), and repatriates part of the product's earnings in the form of a fixed fee and/or royalties to the licensor. The terms of agreement in international product licenses have been considered in other work by Sheldon and Henderson (1991a).

As a form of business activity, the licensing of branded food and related products has existed for many years in both the U.S. and other developed countries. For example, both Coca-Cola and Pepsi-Cola have licensed the domestic canning and distribution of their final products. The activity also crosses national borders. For example, Cadbury-Schweppes and Britvic-Corona own the UK canning and distribution rights to "Coca-Cola" and "Pepsi-Cola" respectively. Numerous examples of U.S. outbound and inbound and third-country licenses have been documented, Sheldon and Henderson (1991b), of which a selection is presented in Tables 14.1-14.3 respectively.

A sample of outbound licenses from the U.S is listed in Table 14.1. These represent a wide range of firms, including both farmer-owned cooperatives such as Land O'Lakes and Welch Foods and proprietary corporations such as Kraft General Foods and Hormel. The products licensed span from variety meats and fruit juices to soup and beer. It is evident that international product licensing from the U.S. involves many well-known brands and many of the foreign licensees are among the world's leading food processors. Interestingly, foreign partners for licensing by farmer cooperatives appear to be proprietary firms rather than cooperatives.

Table 14.2 reports a small number of inbound licenses to the U.S. The products covered are predominantly from the chocolate confectionery and beer sectors and both the licensors and licensees are firms that hold leading positions in their respective markets. It would appear that U.S. firms have not exploited inbound licensing to the same extent as outbound licensing. A number of possible reasons can be forwarded for this: first, there is a perception among some U.S. food manufacturers that U.S. consumers prefer an imported to a licensed foreign product; second, the size of the U.S. market may mean that leading U.S. food manufacturers can exploit most advantages of scale without having to rely on additional production under license; third, many U.S. food manufacturers have purchased rather than licensed the right to use foreign brand names in the U.S.; and fourth, some leading foreign food processors have chosen to invest directly in U.S. production facilities. However, this does not preclude inbound licensing as a potentially profitable strategy for a large number of companies and cooperatives.

Finally, Table 14.3 lists some branded food product licenses involving non-U.S. firms. As with Table 14.1, these cover a broad spectrum of products, firms and markets. For example, Morinaga Foods of Japan licenses its yogurt "Bifidus" to food manufacturers in France and Germany, while the German firm Lutz licenses its ham and sausage brands to the Nichirei group in Japan.

TABLE 14.1 Examples of U.S. Outbound Food Product Licenses

Licensor/Product		Licensee
Anheuser-Busch		
	"Budweiser"	Labatt, Canada
		United Breweries, Denmark
		Guinness, Ireland
		Suntory, Japan
		Oriental Brewery, Korea
		Grand Metropolitan, UK
	"Bud Light"	Labatt, Canada
Adolph Coors		
	"Coors"	Molson, Canada
CPC		
	"Knorr"	Ajinmoto, Japan
Geo. A. Hormel		
	"Spam"	Newforge Foods, UK
		K.R. Darling Downs, Australia
	"Hormel Franks"	Lee Tan Farm Industries, Taiwan
	"Frank 'N' Stuff Franks"	Pure Foods Corp., Philippines
	"Black Label Bacon"	Lee Tan Farm Industries, Taiwan
	"Bacon Bits"	K.R. Darling Downs, Australia
	"Hormel Processed Meats"	Blue Ribbon Products, Panama
Land O'Lakes		
	"Calf Milk Replacer"	Japan
Miller		
	"High Life"	Molson, Canada
	"Miller Lite"	Molson, Canada
Ocean Spray		
	"Ocean Spray"	Pernod-Ricard, France
		SPC, Austria
		Cadbury-Schweppes, Canada
		Pokka, Japan
Phillip Morris/Kraft		
	"Kraft Margarine"	Epic Oil Mills, South Africa
	"Kraft Salad Dressing"	Epic Oil Mills, South Africa
Sunkist Growers		
	"Sunkist"	Morinaga, Japan
		Haitai Beverages, Korea
		Rickertson, Germany
		Cadbury-Schweppes, UK
Welch Foods		
	"Welchs"	Cadbury-Schweppes, Canada

TABLE 14.2 Examples of U.S. Inbound Food Product Licenses

Licensor/Product	Licensee
Cadbury, UK	
"Cadbury Dairy Milk"	Hershey
"Cadbury Fruit and Nut"	Hershey
"Caramello"	Hershey
"Cadbury Creme Eggs"	Hershey
"Roast Almond"	Hershey
Haute Brasserie, France	
"Killian's Red"	Adolph Coors
Löwenbrau, Germany	
"Löwenbrau Pils"	Miller
Nestlé-Rowntree, Switzerland/UK	
"Kit-Kat"	Hershey
"Rolos"	Hershey
Sodima, France	
"Yoplait Yogurt"	Yoplait

Perhaps not surprisingly, several of these are arrangements between firms from industrialized countries and firms in developing countries, e.g., Suchard (U.S./Switzerland) licensing to companies in Indonesia, Mexico, Korea and Malaysia. It is probably the case that many of the latter types of license include supply of process technology as well as use of a brand name. For example Heineken is involved in a number of "technical management" arrangements with licensed brewers in several African countries, Asia and the Caribbean.

It is obvious from this sampling that the international licensing of branded food products is a diverse and widespread phenomenon. This raises the important question as to why it occurs. The rest of this paper is concerned with modelling a number of potential licensing equilibria.

Product Licensing Equilibrium

In light of the above discussion, it is useful to consider brand licensing in a theoretical framework. The following licensing equilibrium, based on a stylized market, is modelled in the context of a simple entry game where

TABLE 14.3 Examples of Third-Country Food Product Licenses

Licensor/Product	Licensee	Licensor/Product	Licensee
Arla, Sweden		Löwenbrau, Germany	
"L+L Dairy Spread"	Morinaga, Japan	"Löwenbrau Special Export"	Allied Lyons, UK
Bond, Australia		Lutz Company, Germany	
"Castlemaine XXXX"	Allied Lyons, UK	"Lutz Ham and Sausage"	Nichieri Group, Japan
"Swan Premium"	Allied, Lyons UK	Morinaga, Japan	
Brasserie Artois, Belgium		"Bifidus Yogurt"	St. Hubert S.A., France
"Stella Artois"	Whitbread, UK		Südmilch AG, Germany
	Molson, Canada		P.T. Enseval, Indonesia
BSN, France		"Morinaga Infant Formula"	
"Kronenbourg"	Courage, UK	Oetker Group, Germany	
Cerveceria Modelo, Mexico		"Oetker Baked Goods"	Podravka, Yugoslavia
"Corona"	Molson, Canada	Phillip Morris/Jacob Suchard Switzerland/US	
Elders, Austalia		"Sugus"	Nestlé Produtos Alimentaros, Portugal
"Fosters"	Beamish and Crawford, Ireland		Beacon Sweets and Chocolates, S. Africa
	Pripps, Sweden		P.T. Super Worldwide Foodstuffs, Indonesia
Guinness, Ireland			Sanborn Hermanos, S.A., Mexico
"Guinness Stout"	Elders, Australia	"Toblerone"	Sanborn Hermanos, S.A., Mexico
Heineken, Holland		"Suchard"	Sanborn Hermanos, S.A., Mexico
"Heineken"/"Amstel"	Whitbread, UK		Tong Yang Confectionary, Korea
	Kirin, Japan		Sanborn Hermanos, S.A., Mexico
	Frydenlund Ringes Bryggerier, Norway	"Milka"	Chocolate Products Manuf., Malaysia
	A.B. Wårby Bryggerier, Sweden	"Van Houten"	General Foods Industries, Indonesia
Kirin, Japan			Sunshine Allied Investments, Singapore
"Kirin"	Molson, Canada	Unilever, Netherlands/UK	
	Sam Miguel, Hong Kong	"Lipton Tea"	Morinaga, Japan
Labatt, Canada		United Breweries, Denmark	
"Labatt"	Vaux Brewery, UK	"Carlsberg"	Photos Photiades, Cyprus
Löwenbrau, Germany			Tou, Norway
"Löwenbrau Pils"	Allied Lyons, UK		Suntory, Japan
		"Tuborg"	Frydenlund Ringes Bryggerier, Norway

product licensing enters explicitly into the strategy space of both a potential licensor and licensee (see Gallini, 1984, and Katz and Shapiro, 1985, for applications to technology licensing). Initially it is assumed that the former firm A, is a monopoly in its own market, producing and selling a single branded food product which it may decide to license in an overseas market. The license is essentially the right to produce the branded product, for which the licensor has property rights. It also includes basic information on how to produce the product, i.e. the "recipe"; the production technology being treated as relatively unsophisticated.

The potential licensee, firm B, is also a monopoly in its own market, selling a branded product which is differentiated from that sold by firm A. Both firms are assumed to have the same cost structures. However, if firm B adds a second product to its portfolio, its unit costs of production may fall due to economies of scope. It is also assumed that there is an aggregate demand for variety, although this is not explicitly modelled.

The licensing decision by firms A and B is examined in terms of a simple game where licensing is an alternative strategy to direct entry by the licensor and an alternative to independent product development by the incumbent firm. The *extensive* form of the game is depicted in Figure 14.1, where firms move sequentially left to right. Initially, it is assumed that the game is only played once and the payoffs to any particular strategy are known to both firms. The equilibrium concept invoked is that of *perfect Nash equilibrium* (Selten, 1975). This rules out non-credible threats by firms in the sense that one firm will attach no credibility to an action threatened by another firm for which it has no incentive to actually carry out.

Equilibrium 1

In this version of the game, the focus is on **node 2** of the game, where firm A moves first. Analyzing the entry/no entry sub-game, the following condition is assumed to hold:

$$\pi_m^B > \pi_d^B > 0 > \pi_w^B \tag{1}$$

The outcome of such a game is well-known (see Dixit, 1982); fighting entry by firm A is not a *credible* threat by firm B as the profits from sharing the market in a Nash equilibrium, π_d^B, are greater than those from fighting, π_w^B. Hence the perfect equilibrium is that of entry by firm A and accommodation by firm B.

Focussing now on the strategy of offering a license at **node 2**, for this to be an equilibrium strategy for firm A, it must also be an equilibrium for firm

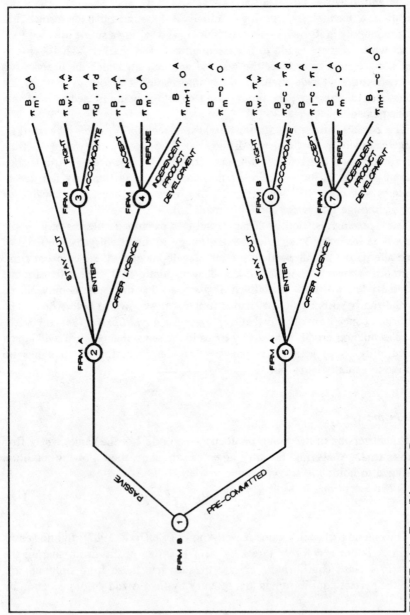

FIGURE 14.1 Entry/Licensing Game

B to accept a license at **node 4**. Clearly, if the following condition holds, firm B will be willing to accept the offer of a license:

$$\pi_l^B > \pi_{m+1}^B > \pi_m^B > \pi_d^B \tag{2}$$

i.e. it is more profitable for the incumbent firm to accept a license, π_1^B, than either developing its own product, π_{m+1}^B, acting as a monopolist, or sharing the market. This condition will be satisfied if firm B's cost function is characterized by economies of scope and high fixed costs of product development. If condition 2 holds, then offering a license will be an equilibrium strategy for firm A if the following holds:

$$\pi_l^A > \pi_d^A \tag{3}$$

where π_1^A are the licensing profits earned by firm A. Assuming (3) holds, (2) must hold, otherwise firm A will simply not offer the license and will enter the market. It can also be noted that if some informational asymmetry is introduced, i.e. firm A does not know the payoffs to firm B from either accepting or refusing the license, then it will always be rational for firm A to follow a strategy of direct entry.

Equilibrium 2

Critical to the above equilibrium is the move sequence in the game, i.e. firm A has been allowed *first-mover* advantage. However, it is possible to allow firm B such an advantage in the sense that it can make irrevocable prior commitments, incurring a sunk cost c, in preparation to fight direct entry by firm A. In the context of branded products, Salop (1979) has suggested that product differentiation and advertising represent examples of such prior commitments.

Therefore, at **node 1** of the game, pre-commitment is a rational strategy for firm B if it is optimal to fight entry i.e.:

$$\pi_w^B > (\pi_d^B - c) \tag{4}$$

Assuming firm A can observe this, it will not enter if firm B is pre-committed, but will do if firm B is passive. Firm B, in turn, will pre-commit if the monopoly profits from doing so exceed the profits from passive market sharing:

$$(\pi_m^B - c) > \pi_d^B \tag{5}$$

Therefore, as long as there is a pre-commitment whose cost satisfies the following condition:

$$(\pi_m^B - \pi_d^B) > c > (\pi_d^B - \pi_w^B) \tag{6}$$

then a credible threat can be employed by firm B such that at **node 5** of the game, entry is no longer an optimal strategy for firm A, i.e. it knows firm B will fight to protect its pre-commitment. Expression (6) is derived by re-arranging (4) and (5) and has the following interpretation. The size of the pre-commitment c must be large enough for firm B to have an incentive to fight entry by firm A, but not so large as to reduce monopoly profits below the profits from market sharing. Essentially, the pre-commitment has to be sufficiently large to be a credible threat against firm A's entry, that is, larger than the opportunity cost to B of fighting entry. If this condition does not hold, the pre-commitment is not credible and it would be more profitable for firm B to accept market sharing.

It is now the case that, at **node 7** of the game, both licensing (π_1^B - c) and developing a new product (π_{m+1}^B - c) are possible outcomes and hence strategies for firm B. Licensing will be an equilibrium if the following conditions hold:

$$(\pi_l^B - c) > (\pi_{m+1}^B - c) > (\pi_m^B - c) \tag{7}$$

$$\pi_l^A > 0 \tag{8}$$

Therefore, in this simple model, the motives for licensing are clear: the licensor aims to extract rents from an imperfectly competitive market overseas that it is unable to enter directly, while the licensee aims to increase monopoly profits via a less costly route than independent product development, taking advantage of economies of scope. However, this model has been constructed using certain simplifying assumptions which can be relaxed, specifically a monopolistic structure in firm B's market, no repetition of the game and complete information on the part of firm A about all of firm B's payoffs.

Relaxation of Assumptions

Incumbent Oligopoly. If it is assumed that firm B operates in a small numbers, non-cooperative oligopoly, then the outcomes of the two games outlined above will not alter substantively, although the incentive structure for the incumbent firms may change due to their strategic interaction. In the case of **equilibrium 1**, direct entry by firm A is again likely to be the dominant outcome of the game, i.e. the following condition holds:

$$\pi_o^A > \pi_l^A \tag{9}$$

where π_o^A are the oligopoly profits accruing to firm A. However, it is important to note that the necessary condition for an incumbent firm, say firm B, to accept a license will now be affected by its conjectures about other firms' licensing decisions, i.e. even though independent product development is more profitable than licensing for firm B, if either a rival firm(s) finds licensing profitable or firm B is uncertain about the profitability of licensing for rival firms, it may desire to pre-empt its rivals from obtaining the license because the loss of future profits from not licensing exceeds the difference between independent product development and purchasing the license:

$$(\pi_o^B - \pi_o^{B'}) > (\pi_{o+1}^B - \pi_l^B) \tag{10}$$

where $\pi_o^{B'}$ are firm B's oligopoly profits if another firm gets a license and π_{o+1}^B are the next-period profits to firm B of independent product development.

Even if (10) is satisfied, it is not sufficient to ensure a licensing equilibrium. In an auction, firm B will only bid up to what it will lose if another firm gains a license, whilst firm A's reservation price will be at least π_o^A, the oligopoly profits it could gain through entry. Hence, the necessary condition for a licensing equilibrium in this case is:

$$(\pi_o^B - \pi_o^{B'}) > \pi_o^A \tag{11}$$

i.e., the amount bid for the license must exceed firm A's oligopoly profits if it chooses to enter.

This seems a particularly strong condition for a licensing equilibrium, hence entry is likely to occur. Turning to **equilibrium 2**, where entry is credibly prevented by the incumbent firms, the licensing equilibrium is now only dependent on it being profitable for both firm A and firm B. However, as just noted, pre-emptive behavior by the incumbent firms now enters the equilibrium where a rival firm(s) finds licensing profitable or there is uncertainty about its profitability. This strategic interaction between firm B and its rivals would seem, *a priori*, to make licensing a more likely outcome in oligopolistic markets than monopolies.

Repetition of Game. Retaining a market structure of monopoly for firm B, repetition of the game in Figure 14.1 can be allowed for. Specifically, the passive incumbent[2] plays against a number of potential foreign entrants A_i, $i=1,...,n$. In the *"one-shot"* version of this game, it has been shown that entry and accommodation is the perfect equilibrium of the game, however, with repetition the question arises as to whether the incumbent can rationally fight early entrants in order to deter future entry.

In the case of infinite repetition of the game, it is possible for firm B to fight entry early on and then enjoy monopoly profits in perpetuity. If r is the rate of interest, fighting occurs when the following condition holds:

$$(\pi_d^B - \pi_w^B) < (\pi_m^B - \pi_d^B)/r(1+r) \tag{12}$$

i.e., the loss from fighting is outweighed by future gains from not sharing the market. In this case, a licensing equilibrium would emerge for the same reasons as described for **equilibrium 2**, that is conditions (7) and (8) hold. However, as noted by Friedman (1977), infinitely repeated games tend to generate multiple equilibria such that the perfect equilibrium of the above game could have entry and accommodation at each play, in which case **equilibrium 1** would hold, i.e., there would be entry and no licensing, unless (2) and (3) hold.

If the game is repeated a finite number of times, there is a unique perfect equilibrium, similar to **equilibrium 1**. Given a finite number of entrants, who have complete information about payoffs and strategies, repetition of the game will generate entry and accommodation each time, assuming of course entry remains profitable, $\pi_o^{Ai} > 0$. The proof of this result is due to Selten, and is commonly known as the *"chain-store"* paradox. Suppose the last round of the game at time t is considered, A_n being the last potential entrant. As there are no more potential entrants, firm B has no incentive to fight, consequently the Nash equilibrium is to share, i.e., **equilibrium 1**. If accommodation is the equilibrium in period t, then fighting entry in t-1 will not be a credible threat, and likewise in t-2. So by backwards induction, the incumbent firm will never fight entry in a finitely repeated game where players have complete information. Consequently, under these assumptions, repetition of the game is unlikely to generate a licensing equilibrium, unless conditions (2) and (3) hold.

Incomplete Information. Suppose now that the game in Figure 14.1 is repeated a finite number of times, but the potential entrants A_i have incomplete information about the payoffs and strategies of the incumbent firm. The focus is on a situation where the potential entrants do not know whether there is a pre-commitment[3] that satisfies condition (6). Clearly a committed firm will always fight entry as sharing will indicate a lack of commitment and hence provoke future entry. However, a passive incumbent may act aggressively in order to be mistaken for a committed firm. As several authors have noted, such incomplete information may be sufficient to allow for reputation-building behavior by incumbent firms which will deter future entry, or at least delay it (see Kreps and Wilson, 1982a, Milgrom and Roberts, 1982, and Easley, Masson and Reynolds, 1985). In this context, it is interesting to introduce licensing into the strategy space of firms.

Suppose the incumbent firm B is drawn from a sample of four types; type 1 is a pre-committed firm who will always fight entry but, with some positive probability v, will accept a license if offered ; type 2 is a passive firm who will accommodate entry whilst types 3 and 4 are non-committed firms that will fight current entry in order to delay/deter future entry and also accept a license(s). Type 3 is an incumbent who fights once and then accommodates and type 4 is an incumbent who fights twice[4] and accepts a license in the third period. There are a number of potential entrants A_i, $i = 1,...,n$, which is sufficient to generate predatory behavior by the incumbent firm. Entrants do not know the type of firm they face but are able to assign probabilities to types 1 to 4 and update these after observing the incumbent's reaction to entry, i.e., firms act in a Bayesian manner. Also, it is assumed that firms incur sunk entry costs f_i and that for firm A_1, $f_1 < f_i$, $i > 1$, which ensures that firm 1 has an entry advantage[5] over the other potential entrants, although with accommodation by the incumbent, the market can sustain further entry until the following condition is met:

$$(\pi_w^{A_i} - f_i)/r \leq 0, \quad i > 1 \tag{13}$$

The equilibrium concept employed is that of *sequential equilibrium* as described by Kreps and Wilson (1982b) and Kreps (1990). Suppose that g is a vector of entrants' and incumbent's strategies and μ is the vector of entrants' expectations about the type of entrant they face. Assuming that all players maximize expected payoffs given the strategies of other players, then an equilibrium to the game can be defined as a set of strategies that satisfy:

$$g \in g*(\mu^{BAY}(g)) \tag{14}$$

and a set of beliefs associated with these strategies satisfies the following:

$$\mu \in \mu^{BAY}(g*(\mu)) \tag{15}$$

Essentially (14) means that firms' strategies are optimal given beliefs and (15) states that beliefs are obtained from strategies and observed actions using Bayes' rule[6]. A sequential equilibrium then is a set of strategies g and beliefs μ where, at any point in the game, a player plays optimally from then on given what has already occurred and their beliefs about what will happen at later nodes of the game.

As Tirole and Easley *et al.* note, finding direct solutions for sequential equilibria can be difficult. However, following a structure suggested by Easley *et al.*, an equilibrium for a specific game can be constructed by working backwards, i.e., assume behavioral rules for the various types of incumbent firm and derive optimal entrant reactions and vice-versa for assumed entry behavior rules.   If there exists a set of parameters that ensure that the incumbent's/entrants' best responses are identical to their assumed behavioral rules, a Nash equilibrium can be derived.

The aim here is to generate an equilibrium where potential entrants are convinced they face a type 1 firm so that they switch strategies to one of offering a license which has a positive probability of being accepted. The sequence of interest is the following:

t=1, firm A_1 enters the market;

t=2, firm A_1 remains in the market even if $\pi_w^{A_1}$ is observed in t=1. If $\pi_d^{A_1}$ is observed in t=1, the other firms, i>1, will enter until $(\pi_o^{A_i} - f_i)/r \leq 0$;

t=3, firm A_1 exits the market if $\pi_w^{A_1}$ is observed in t=1 and 2, and any firm in the sample can offer a license, which the incumbent accepts.

Suppose the initial probability distribution attached to the incumbent firm type is such that, p(type 1)=p_1, p(type 2)=p_2, p(type 3)=p_3, and p(type 4)=p_4, $\Sigma p_i = 1$. Given these probabilities, a set of decision rules can be written down for the entrants that will satisfy the above sequence of events:

$$[p_2 \pi_d^{A_i}/r + (1-p_2)\pi_w^{A_i}/r] - f_i < [v\pi_i^{A_1}/r(1+r) + (1-v)0^{A_i}], \ i>1 \qquad (16)$$

$$(\pi_w^{A_1} - f_1) + [2(p_3 \pi_d^{A_1}/r + (1-p_3)\pi_w^{A_1}/r) - f_1]/(1+r) > 0 \qquad (17)$$

$$[3\{\omega \pi_d^{A_1}/r + (1-\omega)\pi_w^{A_1}/r\} - f_1] - (2\pi_w^{A_1}/r) $$
$$< [v\pi_i^{A_1}/r(1+r) + (1-v)0^{A_1}] \qquad (18)$$

where $\omega = $ p(type 4|$\Omega_{t=3}$), which defines the conditional probability that A_1 faces a type 4 firm given the information set $\Omega_{t=3}$ available at the start of period 3, i.e., $\pi_d^{A_1}$ has not been observed in periods t=1 and 2. Condition (16) shows that for all firms, bar firm 1, the expected profits from entering are less than the expected profits from offering a license at some future date. Condition (17) indicates that for firm 1, the expected profits of entry and remaining in the market for two sequential periods are positive, whilst (18) shows that for firm 1, the expected profits from remaining in the market for a third period are outweighed by the profits from exiting and offering a license, i.e., the odds of the incumbent firm being a type 4 firm are not sufficiently attractive to firm 1 for it remain in the market after period 2.

If the above responses for entrant firms are assumed to be their behavioral rules, a set of optimal reactions can be defined for the incumbent firm that will ensure the sequential equilibrium defined;

Type 1 preys if:

$$
n\{\pi_w^B - (\pi_d^B - c)\}/r < [2\{(\pi_m^B - c)
$$
$$
- (\pi_d^B - c)\}/r + \{(\pi_l^B - c) - (\pi_d^B - c)\}/r(1 + r)]
\tag{19}
$$

where n is the number of times a committed firm fights in order to deter entry[7]. This condition states that the profits from long-run monopoly, inclusive of returns to accepting a license[8], outweigh the short-run losses from fighting.

Type 2 does not prey if:

$$
(\pi_d^B - \pi_w^B) > (\pi_m^B - \pi_d^B)/r(1 + r)
\tag{20}
$$

i.e., the one-period loss from fighting exceeds the difference between monopoly and market sharing in future periods.

Type 3 preys in one period, but not in two if:

$$
(\pi_d^B - \pi_w^B) < (\pi_m^B - \pi_d^B)/r(1 + r)
\tag{21}
$$

and

$$
2(\pi_d^B - \pi_w^B)/r > (\pi_m^B - \pi_d^B)/r(1 + r)
\tag{22}
$$

Type 4 preys twice and accepts a license[9] in the third period if:

$$
2(\pi_d^B - \pi_w^B)/r < [2(\pi_m^B - \pi_d^B)/r + (\pi_l^B - \pi_d^B)/r(1 + r)]
\tag{23}
$$

Conditions (16)-(23) are sufficient to show that an equilibrium with fighting and licensing exists, i.e., neither entrants nor incumbents can improve their payoffs given the others' behavioral rules.

Necessarily the above result is somewhat restrictive and is only one of a number of equilibria that could arise. In particular, the nature of the strategic outcomes could be made much richer if the game did not end at period t=3, i.e., incumbent firms use licensing as a delaying tactic against future entry and entrant firms regard it as a means of revealing information about incumbent firms.   If the bargaining process over the terms of the license is dealt with explicitly, it might be argued, *a priori*, that an incumbent firm, who is no longer willing to fight after period 2, has an incentive to bargain for a license with a long time-horizon in order to delay future entry, whilst firm 1, having incurred losses from entry in periods 1 and 2, will require a license to provide returns over a short time period.   In contrast, a pre-committed firm will care less about the length of the license as future entry will always be fought.   Hence the possible asymmetry between the incumbent's and entrant's time-horizons may reveal information about incumbent type.   However, expanding the potential equilibria to the game does not undermine the basic point that uncertainty about incumbent firms' behavior may generate the offer of a license and a licensing equilibrium.

Summary

In summary, this paper has suggested that the licensing of branded food and related products may become an increasingly important feature of international transactions in the food industry.   Currently, the economic theory of licensing deals predominantly with the transfer of process technology rather than branded products.   Therefore, given the observations on food brand licensing, a conceptual model of a product licensing equilibrium has been presented in order to provide an analytical background to more rigorous empirical work.   This analysis suggests that if licensing is considered as an alternative strategy to entry in a simple game-theoretic structure, then in the simplest type of model, licensing is aimed at extracting rents from imperfectly competitive overseas markets.   In a more complex model, strategic interaction amongst incumbent firms and imperfect information about their payoffs may also be important factors in the decision to license products internationally.

Clearly more research needs to be conducted in this area both in developing the theory and in establishing the quantitative importance of licensing and its determinants.   Also, other licensing issues not addressed in this paper include the notion of an optimal licensing contract, the process of bargaining, the content of brand licensing agreements, and the lifespan of licenses.

Notes

1. The authors would like to acknowledge the Farm Income Enhancement Program at Ohio State University in providing support for some of this research. This paper is based on research conducted as part of North Central Region research project NC-194,

entitled, "Organization and Performance of World Food Systems: Implications for U.S. Policies".

    2. Repeating the game with a pre-committed incumbent is not necessary as the pre-commitment is credible by assumption.

    3. This is probably a reasonable assumption in the case of firms attempting to enter an overseas market.

    4. A non-committed incumbent might be prepared to fight for more than two periods, but this is not necessary for deriving the result of fighting and licensing.

    5. This is a device to allow one firm to do the entering and may be regarded as reasonable in the case of attempting to establish a branded product overseas.

    6. A more formal definition is given in Kreps and Wilson (1982b) and Easley *et al*.

    7. The way the game is structured, the committed firm only has to fight twice.

    8. Condition (7) is assumed to hold.

    9. This assumes that condition (2) holds.

References

Dixit, A. 1982. Recent Developments in Oligopoly Theory. *American Economic Review-Papers and Proceedings* 72(2):12-17.

Easley, D., R. Masson, and R. Reynolds. 1985. Preying for Time. *Journal of Industrial Economics* 33(4):445-460.

Friedman, J. W. 1977. *Oligopoly and the Theory of Games*. Amsterdam: North-Holland.

Gallini, N. T. 1984. Deterrence by Market Sharing: A Strategic Incentive for Licensing. *American Economic Review* 74(5):931-941.

Handy, C. and J. M. MacDonald. 1989. Multinational Structures and Strategies of U.S. Food Firms. *American Journal of Agricultural Economics* 71(5):1246-1254.

Handy, C. and D. R. Henderson. 1990. EC 1992: Implications of a Single EC Market for the U.S. Food Manufacturing Sector. In *Impacts of Europe 1992 on the Processed Food Industries*, Organization and Performance of World Food Systems: NC-194, Economic Studies:Report No.1.

Henderson, D. R. and S. D. Frank. 1990. Industrial Organization and Export Competitiveness of U.S. Food Manufacturers. *Organization and Performance of World Food Systems: NC-194*. Occasional Paper:OP-4.

Katz, M. L. and C. Shapiro. 1985. On the Licensing of Innovations. *Rand Journal of Economics* 16(4):504-520.

Kreps, D. 1990. *A Course in Microeconomic Theory*. Princeton, MA: Princeton University Press.

Kreps, D. and R. Wilson. 1982a. Reputation and Imperfect Information. *Journal of Economic Theory* 27:253-279.

_____. 1982b. Sequential Equilibrium. *Econometrica*. 50:1003-1038.

Milgrom, P. and J. Roberts. 1982. Predation, Reputation and Entry Deterrence. *Journal of Economic Theory* 27:80-312.

Salop, S. C. 1979. Strategic Entry Deterrence. *American Economic Review-Papers and Proceedings* 69(2):335-338.

Selten, R. 1975. Reexamination of the Perfectness Concept for Equilibrium Points in Extensive Games. *International Journal of Game Theory* 4(1):470–486.

Sheldon, I. M. and D. R. Henderson. 1990. Motives for the International Licensing of Branded Food and Related Products. *Organization and Performance of World Food Systems: NC-194*. Occasional Paper:OP-15.

_____. 1991a. Product Licensing as a Competitive Strategy in World Food Markets. *Income Enhancement Study Series*. The Farm Income Enhancement Program, The Ohio State University.

_____. 1991b. The Importance of and Economic Motivation for the International Licensing of Branded Food and Related Products. *Journal of Food Distribution Research* 22(1):101-108.

Tirole, J. 1989. *The Theory of Industrial Organization*. Cambridge MA: MIT Press.

15

Food Processing Industry Structure and Political Influence: Redistribution Through Trade Barriers

Rigoberto A. Lopez and Emilio Pagoulatos

Introduction

The food processing industries in the United States are characterized by relatively low import penetration ratios and export propensities (Connor *et al.*), and by trade barriers that are more restrictive than most manufacturing industries (Lavergne, Ray (1990)). In fact, trade barriers which include nominal tariffs and nontariff barriers (NTBs) act as entry barriers from the standpoint of foreign competitors. The strategic implication is that firms will spend resources to compete in the political arena to influence trade barriers in their favor instead of concentrating their efforts on regular market competition, such as pricing and other nonpricing strategies. The performance implication is that a significant redistribution of wealth occurs as trade barriers increase prices and food processing industry profits. In addition, resources are spent on unproductive activities which may result in a decline of overall economic efficiency. However, to date, there have been no attempts to address the structural and political determinants of trade barriers or their redistributive effects in food processing industries.

The purpose of this paper is to (1) explain the variation of import tariffs across U.S. food processing industries in terms of strategic behavior of firms, as well as industry structure characteristics; and (2) assess the extent of redistribution and potential social costs brought about by import tariffs in U. S. food processing industries. While several of there industries are protected by nontariff barriers, the measurement and analysis of NTBs is more uncertain than that of tariffs, and, therefore, their consideration is beyond the scope of the present paper.

The general framework of analysis is the theory of public choice which analyzes political processes and their interaction with the economy (Mueller, 1979). Of particular interest is the importance of interest group pressures (lobbying) in determining the level of tariffs. In addition to being the first to address specifically the determinants of tariffs in food processing, this paper incorporates a more direct measure of lobbying or political activity than previous studies of total manufacturing, i.e. the number of congressmen supported by the respective industries at the 4-digit SIC level. It also provides a rough estimate of the welfare cost of tariffs for forty-five 4-digit SIC industries, using willingness-to-pay measures.

The next section of the present study presents the conceptual framework for analyzing the economics and politics of tariff formation. The third section presents a general methodology for measuring the redistributive effect and the allocative efficiency loss brought about by tariffs. The fourth section discusses the data and econometric specification, and the fifth section analyzes the empirical results. Brief concluding remarks appear in the final section.

Theory of Tariff Formation

A widely accepted and useful framework to analyze the determinants of tariff levels is one that conceptualizes them within a political market for protection (Anderson and Baldwin, Baldwin (1985, 1989), Pincus, Ray (1981a,b, 1990). As such, domestic food manufacturers can be viewed as the "demanders" of tariff protection, and the government legislators and bureaucrats are the "suppliers" of tariffs.

The rent-seeking approach (recently surveyed by Tollison) provides the arguments for the demand side of protection. Tariffs artificially increase the price of commodities while restricting foreign entry, hence generating rents or avoiding or slowing the demise of a domestic industry. For an industry to exert political pressure, a necessary condition is that it be willing and able to organize its members to obtain contributions needed for lobbying.

Past studies have used several industry characteristics as proxies for political activity that will presumably result in political outcomes such as tariffs. Since trade policy can be viewed as a public good, a common-interest group advocating such a policy needs to overcome the free rider problem typically present in raising lobbying funds.   Thus, lobbying activity is usually approximated by industry organization characteristics such as the number of firms, the concentration ratio, and the size of the industry that are more likely to reduce or eliminate free riding. In other words, due to lack of data, a measure of conduct (lobbying) is indirectly measured by structural characteristics that presumably determine such conduct. More recently, Baldwin (1989), among other economists, has advocated the use of more direct measures such as funding levels and number of employees supported by private interests, as better proxies for lobbying activities.

The supply side of the market is usually characterized by political behavior concerned with re-election, i.e., the maximization of votes (Becker, Pelzman) or concerned with social welfare (principled behavior). If behavior is characterized by vote-seeking, then policymakers or bureaucrats will make choices to balance political support from conflicting-interest groups. Thus, to some extent, they will take into account the interests of consumers and will be concerned with public sympathy or dislike for government protection through import tariffs.

Another important issue is that of the role of comparative advantage in tariff formation. Tariffs usually make sense if an industry has some degree of comparative disadvantage, since only then will lobbying by an industry be more effective in gaining policymakers' and the public's sympathy. Furthermore, tariffs are redundant or ineffective in an industry that has comparative advantage in trade. Comparative advantage is usually measured by growth in employment, output or imports, capital/labor intensity, the degree of labor intensity in production, and skill level of the labor force (Anderson and Baldwin, Lavergne, Ray (1981a, b, 1990)). If an industry is declining, then that industry will more likely be protected by a higher tariff level. Also, lobbying by labor will depend on the labor share of the rents attained through tariffs. Traditionally, U.S. comparative disadvantage has been established in low-skilled, labor intensive industries. Finally, observed tariff levels at any one point in time should depend on historical tariff levels, as tariffs change very slowly over time. Institutional rigidities or lags characterize tariff levels. Only through dramatic shocks to the system, such as international negotiations (GATT), will tariff levels change significantly.

Based on the above discussion, we can broadly hypothesize that tariff formation in food processing industries can be summarized by the following function:

$$T_{kt} = f(Z_{kt}, Z_{ok,t}, CD_{kt}, O_{kt}, T_{k,t-s})$$

where T_k is the tariff level in industry k; t and s are time subscripts; Z_k the lobbying activity by that industry; Z_{ok} the lobbying activity by those adversely affected by the tariff; CD_k a vector of variables describing the degree (if any) of comparative disadvantage; O_{kt} a vector of other variables such as industry structure characteristics; and $T_{k,t-s}$ a historical (previous) level of the tariff.

Redistribution Through Tariffs

We turn next to the redistributive and allocative efficiency aspects of tariffs in food and tobacco manufacturing. Regarding this sector, the most closely related work has focused on estimating the welfare cost of oligopolistic or monopolistic pricing behavior (Gisser (1982), Parker and Connor, Willner). However, those welfare estimates were motivated by the potential presence of market power due to domestic distortions rather than trade barriers and were limited to computing the deadweight loss associated with the misallocation of resources. But, as Tullock and Posner have suggested, the traditional deadweight loss estimates may understate the true cost of monopoly if firms will find it worthwhile to expend resources to obtain and maintain monopoly power through the political process. Furthermore, firms in competition for the monopoly may pay up to the whole amount of the prospective monopoly rent, and the opportunity cost of these expended resources will add to the social cost of monopoly. In other words, the resources used by interest groups to pursue special advantages through government action would be essentially wasted since they reduce rather than create wealth as resources are withdrawn from productive activities.

This socially wasteful competition has been alternatively named "rent seeking" by Krueger and "directly unproductive profit-seeking" (DUP) activity by Bhagwati and is often associated with lobbying to obtain and maintain trade barriers. In this paper we blend recent contributions to the theory of rent seeking with a simple model of trade in order to obtain a measure of the welfare cost of tariffs. The assumptions used are as follows:

1. Cross-price effects are zero (partial equilibrium analysis);
2. The world export supply curve is perfectly elastic (small country assumption);
3. Domestic food manufacturers exhibit perfectly competitive behavior;
4. There is a single homogeneous product and no intraindustry trade; and
5. There are no other impediments to trade (such as, import quotas and health-related restrictions).

Figure 15.1 illustrates equilibria under a tariff and a free trade situation. In the absence of a tariff, arbitrage ensures that the domestic price equals the world price (P_w). Imposition of with an advalorem tariff (T) drives the domestic price up to $P_w(1 + T)$. As a result, domestic production expands from Q_1^s to Q_2^s, domestic consumption contracts from Q_1^d to Q_2^s, imports decline from $(Q_1^d - Q_1^s)$ to $(Q_2^d - Q_2^s)$, and the government collects tariff revenues equal to (G).

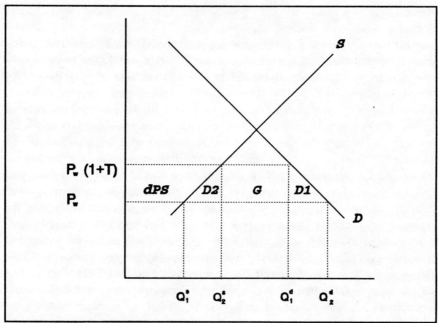

FIGURE 15.1 Redistribution Through Tariffs

In traditional economic analysis, the welfare cost of a tariff is given by triangles $D = D1 + D2$. Similar triangles have been designed as the welfare cost of monopolies (Siegfried and Tiemann, Harberger, Parker and Connor). Following Posner and Krueger, one could argue that the welfare cost of the tariff exceeds the triangles to encompass an area analogous to the loss in consumer surplus depicted by $dCS = dPS + D + G$. That is, the benefit to producers afforded by the tariff (dPS) is dissipated through competitive rent seeking while government revenues (G) are dissipated through administrative costs.

However, one could also argue that firms in the domestic industry are not competing against one another, but rather against foreign food manufacturers. Since domestic firms are better positioned than foreign ones in terms of political lobbying at home, it is less imperative for them to lobby to the point of exhausting all benefits granted by the tariff. Also, tariffs show a high degree of rigidity over time, so that maintaining a tariff might require a lower lobbying cost. Fisher has questioned the validity of Posner's measure of social cost of monopolies and regulation, indicating that the cost of *obtaining* a monopoly need not be wasteful and even if it is, it does not have to match monopoly profits.

Varian introduces rent seeking in a general equilibrium model and concludes that Posner's measure overestimates the social cost of a monopoly. Bhagwati has even sustained that a tariff may be welfare-enhancing in a second best world. In this paper, both the deadweight loss and Posner's social cost are computed as lower and upper bounds, respectively, of the true social cost of tariffs in food manufacturing.

Empirical Procedures

The first objective of our data analysis is to determine the influence of economic and political variables on the inter-industry pattern of tariffs in U.S. food processing. The dependent variable (T72) used in this study was the simple average U.S. nominal tariff rate at the 4-digit SIC level. This variable was obtained for 1972, using a GATT data set, from a study by the U.S. International Trade Commission of the levels of protection in industrial countries following the Kennedy Round of GATT negotiations (USITC (1975)).

Several independent variables were included to account for the major hypothesized factors discussed in section II. First, to account for the slow adjustment of tariffs over time, we included the simple average U.S. nominal tariff rate in 1965 (T65) obtained from the International Trade Commission study (USITC (1975)). It is expected that this variable is positively related to tariffs in 1972.

To control for comparative advantage, we included three variables: LSH, labor's share of value-added; K/L, the capital ratio; and W, the average wage per employee. All three were computed from census data for 1972 (U.S. Bureau of the Census, Census of Manufactures, 1972 (1975)). Labor's share of value-added, defined as payroll/value-added, should be positively related to tariffs, while the capital labor ratio, defined as the gross value of fixed assets per employee, should be inversely related to tariffs. Finally, the average wage for the industry, defined as payroll per employee, serves as a proxy for human capital and should be inversely related to tariffs.

Tariffs are expected to be higher in declining industries. To capture this factor in tariff formation, we included a growth in employment variable (GE) defined as the change in employment from 1963 to 1972 (U.S. Bureau of the Census, Census of Manufacturers, 1972 (1975)). This variable should be inversely related to tariffs.

The impact of market structure characteristics on tariff restrictions is ambiguous. The size of the industry is measured by the log of value of shipments (lnVS) and could reflect the industry's potential political leverage. The four-firm concentration ratio (CR), obtained from the U.S. Bureau of the Census (1981), and the minimum efficient size (MES), obtained from Connor *et al.* (1985), were also included in the model. Their expected influence on tariffs, however, is uncertain since it depends upon the ability of these variables to reflect the effectiveness of lobbying activities.

The final variable included in the model is a proxy for direct lobbying activities by firms in the industry (CONGR) and is defined as the number of congressional candidates (of both parties) supported by food processor organizations and cooperatives (Guither). This variable is expected to positively affect the level of tariffs.

The equation which summarizes the empirical determinants of tariffs in food processing industries discussed above is

$$T72 = f(T65, LSH, K/L, W, GE, lnVS, CR, MES, CONGR)$$
$$\quad\quad + \quad + \quad - \quad - \quad - \quad \pm \quad \pm \quad \pm \quad\quad +$$

The sign below a variable indicates the hypothesized sign of its coefficient; a (+/-) indicates that no *a priori* sign prediction is warranted. The model coefficients were estimated using data for forty five 4-digit SIC food and tobacco manufacturing industries for the year 1972 via ordinary least squares.

To obtain welfare and redistribution measures in terms of observable variables, we adapted the procedures presented by Posner. The derivations of these measures are given in the appendix. The approach is appealing because it requires little data. Our estimates required the following data: the value of shipments, exports and imports, the advalorem tariff level, and estimates of the elasticities of demand and supply. The data used are given in the appendix, Table 15.A.1. The results for 1972 for 45 U.S. food manufacturing industries are presented below.

Empirical Results

The empirical results explaining tariff formation are presented in Table 15.1 while the computed welfare costs and transfers for each of the 45 food processing and tobacco industries are reported in Table 15.2. Note that all estimators were made at the industrial or manufacturer's market level.

Two model specifications are presented in Table 15.1: a complete and a restricted model, denoted Model 1 and 2. In the alternative model (model 2), two insignificant industry-organization variables are excluded. To investigate the relevance of including those variables, we conducted an F-test of the joint hypothesis that the coefficients for industry size (log of value of shipments) and firm size (minimum efficient size) were zero. This hypothesis was rejected as the resultant F-statistic (0.51) with (2,35) degrees of freedom did not exceed the critical F-value of 3.30 at the 5 percent level. Thus, individually or jointly, size

TABLE 15.1 Determinants of Tariffs in U.S. Food Manufacturing

Dependent Variable: Nominal Tariff, 1972 (T72)

Variable	Model 1	Model 2
Constant	21.78[a]	15.03[a]
	(1.82)	(1.70)
Nominal Tariff, 1965 (T65)	.756[c]	.737[c]
	(7.56)	(7.68)
Labor's Share of Value-added (LSH)	16.38	17.38
	(1.21)	(1.320)
Capital Labor Ratio (K/L)	-.211[b]	-.209[b]
	(2.69)	(2.715)
Average Wage Rate (W)	-.713[a]	-.780[b]
	(2.01)	(2.289)
Employment Growth 1963-72 (GE)	-15.82[c]	-16.331[c]
	(3.63)	(3.945)
Log of Value of Shipments (lnVS)	-.815	
	(.733)	
Four-Firm Concentration Ratio (CR)	.104	.159[b]
	(1.11)	(2.393)
Minimum Efficient Size (MES)	.199	
	(.537)	
No. of Congressmen Supported (CONGR)	.047[a]	.041[a]
	(1.85)	(1.700)
R^2 Adjusted	.661	.669
N	45	45
F-Ratio	10.51	13.74

[a] significance at the 90 percent level
[b] significant at the 95 percent level
[c] significant at the 99 percent levels
t-statistics are given in parentheses

of industry and minimum efficient scale did not add any information in explaining tariff levels in food processing.

The models were quite satisfactory in explaining tariffs levels in food and tobacco manufacturing, especially for a cross-sectional, small sample. The adjusted R^2 for Model 1 was .661. (Ray's, and Anderson and Baldwin's work obtain smaller R^2 for significantly larger samples). Of the comparative advantage or labor-related variables, only the share of value labor added variable was not significant at the 10 percent level. The findings indicate that the level of tariffs will be higher, the lower the capital/labor ratio, the lower the wage of labor (less skilled), and the higher the recent loss of employment in a particular industry. Except for the labor share variable, these findings are consistent with those of Anderson and Baldwin and Ray (1981a, b, 1990) for broader interindustry or intercountry tariff samples.

None of the variables used as proxy for industry organization were significant at the 90 percent level. That is, the size of the industry (log of value of shipments), the 4-firm concentration ratio, and the minimum efficient size variables were all insignificant in Model 1. However, the number of congressmen supported by industry was significant at the 10 percent level in both models. Thus, the proxy for direct lobbying activities was more significant in explaining tariffs than the indirect variables of industry organization characteristics. It is important to note that the number of congressmen is a gross measure of political influence, and additional effort is needed to refine this measure in the future. Finally, tariffs in food processing industries appear to be slow in adjusting over time. The coefficient associated with the 1965 (previous) tariff was .756 and highly significant.

Given the lack of significance in industry organization characteristics, Model 2 was estimated without the industry size and minimum efficient scale variables. Since the 4-firm concentration ratio and minimum efficient size were highly correlated, the former was retained in the Model 2 specification. The overall results were qualitatively the same except that the four-firm concentration ratio was now significant at the 5 percent level. Based on Model 2 results, we conclude that more concentrated industries appear to have been more successful in obtaining higher tariffs.

The results in Table 15.2 indicate that the aggregate welfare cost based on price deadweight losses (traditional triangles) are extremely small when compared to the trade-adjusted value of shipments (TAVS) defined as the value of shipments less exports plus imports. TAVS is also referred to as "apparent consumption" at the industrial level. The average deadweight loss amounted to approximately $457 million or 4/10 of a percent of TAVS. However, the transfers from consumers to producers were quite significant. The total transfers amounted to approximately $8.4 billion in 1972 or 7.2 percent of TAVS. Also note, that the government revenues from tariffs were relatively small amounting to $499 million for the whole sector, implying that the main objective of the government was not to raise revenue but to increase domestic prices. If one believes in Posner's approach, the loss in consumer surplus is the social cost since it is dissipated in deadweight losses as well as directly unproductive activities. Posner's loss amounted to approximately $9.3 billion for 1972 which represents close to eight percent of TAVS. Most likely, this

TABLE 15.2 Redistribution Through Tariffs in U.S. Food Manufacturing, 1972

SIC	Industry	Producer Surplus Gain (dPS)	Govt. Revenue (G)	Dead-weight Loss (D)	Consumer Surplus Loss (dCS)	dCS (% of Ap. Consum.) 100%
		Millions 1972 Dollars				
201	**Meat Products**	1633.5	63.7	55.7	1752.9	5.89
2011	Meat Packing	1124.9	61.0	38.4	1224.3	5.49
2013	Saus. & Prep. Meats	223.0	2.2	7.4	232.6	5.94
2016-7	Poultry & Egg Proc.	285.6	0.5	9.9	296.0	8.36
202	**Dairy Products**	1228.4	22.8	63.0	1314.2	8.06
2021	Creamery Butter	11.6	0.0	0.1	11.7	1.48
2022	Cheese	433.0	18.3	34.7	486.0	14.64
2023	Cond. & Evap. Milk	111.9	4.3	4.2	120.4	7.70
2024	Ice Cream	188.7	0.0	14.4	203.1	16.33
2026	Fluid Milk	483.2	0.2	9.6	493.0	5.25
203	**Pres. Fruit & Veg.**	1054.3	38.5	38.7	1131.5	9.77
2032	Canned Specialties	147.5	0.4	3.5	151.4	8.11
2033	Canned Fruit & Veg.	424.2	26.9	17.8	468.9	11.24
2034	Dried Fruit & Veg.	50.4	2.3	1.8	54.5	9.84
2035	Pickled Sauces	80.1	0.5	2.1	82.7	7.11
2037-8	Frozen Fruit & Veg.	352.1	8.4	13.5	374.0	9.79
204	**Grain Mill Products**	475.7	3.7	7.6	487.0	4.21
2041	Flour & Grain Mill	119.1	1.2	1.8	122.1	5.31
2043	Cereal Preparations	29.6	0.1	0.2	29.9	2.72
2044	Rice Milling	20.7	0.1	0.5	21.3	7.29
2045	Blended, Prep. Flour	44.0	0.0	0.8	44.8	6.37
2046	Wet Corn Milling	84.2	1.4	2.8	88.4	11.45
2047-8	Prepared Feeds	178.1	0.9	1.5	180.5	2.82
205	**Bakery Products**	118.7	1.6	1.1	121.5	1.53
2051	Bread & Bakery	48.5	0.1	0.1	48.7	.79
2052	Cookies & Crackers	70.2	1.6	1.0	72.8	4.05

206	**Sugar Confectionery**	421.8	52.0	10.7	484.5	6.25
2061-3	Cane & Beet Sugar	123.2	43.0	1.6	167.8	4.13
2065	Candy & Confection.	249.1	7.0	7.9	264.0	10.52
2066	Chocolate & Cocoa	18.4	1.6	0.2	20.2	2.54
2067	Chewing Gum	31.1	0.4	1.0	32.5	8.50
207	**Fats & Oils**	1079.1	14.9	68.5	1162.5	19.32
2074	Cottonseed Oil Mills	0.4	0.0	0.0	0.4	0.10
2075	Soybean Oil Mills	635.5	0.0	60.7	696.2	24.78
2076	Vegetable Oil Mills	6.3	5.0	0.1	11.4	3.49
2077	Anim. & Marine Fats	19.0	2.4	0.2	21.6	3.48
2079	Lard & Cooking Oils	417.9	7.5	7.5	432.9	22.90
208	**Beverages**	1080.6	223.5	56.9	1361.0	9.17
2082	Malt Liquors	281.0	2.9	7.5	291.4	7.12
2083	Malt	5.9	0.2	0.1	6.2	2.73
2084	Wine & Brandy Sp.	173.7	67.0	17.7	258.4	21.95
2085	Distilled Liquor	365.4	151.5	24.1	541.0	21.94
2086	Soft Drinks	64.4	0.0	0.2	64.6	1.19
2087	Flavor. Extr. & Syrups	190.2	1.9	7.3	199.4	14.02
209	Miscellaneous Food	116.6	39.2	3.0	158.8	2.71
2091	Canned Seafoods	51.1	14.4	2.3	67.8	7.03
2092	Fresh Fish Proc.	22.6	23.2	0.6	46.4	2.27
2095	Roasted Coffee	31.6	1.3	0.0	32.9	1.38
2097	Manufactured Ice	0.0	0.0	0.0	0.0	0.00
2098	Macaroni & Spaghetti	11.3	0.3	0.1	11.7	3.29
21	**Tobacco-Mfrs.**	1170.0	39.3	122.5	1331.8	25.28
2111	Cigarettes	943.6	1.0	102.6	1047.2	29.60
2121	Cigars	25.5	0.7	1.3	27.5	7.99
2131	Chew. & Smok. Tobac.	32.4	5.1	3.1	40.6	24.33
2141	Tobacco, Stemming	168.5	32.5	15.5	216.5	17.77
20-21	**All Food & Tobacco**	8,378.7	499.2	427.7	9,305.6	7.98

Notes: Marginal cost elasticities of 0.5 were assumed for all industries. Demand Elasticity estimates were obtained from Pagoulatos and Sorensen.

figure only provides an upper bound for the social cost of the tariff and it is likely that the social cost of tariff was well below this figure.

Inter-industry comparisons indicate that the greatest relative deadweight losses were concentrated in the milk processing sectors (cheese, ice cream), soybean oil, milk, wine, and tobacco manufacturing industries (cigarette manufacturing, chewing tobacco, and stemming).   In these industries, deadweight losses exceeded 1 percent of TAVS to a maximum of 3 percent in cigarette manufacturing. If one adopts Posner's viewpoint and uses the loss in consumer surplus as the social cost, the same industries involved the largest relative social costs, with some additions: canned fruit and vegetables, wet corn milling, lard and cooking oil, and flour extract syrup, all of which entail social costs that exceeded 10 percent of TAVS.

The results also indicate that meat packing and cigarette manufacturing tariffs induce the greatest transfers from consumers to domestic producers, exceeding 1 billion dollars each in 1972. Other industries that received transfers exceeding $300 million were cheese, fluid milk, canned and frozen fruits and vegetables, soybean oil milling, lard cooking oil and distilled liquor. Together, these eight industries (of a total of 45 industries) accounted for over half of the total tariff-induced transfers accruing to food manufacturers.

Conclusion

This study examined (1) the determinants of the inter-industry pattern of tariff restrictions in forty-five food and tobacco manufacturing industries in 1972 and (2) the redistribution and welfare consequences of these tariffs.   The empirical results support the hypotheses that tariff protection is higher in those industries in which the United States has a long standing comparative disadvantage in trade (as measured by a relatively low capital labor ratio and average wage rate) and those in which employment has been declining over time. The results also indicate that tariffs are associated with direct lobbying activities in the industry and that nominal tariffs are slow in adjusting over time. Only weak support was found for the hypothesis that high concentration in the industry results in increased tariffs.

The redistribution through tariffs accounted for approximately seven percent of the trade-adjusted value of shipments (apparent consumption) of manufactured food and tobacco products. The allocative efficiency loss from tariffs, as measured by the traditional triangles, seemed to have been extremely small and in line with relatively small efficiency losses found for monopoly in the U.S. manufacturing sector (e.g. Harberger, Gisser (1986)). However, the deadweight loss represents a lower bound to the efficiency losses brought about by tariffs because they do not account for administrative costs and directly unproductive activity that may dissipate part of the domestic manufacturers' gain. If all rents are dissipated (Posner) then the social cost of tariffs amounts to approximately eight percent trade-adjusted value of shipments. However this extreme is unlikely. Interindustry comparisons indicate specific "pockets" where tariffs cause the greatest redistribution and perhaps the greatest social costs. These industries include meatpacking, fluid milk, soybean oil, distilled liquor, and cigarette manufacturing. These results should be interpreted with caution as the objective was to provide an illustration of the methodology that might be used if an augmented-social cost model seems appropriate.

Promising extensions and revisions of the present study include updating the data to the most recent years, improving on the measurement of potential political influence, the inclusion of nontariff barriers, the inclusion of possible market failures (e.g. nonmarket effects such as health cost of cigarette consumption) and accounting for possible price leadership behavior in U.S. food manufacturing.

References

Anderson, K. and R. E. Baldwin. 1981. *The Political Market for Protection in Industrial Countries: Empirical Evidence.* Washington D.C.: World Bank Staff Working Paper No. 492.

Baldwin, R. E. 1985. *The Political Economy of U.S. Import Policy.* Cambridge, MA: The MIT Press.

Baldwin, R. 1989. The Political Economy of Trade Policy. *Journal of Economic Perspectives* 3: 119-135.

Becker, G. S. 1983. A Theory of Competition Among Pressure Groups for Political Influence. *Bell Journal of Economics* 19:371-400.

Bhagwati, J. N. 1982. Directly Unproductive, Profit-Seeking (DUP) Activities. *Journal of Political Economy* 90:988-1002.

Connor, J. M., R. T. Rogers, B. W. Marion, and W. F. Mueller. 1985. *The Food Manufacturing Industries: Structure, Strategies, Performance, and Policies.* Lexington, Mass.:Lexington Book.

Fisher, F. M. 1985. The Social Cost of Monopoly and Regulation: Posner Reconsidered. *Journal of Political Economy* 93: 410-416.

Gisser, M. 1982. Welfare Implications of Oligopoly in U.S. Food Manufacturing. *American Journal of Agricultural Economics* 64: 616-624.

_____. 1986. Price Leadership and Welfare Losses in U.S. Manufacturing. *American Economic Review* 76:756-67.

Godek, P. E. 1985. Industry Structure and Redistribution through Trade Restrictions. *Journal of Law and Economics* 28:687-703.

Guither, H. D. 1980. *The Food Lobbyists.* Lexington, Mass.: Lexington Books.

Harberger, A. 1954. Monopoly and Resource Allocation. *American Economic Review* 44: 77-87.

Krueger, A. O. 1974. The Political Economy of the Rent-Seeking Society. *American Economic Review* 64:291-303.

Lavergne, R. P. 1983. *The Political Economy of U.S. Tariffs: An Empirical Analysis.* New York: Academic Press.

Mueller, D. C. 1979. *Public Choice.* Cambridge: Cambridge University Press.

Nelson, D. 1988. Endogenous Tariff Theory: A Critical Survey. *American Journal of Political Science* 32(1988): 796-837.

Pagoulatos, E. and R. Sorensen. 1986. What Determines the Elasticity of Industry Demand? *International Journal of Industrial Organization* 4:237-250.

Parker, R. C. and J. M. Connor. 1979. Estimates of Consumer Loss Due to Monopoly in the U.S. Food Manufacturing Industries. *American Journal of Agricultural Economics* 61: 626-38.

Peltzman, S. 1976. Toward a More General Theory of Economic Regulation. *Journal of Law and Economics* 19:211-240.

Posner, R. A. 1975. The Social Cost of Monopoly and Regulation. *Journal of Political Economy* 83: 807-27.

Ray, E. J. 1981a. The Determinants of Tariff and Nontariff Restrictions in the United States. *Journal of Political Economics.* 89:105-121.

_____. 1981b. Tariff and Nontariff Barriers to Trade in the United States and Abroad. *Review of Economics and Statistics* 63: 161-168.

_____. 1990. Protection of Manufactures in the United States. In *Global Protectionism: Is the U.S. Playing on a Level Field* ed. D. Greenaway. New York, N.Y.: MacMillan.

Saunders, R. S. 1980. The Political Economy of Effective Tariff Protection in Canada's Manufacturing Sector. *Canadian Journal of Economics* 13:340-348.

Siegfried, J. J. and T. K. Tiemann. 1974. The Welfare Cost of Monopoly: An Inter-Industry Analysis. *Economic Inquiry* 12:190-202.

Tollison, R. D. 1982. Rent Seeking: A Survey. *Kyklos* 35:575-602.

Tullock, G. 1967. The Welfare Cost of Tariffs, Monopolies, and Theft. *Western Economic Journal* 5:224-32.

U.S. Bureau of the Census. U. S. Department of Commerce, *Annual Survey of Manufactures*, Washington, D.C., various years.

_____. 1975. *Census of Manufactures, Industry Statistics, 1972,* Washington, D.C.

_____. 1981. *Census of Manufactures, 1977: Concentration Ratios in Manufacturing,* Washington, D.C.

_____. 1979. *U.S. Commodity Exports and Imports as Related to Output, 1972 and 1971,* Washington, D.C.

U.S. Bureau of Economic Analysis. 1979. U. Department of Commerce, *The Detailed Input-Output Structure of the U.S. Economy: 1972* , Washington, D.C.

U.S. International Trade Commission. 1975. *Protection in Major Trading Countries.* USITC Publication No. 737, Washington, D.C.

Varian, H. R. 1989. Measuring the Deadweight Costs of DUP and Rent-Seeking Activities. *Economics and Politics* 1:81-95.

Willner, J. 1989. Price Leadership and Welfare Losses in U. S. Manufacturing: Comment. *American Economic Review* 79:604-609.

Appendix

Derivation of Redistribution and Deadweight Measures

For simplicity, let $P_1 = P_w$ the world price and $P_2 = P_w(1 + T)$, the domestic price. Define $K = P_1/P_2 = 1/(1 + T)$, where T is an advalorem tariff rate, and $Q^d = \alpha P^{-\epsilon}$, where Q^d is domestic demand at the manufacturing level, α a constant, P is domestic manufacturer's price, and ϵ is the absolute value of the price elasticity of demand.

Define trade-adjusted value of shipments (or industry apparent consumption) as $TAVS = VS - VX + VM(1 + T)$, where VS is domestic value of shipments, VM is value of imports, and VX is value of exports. Let $L_1 = dPS + D2 + G$ in Figure 15.1. Note that

$$DI = 1/2(Q_1^d - Q_2^d)(P_2 - P_1).$$

Thus, using our definition of demand above:

$$\frac{D1}{L1} = \frac{(KP_2)^{-\epsilon} - P_2^{-\epsilon}}{2P_2^{-\epsilon}} = \frac{K^{-\epsilon} - 1}{2}.$$

The total loss of consumer surplus (Posner's social cost) is given by $dCS = D1 + L1$. Thus,

$$dCS = D1 + \frac{2D1}{K^{-\epsilon} - 1} = TAVS\,(1 - K)\,\frac{(K^{-\epsilon} + 1)}{2}.$$

The deadweight loss associated with consumption is given by

$$D1 = \frac{dCS(K^{-\epsilon} - 1)}{(K^{-\epsilon} + 1)}.$$

Define domestic supply as $Q^s = \beta P^\eta$. Let $L2 = dPS + D2$. Similarly, the relative deadweight loss associated with the increased cost caused by additional domestic output is given by

$$\frac{D2}{L2} = 1/2(1 - K^\eta)$$

Also,

$$L2 = P_2 Q_2 T/(1 + T) = (VS - VX)T/(1 + T),$$

TABLE 15.A.2 Data Used for Computations of Table 15.2

SIC	Industry	Tariif Rate (Tx100%)	Value of Shipments (VS) $ Million	Value of Exports (VX) $ Million	Value of Imports (VM $ Million	Price Elasticity of Demand
2011	Meat Packing	5.7	21637.0	487.0	1069.5	0.67
2013	Saus. & Prep. Meats	6.2	3905.0	27.4	35.2	0.58
20167	Poultry & Egg Proc.	9.0	3584.0	50.0	5.2	0.30
2021	Creamery Butter	1.5	808.0	23.0	0.5	0.52
2022	Cheese, Natural	16.4	3195.0	5.2	111.5	0.50
2023	Cond. & Evap. Milk	8.2	1668.0	162.7	52.0	0.42
2024	Ice Cream, Frozen	18.8	1245.0	1.5	0.2	0.36
2026	Fluid Milk-Mfrs.	5.5	9396.0	2.9	3.6	0.24
2032	Canned Specialties	8.8	1877.0	14.8	4.6	0.056
2033	Canned Fruit & Veg.	12.5	4044.0	114.1	215.6	0.19
2034	Dried Fruit & Veg.	10.8	607.0	76.2	21.7	0.18
2035	Pickled Sauces	7.6	1167.0	12.2	7.1	0.20
2037-8	Fozen Fruit & Veg.	10.7	3784.0	48.7	78.4	0.24
2041	Flour & Grain Mill	5.6	2380.0	104.0	21.1	0.04
2043	Ceral Preparations	2.8	1126.0	32.9	3.5	0.03
2044	Rice Milling	7.8	681.0	389.9	1.9	0.18
2045	Blended, Prep. Flour	6.8	705.0	2.1	0.2	0.037
2046	Wet Corn Milling	12.9	832.0	73.0	10.9	0.037
2047-8	Pet Foods, Pr Feeds	2.9	6439.0	73.6	30.8	0.076
2051	Bread, Bakery Prod.	0.8	6132.0	11.0	4.7	0.210
2052	Cookies, Crackers	4.2	1764.0	4.0	37.6	0.170

SIC	Industry				
206123	Cane & Beet Sugar	4.3	3030.0	9.4	1000.4	0.094
2065	Candy, Confection	11.7	2473.0	28.7	59.9	0.070
2066	Chocolate, Cocoa	2.6	736.0	4.6	60.8	0.310
2067	Chewing Gum-Mfrs	9.2	383.0	5.2	4.0	0.210
2074	Cottonseed Oil	0.1	459.0	66.2	2.2	0.014
2075	Soybean Oil Mill	31.9	3357.0	547.6	0.1	0.17
2076	Vegetable Oil	3.6	261.0	79.0	137.5	0.31
2077	Anim. Marine Fats	3.6	765.0	213.8	65.3	0.070
2079	Lard, Cooking Oil	28.8	2068.0	81.0	26.0	0.19
2082	Malt Liquors	7.6	4054.0	2.7	37.5	0.22
2083	Malt-Mfrs	2.8	226.0	7.0	6.6	0.18
2084	Wine, Brandy	27.1	865.0	1.7	247.0	0.24
2085	Distilled Liquor	28.0	1798.0	24.8	541.0	0.023
2086	Bottled Soft Drink	1.2	5454.0	2.8	1.3	0.042
2087	Fl Extract Syrup	16.3	1472.0	63.9	11.4	0.006
2091	Canned Seafoods	7.4	810.0	55.4	194.3	0.559
2092	Fresh Fish Proc	2.3	1084.0	74.5	1009.2	0.848
2095	Roasted Coffee	1.4	2329.0	30.1	95.6	0.11
2097	Manufactured Ice	0.0	116.0	0.1	0.1	0.029
2098	Macaroni, Spaghetti	3.4	348.0	0.4	8.1	0.096
2111	Cigarettes-Mfrs	40.8	3745.0	210.6	2.5	0.12
2121	Cigars-Mfrs	8.4	339.0	3.0	8.8	0.74
2131	Chew, Smokeless Tob.	31.2	180.0	34.6	16.2	0.17
2141	Tobacco, Stemming	20.7	1657.0	628.0	157.1	0.373

$$D2 = 1/2 L2 (1 - K^{\eta}).$$

Government revenues are given by $G = VM \cdot T$. The dollar value of shipments and trade data *(VS, VM, VX)* came from *Census of Manufacturers*, 1972(1975). The tariff data *(T)* were obtained from the *International Trade Commission Study* (USITC (1975)). The elasticities of demand (ϵ) came from Pagoulatos and Sorensen, while the elasticity of supply (η) were assumed to be equal to 0.5 for all industries. The data are given in Table 15.A.1.

PART SEVEN

Private and Public
Decision Making

16

The Comprehensiveness of Strategic Decision Making and Its Relationship to Business Unit Performance

Randall E. Westgren, Steven T. Sonka,
and Gunta S. Vitins

In an era of increasing structural change among agricultural industries in the US agribusiness sector, managers of farms and other firms are faced with planning for uncertain, turbulent futures (King and Sonka, 1986). In an industrialized agriculture sector characterized by myriad interrelationships across products and market channel participants (including imports) and unstable government intervention, the synapse between firms and their business environment is complex. Strategic management is the long term process of managing that synapse (Westgren and Cook, 1986).

Strategic Management: Content and Process

Measuring the effect of strategic management on firm performance is difficult due to the time dimension; by definition, strategic decisions are long-run and performance measures such as profitability which are made at a single point in time cannot capture the cumulative performance effects of strategic choice (Hambrick, 1980; Fahey and Christensen, 1986). As well, some management scholars maintain that the *content* (e.g., diversification, cost leadership, product innovation) of the strategy may not be as important to performance as the *process* of devising the strategy (Quinn, 1980). Strategic management research has thus evolved into two thrusts: content-performance research and process-performance research (Robinson and Pearce, 1988).

However, theorists are divided over the process of strategic management. One body of thought arises from the traditions of long-range planning as the precursor to the more broadly defined strategic management (Westgren and Cook, 1986). Planning models describe a process of setting long term goals, identifyingmarketing and production strategies to achieve the goals, and structuring the resources of the firm to pursue the strategies. This *synoptic* ("ends-ways-means") process is typified by work of Porter (1980), Hofer and Schendel (1978), and Robinson and Pearce (1988). A second body of theory treats the bounded rationality of management and the impact of business environment as constraints to the rational planning process. Variously described as *emergent, contingent,* or *incremental,* this process is reactive to the environment and explicitly relies on idiosyncratic organizational structure and behavior to yield episodic and myopic strategies (Mintzberg, Raisinghani, and Théorêt, 1976).

Fredrickson (1984,1986) summarizes the literature on synoptic and incremental strategic processes and distinguishes between them on the basis of six characteristics. Rational, synoptic processes

1. are proactively initiated as a result of constant environmental surveillance,
2. attempt to achieve specific planning goals,
3. identify goals prior to evaluating strategies to achieve them,
4. objectively choose among alternatives based on ability to attain prestated goals,
5. exhibit comprehensiveness in analytical and choice activities, and
6. comprehensively integrate the decisions into an overall strategy.

Incremental, or emergent, strategy processes tend to be antithetical to the synoptic model across the six characteristics. The process tends to be

1. reactive to problems,
2. remedial to the current state rather than goal-directed,
3. contemporaneous in combining goals and strategies in choice,
4. subjectively based on the intertwined goals and alternatives,
5. restricted in analysis of alternatives and simple in decision heuristics, and
6. characterized by little attempt to make decisions integral to the overall company.

Fredrickson argues that comprehensiveness of the decision process may be the most important characteristic. Since the synoptic models of Porter and others require comprehensiveness in their normative paradigms of strategic management, and the incremental models of Quinn and others preclude comprehensiveness, this characteristic is useful in discriminating between the two regimes in empirical analysis. In an industry with an unstable environment, there will be a negative relationship between comprehensiveness and firm performance. In an industry with a stable environment, comprehensiveness will be positively correlated with firm performance. Fredrickson tested these hypotheses with an unstable industry (sawmilling and planing) and a stable industry (paints and coatings) and found support for the hypotheses.

Study Sample and Methods

The sample of firms is drawn from the membership of the California League of Food Processors, the primary trade association. The CEO of each of the 58 members was contacted. Twenty four firms contributed to the first step of the research design. The sample reflects the diversity of the total membership. Sample members include small single-product firms, large multinational firms, cooperatives, proprietary firms, firms with branded products, and industrial manufacturers.

Comprehensiveness Measures

Building a useful construct for comprehensiveness is no easier than for any of the other characteristics of planning processes. Where economics relies on rational, optimizing behavior that subsumes cognitive processes of individual managers, the organizational behavior discipline that undergirds process research requires explicit behavioral constructs for managerial decision-making. However, with a working definition of comprehensiveness, such as "the extent to which an organization attempts to be exhaustive and inclusive in making and integrating strategic decisions" (Fredrickson and Mitchell), a survey instrument can be designed to measure this multiattribute construct.

The method used in this paper follows Fredrickson (1984, 1986) and Fredrickson and Mitchell (1984). It consists of a two step survey technique of managers within a single SIC industry. Fredrickson's prior studies of the sawmill and planing industry (SIC 2421) and paints and coatings (SIC 2851) follow the same design as this study of SIC 2033, fruit and vegetable canning.

The first step is to personally interview CEOs or senior planning officers from a sample of firms in the industry. Respondents are asked to identify priority (strategic) issues facing the industry and rate their importance to the firm over the next five years. In addition, the respondents described the nature of the planning process in the firm; i.e. did they have a written long-range plan? Results of a pretest of this survey instrument for agribusinesses can be found in Westgren, Sonka, and Litzenberg. From the interviews, a decision scenario is written for a hypothetical firm in the packing industry. The scenario follows the decision process of the management in a sequence of four explicit steps: (1) determining the problem behind recent performance decline, (2) generating alternatives to solve the problem, (3) evaluating the alternatives, and (4) integrating the decision. The scenario is not based on normative behavior; it reflects typical behavior from the first survey.

The scenario and a second survey instrument are sent to the first respondent and up to four other managers designated by the CEO. Each respondent in the second step survey is asked to read the scenario and fill out the survey as if their firm faced the same problem. Respondents are prompted that the resolution of the problem follows the four step process described above and are asked questions about the process they would follow in the firm.

For each step, the following measures of comprehensiveness are elicited:

1. How large is the cadre of managers that would be involved in the step? (i.e., One specific individual, existing committee, ad hoc working group);
2. How much reliance would be placed on persons outside the firm? (e.g. consultants, customers, growers, equipment manufacturers);
3. How large an outlay of direct expenses would be spent?

There are 10 questions for which 4 to 9 separate responses are elicited. Table 16.1 illustrates two of these multi-item questions. For the purpose of building index scores of the comprehensiveness of each step in the decision process, composite scores of the multi-item questions need to be constructed to maintain the implicit equal weighting of the 34 total questions. The multi-item composites are simply the arithmetic means of the separate responses.

In addition to measures of comprehensiveness common to each step, indications of the breadth of analysis and reporting relevant to a given step are elicited. As well, the respondent is prompted for the scope of membership of committees and study groups, how widely information is shared beyond the decision-making group, and what techniques of analysis would be employed.

Numerical scores for the ninety five responses are coded. Scores are aggregated for indexes of comprehensiveness for each step in the decision process and overall. Comprehensiveness scores for each of the four steps in the decision-making process and for the overall process are tested for correlation with seven measures of effectiveness of the planning system and with seven measures of firm performance.

Twenty firms completed the second part of the research design. A total of 33 managers filled out the questionnaire on the scenario. Most of the small family-owned companies had a single respondent, as there was effectively no professional management cadre from which to draw multiple observations.

Performance Measures

To measure performance of California food processors, two sets of subjective measures are used. Following Dess and Robinson, respondents rated

TABLE 16.1 Examples of Multi-Item Questions

(Scale from 1 = very unlikely to 7 = very likely)

How likely is it any of the following *outsiders* would be contacted to provide information regarding construction of a new plant:

    a. Individuals from other food processing firms
    b. Consultants in the food processing industry
    c. Growers
    d. Customers
    e. Equipment manufacturers
    f. Financial consultants
    g. Individuals from firms in other industries
    h. Management consultants

In determining whether to build the plant, how likely is it that your firm's *analysis* would:

    a. Analyze markets for the plant's existing products
    b. Analyze the future supply and costs of raw materials
    c. Assess the plant's flexibility in producing other products
    d. Evaluate alternative levels of automation
    e. Include alternative equipment configurations
    f. Assess the impact of future changes in transportation costs

their firm's performance on a five point interval scale (top 20% of firms in the industry, second 20%, . . ., bottom 20% of firms in the industry) for six performance variables during the last 5 years: total sales growth, earnings growth, return on investment, employee satisfaction, promotion of firm image, and overall performance. Subjective measures are especially useful for researching privately held firms and conglomerate business units, without sacrificing validity (Dess and Robinson, 1984). Privately held firms do not issue public statements of financial performance and the performance of strategic business units with conglomerates is usually buried in consolidated statements. The fruit and vegetable packing industry in California is dominated by these two types of firms, making a subjective measurement regime necessary.

A second set of subjective measures developed by Ramanujam, Venkatraman, and Camillus (1986) is used to elicit managers' perceptions of the effectiveness of their firm's planning process. On a seven point scale (7 = very effective, 1 = not effective), respondents rated effectiveness of the planning system in

a)   predicting future industry trends,
b)   generating information for evaluating alternative strategies,
c)   identifying and avoiding problems,
d)   enhancing managerial planning ability,
e)   improving short term performance,
f)   improving long term performance, and
g)   supporting overall decision-making.

These measures are designed to examine the link between process and performance. While all measures are tied to planning, the process being measured is not specifically the planning, or synoptic, mode.

Research Results

The analysis of survey responses occurs at three levels of response aggregation:

1.   The average response of all managers in the business unit for each single-response question and for the mean response for multi-item questions,
2.   An index of comprehensiveness for each of the four steps in the decision process, and
3.   An index of comprehensiveness for the entire decision process.

Thus, there exists one score for each respondent firm for each question. Multi-item questions are collapsed into a single score. The index of comprehensiveness for each of the four steps in decision making—situation diagnosis, generating alternatives, evaluating alternatives, and integrating the decision into operations—is calculated as the unweighted mean of the individual question scores. The overall index of comprehensiveness is the simple unweighted mean of the four step indices. This method of aggregation replicates the indices computed by Fredrickson and Fredrickson and Mitchell.

Because the behavioral construct scores for comprehensiveness used in the analysis were constructed from responses across managers within firms and across different questions, a reliability test was performed on each dimension of aggregation. Cronbach's alpha is computed for the aggregation among managers within each respondent firm, for the aggregation of multi-item question responses into a single score, for each of the four steps of the theoretical model using the responses elicited for the questions in each step, and for the aggregation of the four step scores into the overall index. Chronbach's alpha is a common reliability coefficient for the internal consistency of a test. Thus, in the case of the aggregation of single questions into a composite score for a decision step, the alpha measures the correlation between individual question scores and the calculated composite scores.

The α score is computed as

$$\alpha = \frac{k\overline{cov/var}}{1+(k-1)\overline{cov/var}}$$

where k is the number of items in the index, *cov* is the average covariance between items, and *var* is the average variance of the items.

Validity Checks

Respondents are asked four questions about their perceptions of the comprehensiveness of the four steps in the decision-making processes in their own firm. These questions are designed as validity checks for the composite scores computed from the 34 single- and multiple-response questions. To test the assumption that the characteristics of comprehensiveness are consistent across strategic decisions, the relationship between the step composites and the four construct validity questions is assessed. Responses to the construct validity questions exhibit high correlations with the composites for the situation diagnosis ($N=20$, $r=0.79$, $p=.000$), alternative generation ($N=20$, $r=0.72$, $p=.000$), alternative evaluation ($N=20$, $r=0.65$, $p=.001$), and decision integration ($N=20$, $r=0.63$, $p=.002$) sections of the questionnaire. When a single measure of comprehensiveness is computed as the mean of the four construct validity questions, it exhibits a strong relationship ($N=20$, $r=0.84$, $p=.000$) with the overall measure calculated from the scenario-based questionnaire. "These results suggest that strategic process characteristics such as comprehensiveness tend to be consistent across decisions and that the profiles developed in response to the scenario were representative of the firms' strategic decision processes" (Fredrickson & Mitchell, 1984).

The relationship between a composite measure of effectiveness of the firm's current planning system and the response to the question on overall effectiveness is also assessed and shows a strong correlation ($N=20$, $r=.89$, $p=.000$). The composite measure of effectiveness is computed as an unweighted mean of the scores for six individual measures of effectiveness (e.g. predicting future trends in the industry, improving long term performance).

A strong relationship is also exhibited between a computed composite measure of firm performance and the response to the single-item question on overall performance of the firm ($N=20$, $r=.87$, $p=000$). The composite is the unweighted mean of scores for the five performance items: sales growth, earnings growth, ROI, enhancing employee satisfaction, and promoting a positive public image.   These results, and the computed alpha coefficients, suggest that the composite measures of performance and effectiveness of planning system constructed from multiple survey questions are reliable and consistent.

Additional analysis indicated that accumulated experience in the company or in the food industry are not confounding factors. Firm seniority exhibits no significant relationship with any of the step composites, or the overall measure of comprehensiveness. Similarly, industry seniority does not exhibit significant relationships with any of the step composites or when correlated with the overall measure of comprehensiveness.

Hypothesis Testing

The hypothesis that comprehensiveness of strategic decision making is positively related to performance was tested. Due to limitations in sample size relative to the large number of individual and composite measures of comprehensiveness and performance, partial correlation analysis was used to test the hypothesis. The effectiveness of the planning system, as measured subjectively by the management team, is also tested for correlation with comprehensiveness measures. The perceived effectiveness of the decision making system is an intermediate construct between comprehensiveness and performance. That is, one would expect that the relationship between a comprehensive decision process and high performance would be mitigated if the process is ineffective.

Partial correlations were computed between the 39 measures of comprehensiveness (24 single response questions + 10 multi-item composites + 4 step scores + overall score) and the 9 measures of the effectiveness of the planning system (7 single measures + 1 overall measure + computed composite score) as well as the 7 measures of performance (5 single-item measures + 1 overall measure + computed composite score).

The results of the partial correlations are contained in the following tables. The significance level is based on a one-tailed test for all partial correlations. Only variables which are significantly correlated with the chosen performance or effectiveness measure are shown. That is, a blank next to a comprehensiveness measure indicates a statistically insignificant relationship. Tables 16.2 through 16.5 show results for four measures of performance. Each table shows the 34 individual measures of comprehensiveness grouped into the four steps of the decision process. Each step in the decision process has a measure of the size of the management cadre with responsibility for the step ("primary responsibility"), the willingness to use outsiders ("rely on outsiders"), number of employees involved in the step ("# of employees involved), the value of out of pocket expenses that would be committed to the decision step ("out of pocket expenses"), and the number and range of outsiders used in the decision

TABLE 16.2 Partial Correlations Between Comprehensiveness Measures and Comparative Total Sales Growth (n=20)

Situation Diagnosis		Alternative Generation		Alternative Evaluation		Decision Integration	
1. Primary Responsibility	.58[c]	1. Primary Responsibility		1. Primary Responsibility	.33[a]	1. Primary Responsibility	
2. Rely on Outsiders		2. Rely on Outsiders	.39[b]	2. Rely on Outsiders		2. Rely on Outsiders	
3. Problem Identified		3. Alternative Identified	.48[b]	3. Alternatives Determined		3. Integrating the Decision	
4. # of Employees Involved		4. Dropped Alternative	.50[b]	4. # of Employees Involved		4. Depts. Affected	.46[b]
5. Out of Pocket Expenses		5. # of Employees Involved		5. Out of Pocket Expenses		5. # of Employees Inv.	.49[b]
6. Years of Historical Data		6. Out of Pocket Exp.	.38[a]	6. Years of Historical Data		6. Out of Pocket Expenses	
7. Outsiders Contacted		7. Outsiders Contacted		7. Outsiders Contacted		7. Outsiders Contacted	
8. Analysis of the Problem		8. Techniques Used		8. Criteria Used		8. Techniques Used	
				9. Written Summary or Report			
				10. New Plant Analysis			
Situation Diagnosis Composite	.37[a]	Alternative Generation Composite.	.58[c]	Alternative Evaluation Composite	.42[b]	Decision Integration Composite	

Overall Process Composite .42[b]

Level of significance: [a] p ≤ .10
  [b] p ≤ .05
  [c] p ≤ .01

Note: In tables 16.2 through 16.7, only significant correlation coefficients are shown.

TABLE 16.3 Partial Correlations Between Comprehensiveness Measures and Comparative Growth in Earnings (n=20)

Situation Diagnosis		Alternative Generation		Alternative Evaluation		Decision Integration	
1. Primary Responsibility		1. Primary Responsibility		1. Primary Responsibility		1. Primary Responsibility	
2. Rely on Outsiders	.38[a]	2. Rely on Outsiders	.33[a]	2. Rely on Outsiders		2. Rely on Outsiders	.55[c]
3. Problem Identified		3. Alternative Identified		3. Alternatives Determined		3. Integrating the Decision	
4. # of Employees Inv.	.47[b]	4. Dropped Alternative		4. # of Employees Involved		4. Depts. Affected	
5. Out of Pocket Exp.	.38[a]	5. # of Employees Inv.	.33[a]	5. Out of Pocket Expenses		5. # of Employees Inv.	
6. Yrs. of Historical Data	.53[b]	6. Out of Pocket Exp.	.41[b]	6. Years of Historical Data		6. Out of Pocket Exp.	.38[a]
7. Outsiders Contacted	.33[a]	7. Outsiders Contacted	-.40[b]	7. Outsiders Contacted	-.36[a]	7. Outsiders Contacted	
8. Analysis of Problem	.35[a]	8. Techniques Used		8. Criteria Used		8. Techniques Used	
				9. Written Summary or Report			
				10. New Plant Analysis			
Situation Diagnosis Composite	.49[b]	Alternative Generation Composite		Alternative Evaluation Composite	.32[a]	Decision Integration Composite	
		Overall Process Composite .32[a]					

Level of significance: [a] p ≤ .10 [b] p ≤ .05 [c] p ≤ .01

Note: In tables 16.2 through 16.7, only significant correlation coefficients are shown.

TABLE 16.4 Partial Correlations Between Comprehensiveness Measures and Comparative Overall Performance (n=20)

Situation Diagnosis		Alternative Generation		Alternative Evaluation	Decision Integration	
1. Primary Responsibility		1. Primary Responsibility	.45[b]	1. Primary Responsibility	1. Primary Resp.	.42[b]
2. Rely on Outsiders		2. Rely on Outsiders		2. Rely on Outsiders	2. Rely on Outsiders	.51[b]
3. Problem Identified		3. Alternative Identified		3. Alternatives Determined	3. Integrating the Decision	
4. # of Employees Inv.		4. Dropped Alternative		4. # of Employees Involved	4. Depts. Affected	.37[a]
5. Out of Pocket Exp.		5. # of Employees Inv.		5. Out of Pocket Expenses	5. # of Employees Inv.	
6. Yrs. of Historical Data	.41[b]	6. Out of Pocket Exp.	.31[a]	6. Years of Historical Data	6. Out of Pocket Exp.	
7. Outsiders Contacted		7. Outsiders Contacted		7. Outsiders Contacted	7. Outsiders Contacted	-.37[a]
8. Analysis of Problem		8. Techniques Used		8. Criteria Used	8. Techniques Used	
				9. Written Summary or Report		
				10. New Plant Analysis		

Situation Diagnosis Composite	.34[a]	Alternative Generation Composite		Alternative Evaluation Composite	Decision Integration Composite	

Overall Process Composite

Level of significance: [a] p ≤ .10   [b] p ≤ .05   [c] p ≤ .01

Note: In tables 16.2 through 16.7, only significant correlation coefficients are shown.

TABLE 16.5 Partial Correlations Between Comprehensiveness Measures and Composite Measure of Firm's Performance (n=20)

Situation Diagnosis		Alternative Generation		Alternative Evaluation		Decision Integration	
1. Primary Responsibility	.32[a]	1. Primary Responsibility		1. Primary Responsibility	.34[a]	1. Primary Responsibility	
2. Rely on Outsiders	.32[a]	2. Rely on Outsiders	.40[b]	2. Rely on Outsiders		2. Rely on Outsiders	.49[b]
3. Problem Identified		3. Alternative Identified		3. Alternatives Determined		3. Integrating the Decision	
4. # of Employees Inv.		4. Dropped Alternative		4. # of Employees Involved		4. Depts. Affected	.48[b]
5. Out of Pocket Exp.		5. # of Employees Inv.		5. Out of Pocket Expenses		5. # of Employees Inv.	.41[b]
6. Yrs. of Historical Data	.36[a]	6. Out of Pocket Exp.	.41[b]	6. Years of Historical Data		6. Out of Pocket Exp.	
7. Outsiders Contacted		7. Outsiders Contacted		7. Outsiders Contacted		7. Outsiders Contacted	
8. Analysis of Problem		8. Techniques Used		8. Criteria Used		8. Techniques Used	
				9. Written Summary or Report			
				10. New Plant Analysis			
Situation Diagnosis Composite	.43[b]	Alternative Generation Composite	.39[b]	Alternative Evaluation Composite	.40[b]	Decision Integration Composite	.32[a]

Overall Process Composite .40[b]

Level of significance: [a] p ≤ .10
                        [b] p ≤ .05
                        [c] p ≤ .01

Note: In tables 16.2 through 16.7, only significant correlation coefficients are shown.

step ("outsiders contacted"). The other individual measures are specific to the comprehensiveness of the particular step in the process, such as the sophistication of the criteria used in evaluating alternatives.

Table 16.2 illustrates the relationships between the 39 comprehensiveness scores and the performance measure of total sales growth. Sales growth performance is correlated with at the 5% significance level with the overall composite score for comprehensiveness shown at the bottom of the table. Two of the step scores for comprehensiveness, situation diagnosis and alternative generation, are positively correlated with sales growth (seen in next-to-last line in table). The step score for alternative generation shoes the stronger relationship (significant at the $p \leq .01$ level), which follows from the significance of four of the constituent measures of this step. Comprehensive planning, as measured by the extent of reliance on outsiders, the extensiveness of the process to identify alternatives (shown as " alternative identified"), the extensiveness of the decision process to drop an alternative (shown as "dropped alternative"), and the value of out of pocket expenses for this step in the process, is correlated to total sales growth.

Only one of the constituent comprehensiveness measures in the situation diagnosis composite is significant, despite the composite significance of this step in the decision process. The size of the management cadre with responsibility for diagnosing the strategic situation is significant at the 1% level, but no other individual measures are significantly correlated to sales growth (performance).

The results shown in table 16.3 follow roughly the same pattern as the prior table. The performance measure is comparative earnings growth. Again, the overall composite measure of comprehensiveness is positively correlated with performance. Six of the individual comprehensiveness measures, as well as the composite step score for situation diagnosis, show positive correlations with performance. The comprehensiveness of the last two steps in the planning process, evaluating alternatives and integrating the decision, shows no correlation to this measure of performance. It is possible this reflects that responsibility for these two steps is more closely held in the management cadre, which is likely in the closely held firms in the sample.

Within the decision step of situation diagnosis, six of the eight individual measures show significant correlation to earnings growth: the willingness to rely on outsiders, the extensiveness of the management cadre with some responsibility in the analysis, dollar value of out-of-pocket expenses for this step, the number of years of historical data gathered, the number of outsiders contacted, and the range of analyses included in the situation diagnosis (e.g. historical product costs, customer satisfaction data, estimates of competitor's costs).

There are two measures of overall performance that can be tested for relationship to comprehensiveness of planning. The first is an overall performance criterion elicited from questionnaire respondents in addition to the five specific measures of performance, such as total sales growth and earnings growth. The second is the composite score computed as the unweighted average of the scores for the five specific measures. Tables 16.4 and 16.5 compare the results of testing these two performance measures against comprehensiveness. Clearly, the composite measure (table 16.5) shows a stronger relationship to comprehensiveness constructs, particularly for the step scores and the overall process scores. This result can be attributed to the nature of the complex behavioral constructs. The composite scores for the four steps in the decision making process and the overall process contain all the information on the individual measures of comprehensiveness, many of which may be individually inappropriate to a given firm's decision process. The more inclusive measures of comprehensiveness will capture information on low- and medium-level scores for items that may be individually statistically insignificant, but will contribute to an overall higher composite score. The same argument might be made for the composite performance score, which contains information from all five of the specific performance criteria, rather than a "whitened" overall assessment by the managers.

In addition to the hypothesis that comprehensiveness is positively related to performance, a second hypothesis is tested. Is the comprehensiveness of the planning process positively related to managers' perceptions of the effectiveness of the planning process? Six specific measures and one overall measure of effectiveness are elicited from respondents. A composite score computed from the unweighted average of the six specific measures is also tested. In contrast to the tested relationships with performance, comprehensiveness is not highly correlated with the composite measure of effectiveness. In fact, the *content*-oriented measures of effectiveness (prediction of trends, generating information, identifying and avoiding problems) show virtually no correlation with specific or composite measures of comprehensiveness. The *process*-oriented measure of effectiveness (enhancing managerial planning ability) is strongly correlated for *all* composite measures of comprehensiveness and 24 of 34 of the specific measures of comprehensiveness (Table 16.6). Thus, the managers that feel their planning system is effective in enhancing the planning abilities of the top management team also identify high levels of comprehensiveness in the process. In contrast, Table 16.7 shows the relationships between the composite measure of effectiveness and the comprehensiveness measures . The inference is that effectiveness of the planning system is correlated to comprehensiveness of the first step, situation diagnosis, and not the rest of the process.

Conclusions and Discussion

The results of the data analysis support the hypothesis that in this industry, comprehensiveness in the planning process is correlated with higher performance. The more synoptic is the planning, the more able is the firm at

TABLE 16.6 Partial Correlations Between Comprehensiveness Measures and Effectiveness in Enhancing Managerial Planning Ability (n=20)

Situation Diagnosis		Alternative Generation		Alternative Evaluation		Decision Integration	
1. Primary Responsibility	.34ᵃ	1. Primary Responsibility	.41ᵇ	1. Primary Responsibility	.37ᵃ	1. Primary Responsibility	.39ᵇ
2. Rely on Outsiders	.43ᵇ	2. Rely on Outsiders	.43ᵇ	2. Rely on Outsiders	.39ᵇ	2. Rely on Outsiders	.39ᵇ
3. Problem Identified	.34ᵃ	3. Alternative Identified	.33ᵃ	3. Alternatives Determined	.53ᶜ	3. Integrating the Dec.	.53ᶜ
4. # of Employees Inv.	.49ᵇ	4. Dropped Alternative	.38ᵃ	4. # of Employees Involved	.50ᵇ	4. Depts. Affected	.37ᵃ
5. Out of Pocket Exp.	.68ᶜ	5. # of Employees Inv.	.49ᵇ	5. Out of Pocket Expenses	.60ᶜ	5. # of Employees Inv.	.43ᵇ
6. Yrs. of Historical Data	.45ᵇ	6. Out of Pocket Exp.	.53ᵇ	6. Years of Historical Data		6. Out of Pocket Exp.	.53ᶜ
7. Outsiders Contacted	.47ᵇ	7. Outsiders Contacted		7. Outsiders Contacted		7. Outsiders Contacted	
8. Analysis of Problem	.58ᶜ	8. Techniques Used		8. Criteria Used	.34ᵃ	8. Techniques Used	
				9. Written Summary or Report			
				10. New Plant Analysis			
Situation Diagnosis Composite	.71ᶜ	Alternative Generation Composite	.58ᶜ	Alternative Evaluation Composite	.62ᶜ	Decision Integration Composite	.61ᶜ

Overall Process Composite .73ᶜ

Level of significance: ᵃ p ≤ .10
 ᵇ p ≤ .05
 ᶜ p ≤ .01

Note: In tables 16.2 through 16.7, only significant correlation coefficients are shown.

TABLE 16.7 Partial Correlations Between Comprehensiveness Measures and Composite Measure of Effectiveness of Firms' Current Planning System (n=20)

Situation Diagnosis	Alternative Generation	Alternative Evaluation	Decision Integration
1. Primary Responsibility .59[c]	1. Primary Responsibility	1. Primary Responsibility	1. Primary Responsibility
2. Rely on Outsiders	2. Rely on Outsiders	2. Rely on Outsiders .40[b]	2. Rely on Outsiders
3. Problem Identified	3. Alternative Identified	3. Alternatives Determined .35[a]	3. Integrating the Dec.
4. # of Employees Inv.	4. Dropped Alternative	4. # of Employees Involved	4. Depts. Affected
5. Out of Pocket Exp.	5. # of Employees Inv.	5. Out of Pocket Expenses .31[a]	5. # of Employees Inv. .34[a]
6. Yrs. of Historical Data .35[a]	6. Out of Pocket Exp. .51[b]	6. Years of Historical Data	6. Out of Pocket Exp.
7. Outsiders Contacted	7. Outsiders Contacted	7. Outsiders Contacted	7. Outsiders Contacted
8. Analysis of Problem .39[b]	8. Techniques Used	8. Criteria Used	8. Techniques Used
		9. Written Summary or Report	
		10. New Plant Analysis	
Situation Diagnosis Composite .50[b]	Alternative Generation Composite	Alternative Evaluation Composite .62[c]	Decision Integration Composite .61[c]

Overall Process Composite .73[c]

Level of significance:
[a] p ≤ .10
[b] p ≤ .05
[c] p ≤ .01

Note: In tables 16.2 through 16.7, only significant correlation coefficients are shown.

finding a proper strategic fit with the environment. As one would expect with a complex behavioral construct such as comprehensiveness, the most significant measures were the composites of the various components of comprehensiveness. The aggregation of the breadth of management involvement, number of outsiders, amount of money spent, number of analyses and reports generated, and level of process structure represents a true comprehensiveness construct which contains more information on planning behavior than any of the individual measures. Comprehensiveness is also more highly correlated to a composite measure of performance than it is to more specific measures of performance. This follows from the fact that firms are pursuing different strategies (private labels, niche marketing, etc.) that will have different performance objectives. Thus, any specific measure of performance may not be particularly relevant to a given firm.

It is also shown that comprehensiveness is highly correlated to the effectiveness of the planning system at enhancing managerial planning abilities. While this may seen tautological on the surface, it is not. While comprehensiveness may not be related to the effectiveness of the planning system at generating information related to the *content* of strategies, it appears to promote a better *process* of strategic decision-making. Managers seem to be more satisfied with decision making in the synoptic process, compared to an incremental process, even though they may not be able to relate the process to superior outcomes. That is, satisfaction with the process of making a decision is associated with a high degree of comprehensiveness in the process, even if superior performance is not a concomitant result.

Given the number of firms in the sample, it is not feasible to segregate the firms by size to statistically test for variation in comprehensiveness, performance, or planning system effectiveness based upon organization size. Correlation analysis does not revel any significant relationships between the two measures of size—number of permanent employees and number of peak season employees—and comprehensiveness, performance, or planning system effectiveness. Scanning the responses from the first step of the survey design, there appears to be no relationship between size or scope of operations and whether a formal planning process is in place.

It is also impossible to control for the content of organizational strategy. This problem is not limited to the sample of firms in this study, but is endemic to strategy research. The diversity of markets served, breadth of product lines, ownership, and other organizational characteristics precludes the investigation of performance relationship to strategic content. Porter (1985) highlights the dimensionality of these choices in his extensive lists of cost and uniqueness drivers. The dimensionality of the strategy set available to participants in the food industries and the time lags between strategic choice and the subsequent performance are nearly intractable problems in testing the strategy—performance relationship. Aside from the treatment of strategic variables in industrial organization economics, there is no published literature where content—performance analysis has been accomplished for the food industries. However, the industrial organization literature uses a limited set of strategy variables (i.e. price, dollars of advertising expenditure, new product introductions) and a has limited ability to capture the dynamics of strategy—performance feedback.

The study of the process—performance relationship is important to the portfolio of research on strategic management in the food industries. Where industrial organization research relates strategy content to profit (or other) performance, the process by which strategies are chosen is ignored. An implied optimizing strategy is presumed in economics that cannot be observed at the firm level. Organizational behavior theory, which drives much of the strategic process literature, is concerned with the behavior of individuals and coalitions within the firm. This research tradition should be useful in future research into food industry performance, because a smaller and smaller number of organizations are dominating these markets and by implication a smaller and smaller number of coalitions of managers are setting strategy. By understanding the decision processes being used, researchers may be better able to understand the dynamics of strategy and counterstrategy in the food industries.

This research follows the experimental design of Fredrickson, who examined the relationship between comprehensiveness in planning and organizational performance in a stable (paints and coatings) and an unstable (sawmilling) industry. By the same criteria used in Fredrickson's studies, the fruit and vegetable canning industry is stable. The hypothesis that under a stable environmental regime, comprehensiveness of the planning process leads to higher performance is tested and accepted. This research extends the analysis to include subjective measurement of performance, in the absence of financial data for the business units in the sample. Also, the research investigates the relationship between comprehensiveness and effectiveness of the planning process. The important result is that effectiveness, as measured by the planning ability of the management, is enhanced in a comprehensive process. Given that comprehensiveness is a major component of a synoptic process, this implies that a comprehensive process leads to an effective process in this industry environment. One can draw the conclusion that repeated cycles of effective strategic planning will lead to higher performance in the long run as a superior fit between the organization and the environment is attained.

References

Aldrich, H. E. 1979. *Organizations and Environments*. Englewood Cliffs, NJ: Prentice-Hall.

Ansoff, H. I. 1965. *Corporate Strategy*. New York: McGraw-Hill.

Bourgeois, L. J. 1980. Strategy and Environment: A Conceptual Integration. *Academy of Management Review* 5(1):25-39.

Dess, G. G. and D. W. Beard. 1984. Dimensions of Organizational Task Environments. *Administrative Science Quarterly* 29:52-73.

Dess, G. G. and R. B. Robinson, Jr. 1984. Measuring Organizational Performance in the Absence of Objective Measures: The Case of the Privately-held Firm and Conglomerate Business Unit. *Strategic Management Journal* 5:265-273.

Fahey, L. and H. K. Christensen. 1986. Evaluating the Research on Strategy Content. *Journal of Management* 12(2):164-184.

Fredrickson, J. W. and T. R. Mitchell. 1984. Strategic Decision Process: Comprehensiveness and Performance in an Industry with an Unstable Environment. *Academy of Management Journal* 27(2):399-423.

Fredrickson, J. W. 1984. The Comprehensiveness of Strategic Decision Process: Extension, Observations, Future Directions. *Academy of Management Journal* 27 (3):445-466.

_____. 1986. An Exploratory Approach to Measuring Perceptions of Strategic Decision Process Constructs. *Strategic Management Journal* 7:473-483.

Hambrick, D. C. 1980. Operationalizing the Concept of Business Level Strategy in Research. *Academy of Management Review* 5:567-576.

Hofer, C. W. and D. E. Schendel. 1978. *Strategy Formulation: Analytical Concepts*. St. Paul, MN: West Publishing.

King, R. P. and S.T. Sonka. 1986. Management Problems of Farms and Agricultural Firms. Discussion paper No. 44, Strategic Management Research Center, University of Minnesota.

Mintzberg, H. 1978. Patterns in Strategy Formation. *Management Science*: 934-948.

Mintzberg, H., H. Raisinghani, and A. Théorêt. 1976. The Structure of Unstructured Decision Processes. *Administrative Science Quarterly* 21:246-275.

Pearce, J. A. and R. B. Robinson. 1982. *Strategic Management*. Homewood, IL: Irwin.

Porter, M. E. 1980. *Competitive Strategy: Techniques for Analyzing Industries and Competitors*. New York: Free Press.

_____. 1985. *Competitive Advantage*. New York: Free Press.

Prescott, J. E. 1986. Environments as Moderators of the Relationship Between Strategy and Performance. *Academy of Management Journal* 29(2):329-346.

Quinn, J. B. 1980. *Strategies for Change: Logical Incrementalism*. Homewood, IL: Irwin.

Ramanujam, V., N. Venkatraman, and J. C. Camillus. 1986. Multi-Objective Assessment of Effectiveness of Strategic Planning: A Discriminant Analysis Approach. *Academy of Management Journal* 29(2):347-372.

Rhyne, L. C. 1986. The Relationship of Strategic Planning to Financial Performance. *Strategic Management Journal* 7:423-436.

Robinson, R. B. and J. A. Pearce. 1988. Planned Patterns of Strategic Behavior and Their Relationship to Business-Unit Performance. *Strategic Management Journal* 9:43-60.

Westgren, R. E. and M. L. Cook. 1986. Strategic Management and Planning. *Agribusiness: An International Journal* 2(4):477-489.

Westgren, R. E., S. T. Sonka, and K. K. Litzenberg. 1988. Strategic Issue Identification Among Agribusiness Firms. *Agribusiness: An International Journal* 4(1):25-38.

17

An Economic Investigation of Federal Antitrust Enforcement in the Food System

Warren P. Preston and John M. Connor

In the century-long history of antitrust enforcement in the United States, the 1980s stand out as a distinctive and perhaps even unique era. Following an activist era of antitrust enforcement that peaked in the mid-1970s, enforcement patterns changed markedly under the Reagan administration (Connor *et al.*, 1985; Gallo, Craycraft and Bush, 1985; Langenfeld and Scheffman, 1990; Subcommittee on Commerce, Consumer and Monetary Affairs, 1983). Particularly in the case of the Federal Trade Commission (FTC), commentators widely ascribe changes in enforcement to the ascendence of economic efficiency as a philosophical foundation for antitrust policy. Importantly, the role of economics grew from mainly *ex post* litigation support to *ex ante* case selection and development (Langenfeld and Scheffman, 1990).

The economic philosophy embraced by the Reagan FTC reflected a more purely market efficiency-oriented criterion for enforcement (Kovacic, 1988). Less concern for equity or firm size was evidenced. Reagan's first appointee to chair the FTC, James C. Miller III, expeditiously replaced existing bureau heads with administrators willing to implement market-oriented reforms (Walton and Langenfeld, 1989). Chicago School tenets of antitrust policy (Posner 1979) formed the intellectual basis for reforms sought during the Reagan administration. As a result, merger enforcement loosened and broad pioneering cases that had been initiated in earlier years were settled or dropped (Tollison, 1983).

The permissive antitrust enforcement policy of the 1980s facilitated a merger wave that shaped a sea-change in industrial structure (Baldwin, 1990). In the food system, acquisitions among food and tobacco manufacturing firms swelled during the 1980s (Connor and Geithman, 1988).   Comparable restructuring occurred   in food wholesaling and retailing (U.S. Dept. of Agriculture).   Thus, previous studies have described both the general declines in antitrust enforcement activity and the consequent restructuring that occurred in the food system during the past decade.   Comparatively little detailed analysis, however, has been conducted to document recent changes in antitrust enforcement efforts directed at the food industries.

In a classic study of antitrust activity, Posner (1970) argued that statistical analysis of antitrust enforcement can serve as a basis for sound policy planning by the antitrust agencies.   The present research addresses this need by conducting such an analysis of FTC enforcement in the food system. Specifically, the primary research objective is to document and describe the level and distribution of FTC antitrust activity directed toward the food manufacturing industries during the 1980s.

Data

This study analyzes a unique data set obtained by the authors under Freedom of Information Act requests.   The data represent reasonably precise measures of resources devoted to antitrust, namely, professional hours charged to antitrust matters by the Federal Trade Commission.[1]   For purposes of the present research, matters encompass all investigations and cases on which antitrust enforcement effort was expended.   Cases refer only to matters of public record, namely matters that proceeded beyond the investigation stage and received docket numbers.   Professional hours include time expended by lawyers and economists on antitrust enforcement but not on other matters such as consumer protection.   The data include all effort reported for matters closed during calendar years 1981 through 1989.

Measuring antitrust inputs confers several advantages over the more typical way of measuring antitrust enforcement, which is numbers of cases brought tabulated in various ways such as by industry and violation alleged (e.g., Gallo, Craycraft and Bush, Posner 1970).   Although bringing cases is certainly an important and probably the most visible aspect of the antitrust agencies' duties, professional time spent on cases does not account for all of the agencies' budgets.   Much effort also is expended on preliminary investigations and other enforcement activities that are not reported publicly through channels such as the *Trade Regulation Reporter* published by Commerce Clearing House and the *FTC Docket of Complaints*.   Hence, case counts neglect the important deterrent effect of "spotlighting" by the antitrust agencies, even though investigation may not proceed beyond the preliminary phase.   Such activities are included in the accounting of professional hours.   Additionally, tabulation of cases brought is a scaleless measure of antitrust effort in the sense that minor lawsuits count the same as major, precedent-setting landmark cases.   Conversely, the extent and precedent-setting importance of an antitrust matter presumably dictates the level of professional resources devoted to it. Measuring antitrust activity by professional hours expended rather than numbers of cases brought is thus an important innovation of this study.

The current study develops detailed cross-tabular analysis of professional effort expended on FTC antitrust enforcement in the food system. The data are categorized by mainly four-digit Standard Industrial Classification (SIC) industries and by federal fiscal years. Included is all antitrust activity reported among any of the four-digit SIC 20 industries related to food processing. Classification of professional resources expended by violations alleged also are reported. Names of individual companies under investigation are highlighted as appropriate to the analysis of prominent cases.

Cumulative Measures of Antitrust Effort

Table 17.1 shows various measures of the cumulative effort expended by the FTC on food manufacturing industry matters closed during 1981 through 1989. One measure of the both the number and size of matters pursued by the FTC is cumulative total effort. Total effort shows how much professional time was expended on matters from initiation to closure (Figure 17.1). For example, 33 professional years were devoted to the matters that closed in 1981. The early

TABLE 17.1 Cumulative Federal Trade Commission Antitrust Effort Expended in the Food Manufacturing Industries, 1981-1989

	Matters Closed in Years				
Item	1981	1982-83	1984-85	1986-89	1981-89
	Professional Years per Year				
Cumulative Total Effort	33.24	105.06	50.94	10.69	43.19
Effort Excluding "Big 3"[a]	33.24	42.01	12.41	10.69	20.62
	Professional Years per Matter				
Matter Type:					
Investigations (A,B)	1.07	0.56	0.57	1.33	0.77
Enforcement (E,F,G,I)	3.17	16.09	27.39	0.24	12.82
Projects (D)	0	2.26	2.09	0.80	1.64
	Number of Matters Closed per Year				
Matter Type:					
Investigations	14	13.5	15.5	9.75	12.33
Enforcement	6	6	1.5	0.5	2.56
Projects	0	1	0.5	0.5	0.56

[a]Kellogg *et al.*, ITT-Continental Baking, and General Foods.

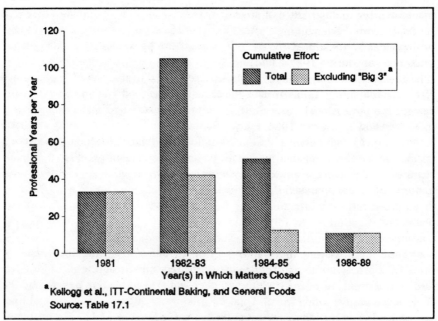

FIGURE 17.1 Cumulative Federal Trade Commission Antitrust Effort Expanded in the
Food Manufacturing Industries, Matters Closed 1981-1989

Reagan years, 1982 and 1983, represent high water marks for termination
oflarge cases. In each of those two years, more than 100 professional years of
FTC effort were accumulated in the matters brought to closure. Conversely,
less than 11 professional years of effort were accrued in the food manufacturing
industry matters that closed during each of the four latter years of the Reagan
administration.   Three disproportionately large cases initiated in the early 1970s
accounted for much of the antitrust enforcement effort amassed in matters closed
during the study period. The cases were Kellogg *et al.*, which closed in 1983,
and ITT-Continental Baking and General Foods, both of which closed in 1984.
All three cases involved monopolization violations, and two of them alleged
price discrimination violations. Dropping the "Big Three" food cases was part
of a larger pattern of change in antitrust enforcement at the beginning of the
Reagan administration. Among its first acts was to dismiss almost all of the big
monopoly cases that were under investigation: the IBM, ATT, petroleum, and
automobile matters are examples. "In their place the FTC sued a small group
of attorneys who represent indigent criminal defendants in Washington, D.C.,
and started investigations of state boards that license taxicabs . . . and of labor
unions that represent actors" (Davidson, 1985, p. 125). Because the Big Three
cases towered over the remaining enforcement workload among food

manufacturing industries, a different picture emerges when their impact is removed from the remaining data. First, the sheer size of the Big Three cases is dramatically illustrated by the 60 percent drop in cumulative professional years per year during 1982-83 and the 75 percent drop during 1984-85. Second, it becomes apparent that the Reagan reforms in antitrust enforcement among the food manufacturing industries were fully implemented by 1984. Namely, cumulative professional years per year dropped to a lower plateau at about a third of its level prior to 1984.

Changes in cumulative professional effort on matters closed in a given period can be disaggregated into changes in the number of matters closed and changes in the average professional effort expended per matter. The great majority of matters pursued by the FTC can be classified as either investigations or enforcement. Investigations may be initial phase or full phase. An initial phase investigation might follow from a complaint filed by a competitor or a consumer. If an initial phase investigation produces sufficient cause for continued investigation, the matter is brought before the Commission for a decision to proceed with a full phase investigation. Investigations that ultimately lead to consent decrees, internal FTC adjudication, external courts, or compliance are classified herein as enforcement matters. Projects are another, less common type of matter. Projects may involve investigation or analysis of, say, a common industry practice that may support subsequent investigation of particular firms or groups of firms.

The number of matters closed per year dropped from about 20 during the Carter and early Reagan years to less than 11 during the latter Reagan years (Figure 17.2). Much of the decline can be attributed to attrition of enforcement, which dropped from six to less than one matters closed per year. Investigations experienced a similar decline, from about 14 or 15 closed per year to fewer than 10 closed per year. No projects were closed in 1981, and these matters remained few in number throughout the rest of the decade.

As would be expected, matters that terminate at the enforcement stage require much more professional time relative to those that close as investigations (Figure 17.3). On average, enforcement matters among the food manufacturing industries consumed almost 17 times more professional time than did investigations. The average food manufacturing industry enforcement matter entailed almost 13 years of professional effort, but this number is heavily influenced by the extensive effort expended on the Big Three cases. Professional years per investigation dipped during the first half of the 1980s, but rebounded to 1.3 professional years per matter during the latter half of the decade. This change may be attributed to an increase in the number of matters closed as initial phase investigations from 1982 through 1985, followed by a move toward fewer initial phase and more full phase investigations closed during 1986 through 1989. The few projects that were terminated required about one or two professional years each.

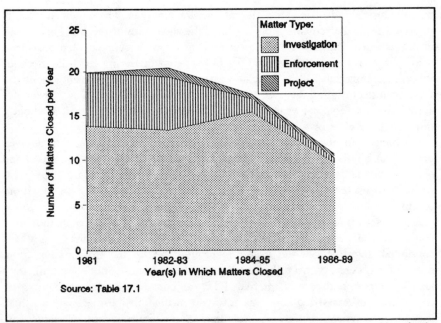

FIGURE 17.2 Number of Federal Trade Commission Matters Closed per Year in the
Food Manufacturing Industries, 1981-1989

An alternative measure of FTC activity is the duration of antitrust matters,
or the length of time from initiation to closing of matters. Table 17.2 presents
the mean duration of food manufacturing industry antitrust matters closed
between 1981 and 1989. These numbers must be interpreted with caution and
are reported solely as rough guides to the comparative durations of various types
of matters. The duration of a given matter is computed as the difference
between the year in which the matter closed and the first year in which
professional time was expended on the matter, plus one year. Thus, a matter
that began and ended in the same fiscal year would have a duration of one year,
while a matter that spanned two fiscal years would be credited with a duration
of two years. For matters begun prior to 1978, the year of initiation is
unavailable. Hence, the lower bound of the duration for such a matters was
calculated as the year in which the matter closed minus 1978, plus one year.

Initial phase investigations among the food manufacturing industries lasted
less than two years on average, and this represents an upper bound estimate
because of the manner in which duration was computed. By reason of FTC
management practices, within a few months initial phase investigations are either
terminated or boosted to full phase investigations. There was a substantial
change in the number of matters closed as preliminary investigations over the
course of the study period. About 10 preliminary investigations were closed

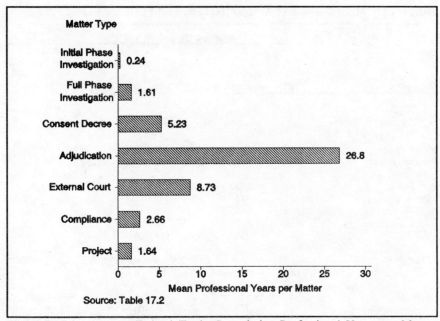

FIGURE 17.3   Average Federal Trade Commission Professional Years per Matter, Matters Closed in the Food Manufacturing Industries, 1981-1989

each year from 1981 through 1985, but only 15 preliminary investigations in total were closed during the final four years of the Reagan administration.

Virtually by definition, full phase investigations took longer to complete than did initial phase investigations. The average duration for full phase investigations was less than three years, with a generally declining trend over the study period. The average duration fell from five years for matters concluded in 1981 to just over two years for matters closed during 1986 to 1989. There was a concurrent increase in the number of preliminary investigations closed each year, up from four in 1981 to six per year during the latter Reagan years.

For the four consent decrees that were secured from food manufacturing firms, about five years elapsed from the years in which the matters were initiated until the consent decrees were obtained. From 1984 to 1989, no consent decrees were entered into with food manufacturers. Additionally, no food manufacturing cases were adjudicated after 1985. Because the bulk of the adjudicated matters were initiated prior to 1978, estimates of the duration of these matters are truncated. Nonetheless, mean duration for internally adjudicated matters was at least four years and in actuality probably closer to five to seven years.

TABLE 17.2 Mean Duration of Antitrust Matters, Matters Closed During 1981-1989

Matter Type	1981	1982-83	1984-85	1986-89	1981-89	Average Professional Years per Matter
			Matters Closed in Years			
			Years (No. of Matters)			
Initial Phase Investigation	1.70 (10)	1.62 (21)	2.09 (22)	1.73 (15)	1.78 (68)	0.24
Full Phase Investigation	5.00 (4)	3.00 (6)	3.33 (9)	2.13 (24)	2.77 (43)	1.61
Consent Decree	8.00 (1)	4.33 (3)	-- (0)	-- (0)	5.25 (4)	5.23
FTC Adjudication	>2.25[a] (4)	>5.00[a] (3)	>7.00[a] (2)	-- (0)	>4.22[a] (9)	26.80
External Adjudication	-- (10)	>11.00[a] (1)	-- (0)	-- (0)	>11.00[a] (1)	8.73
Compliance	>2.00[a] (1)	>3.60[a] (5)	>6.00[a] (1)	1.00 (2)	>3.11[a] (9)	2.66
Projects	-- (0)	4.00 (2)	4.00 (1)	2.50 (2)	3.4 (5)	1.64

[a] Lower bound for mean duration. Year of initiation unavailable for adjudicated matters opened prior to 1978.

Industries Investigated

Current year effort by the FTC on matters related to the food manufacturing industries varied considerably during the 1980s (Table 17.3). A total of 139 matters were closed from 1981 through 1989. Annual professional effort expended on these matters dropped by half after the Reagan administration was seated, from 41 professional years per year during 1978-1981 to 22 professional years per year during 1982-1983 (Figure 17.4). The decline in enforcement among food manufacturing firms is even more dramatic considering the fact that the data for the 1978-1981 period omit effort devoted to any matters closed prior to 1981. Professional effort declined further under the Reagan White House, reaching its nadir during 1984-1985. Only a negligible six professional years per annum were devoted to antitrust enforcement among the food manufacturing firms during that period. Antitrust activity rebounded

TABLE 17.3   Current Federal Trade Commission Antitrust Effort by Food
Manufacturing Industry, Federal Fiscal Years 1978-1989

Industries (SIC)	1978-81	1982-83	1984-85	1986-89
	Professional Years per Year[a]			
Big Three[b]	29.42	4.24	0.01	0.00
Meat and Poultry (201)	0.28	0.08	0.00	0.00
Fluid Milk (2026)	1.95	4.61	0.94	0.19
Other Dairy (202 other)	0.25	0.02	1.08	0.01
Fruits & Vegetables (203)	5.37	0.13	0.00	0.19
Other Grains (204 other)	0.04	2.86	0.07	0.59
Bakery Products (205)	0.69	3.67	0.12	0.00
Confectionery (2065-67)	0.49	0.15	0.00	0.00
Oils (207)	0.00	0.00	0.00	0.52
Soft Drinks (2086-87)	1.98	2.33	1.54	5.05
Beer/Other Alch. Bevs. (2082-85)	0.00	1.98	1.10	0.23
Seafood (2091-92)	0.38	0.97	0.17	0.15
Miscellaneous Foods (2098-99)	0.27	0.89	0.94	1.32
Diversified[c]	0.00	0.41	0.12	0.79
total	41.11	22.35	6.16	10.79

[a]Effort on matters closed during 1981-1989 only.

[b]Kellogg, *et al.*, ITT-Continental Baking, and General Foods, which otherwise
would be classified in the breakfast cereals, bread, and "diversified" industries.

[c]Firms classified by more than one three-digit SIC code.

during the last four years of the Reagan administration, when effort climbed
back to almost 11 professional years per year. The recovery likely was greater
given the fact that current year antitrust activity is subject to under counting,
especially during the latter years of the study period. The reason is that the data
omit effort on open matters. That is, data collected for the present study
exclude activity attributed to any matters active during the study period that
remained open after 1989. The latter years of the study period include a higher
proportion of such unresolved matters relative to earlier years.

In addition to changes in the level of antitrust enforcement among food
manufacturers, the portfolio of industries subject to investigation changed (see
Figure 17.5 for a portfolio of matters closed during the entire study period).
During the Carter years, the Big Three monopolization cases accounted for more
than two-thirds of the FTC's annual professional activity directed at the food
manufacturing industries. The Big Three cases involved the cereal breakfast
foods, bread, and roasted coffee four-digit SIC industries. Except for these
three cases, there was little activity allocated to other matters within the relevant
parent three-digit industries (grain mill products, bakery products, miscellaneous
foods and kindred products). More than five professional years per annum were

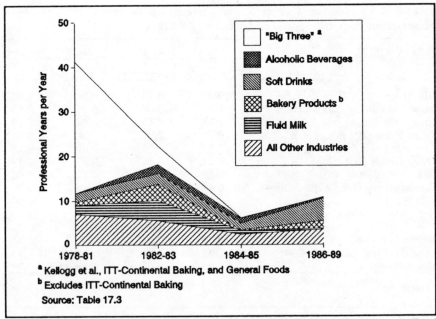

FIGURE 17.4   Current Federal Trade Commission Antitrust Effort Among Food
Manufacturing Industries, Matters Closed 1981-89

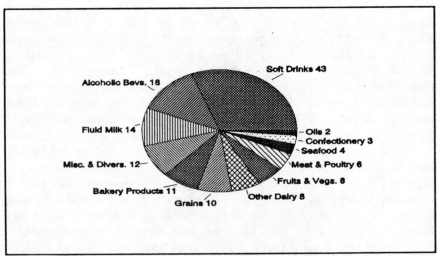

FIGURE 17.5 Distribution of Federal Trade Commission Antitrust Matters Among Food
Manufacturing Industries, Matters Closed 1981-89

directed toward the preserved fruits and vegetables industry during the 1978-1981 period. Other industries in which there was significant enforcement activity under the Carter-influenced FTC budget include fluid milk and soft drinks, with effort of about two professional years per annum each.

As the Reagan administration came into control, the Big Three cases were rather quickly brought to a close. Hence, annual professional effort on these cases during 1982-1983 dropped seven-fold from the previous four years. Some of the decline in resources devoted to the Big Three cases was offset by increased enforcement directed toward other food manufacturing firms. Foremost among the industries in which enforcement increased were fluid milk, bakery products (other than bread), grain mill products (other than cereal breakfast foods), soft drinks, and alcoholic beverages. Nonetheless, the increased effort among these industries was insufficient to compensate for the retreat in antitrust enforcement resulting from the termination of the Big Three cases.

By 1984-1985, FTC antitrust oversight of the food manufacturing industry nearly ground to a halt. A bit of professional time was spent in closing the last of the Big Three cases during 1984. Compared to the previous two years, during 1984-1985 each and every three-digit SIC industry within food and kindred products manufacturing experienced declining FTC professional effort. Effort of about a professional year or more per annum was reported in dairy products (both fluid milk and all other dairy products), soft drinks, alcoholic beverages, and miscellaneous foods.

Most of the recovery in antitrust effort that occurred in the food manufacturing industries during the latter part of the 1980s can be attributed to additional enforcement directed in two areas: bakery products and soft drinks. Professional activity nonetheless remained low among the food manufacturing industries. In addition to bakery products and beverages, the only other three-digit industry subject to at least one professional year of FTC activity per annum was miscellaneous foods and kindred products.

Violations Alleged

An additional way to characterize changes in antitrust activity is to examine the allocation of enforcement effort by types of violations alleged. Table 17.4 shows current year allocations of FTC professional activity among the food manufacturing industries categorized by sections of the law investigated. Several patterns of change are worth noting. Through the first half of the Reagan administration, professional effort per annum on price fixing matters increased from the Carter-determined budget. Such a reorientation is consistent with the Reagan administration's commitment to focus antitrust enforcement toward conduct-related matters rather than structural issues.[2] Interestingly, effort on

TABLE 17.4   Federal Trade Commission Antitrust Effort Expended in the Food Manufacturing Industries, by Type of Violation Alleged, Fiscal Years 1978-1989

Type of Violation[a]	Carter Admin. 1978-81	Reagan Administration		
		1982-83	1984-85	1986-89
	Professional Years per Year			
Price Fixing (Sherman §1)	0.32	2.86	2.33	0.97
Monopolization (Sherman §2)	31.88 (2.46)[b]	4.86 (0.68)	0.37 (0.30)	0.61
Price Discrimination (Clayton §2)	16.45 (0.31)	4.20 (0.60)	0.26 (0.20)	0.26
Exclusive Dealing and Tying (Clayton §3)	2.10	3.89	2.18	0.30
Mergers (Clayton §7)	7.06	13.25	2.92	9.88
Horizontal Interlocks	0.01	0.01	0	0
Other	0.09	0.03	0.59	0.22
total[a]	57.91 (12.30)	29.11 (21.33)	8.64 (8.52)	12.23 --

[a]If the same matter involves two violations, the effort is double counted.

[b]Numbers in parentheses remove the effort expended on three huge monopolization cases begun before 1987, viz., Kellogg, *et al.*; ITT-Continental Baking; and General Foods.

price fixing matters declined during the latter Reagan years even as overall enforcement among the food manufacturing industries rose.

Monopolization matters showed the most drastic decline in effort from the beginning to the end of the study period. Annual effort on food manufacturing monopolization matters dropped incredibly from almost 32 professional years during 1978-1981 to less than two-thirds of a professional year during each of the last four years of the study period. Price discrimination matters exhibited similar proportional declines in enforcement effort, but at about half the scale.[3] During the late Carter and early Reagan years, the Big Three cases dominated the FTC antitrust budget directed at monopolization and price discrimination enforcement among food manufacturing firms. Removing the effort attributed to the Big Three cases shows that other potential monopolization and price discrimination violations by food manufacturers were not vigorously pursued by the FTC.

Exclusive dealing and tying arrangements (Clayton Section Three violations) involving food manufacturing firms received at least two years per annum of professional time from 1978 through 1985. Indeed, the annual allocation toward these types of violations nearly doubled to almost four professional years per year during the first two Reagan antitrust budgets. During the second term of the Reagan administration, however, effort directed toward exclusive dealing and tying violations dropped to a scant one-third of a professional year per annum.

Public commentary and criticism of enforcement policy reached a crescendo during the tenure of the initial Reagan appointees to head the federal antitrust agencies. Perhaps no other segment of antitrust enforcement authority received more attention from the public than did merger enforcement. The relaxed merger enforcement climate under the Reagan administration has been cited as an important contributing factor for the wave of mergers and acquisitions that swept over the food manufacturing industry during the 1980s (Connor and Geithman). While the *disposition* of merger investigations may have fostered a more hospitable climate for consolidation of firms, the same conclusion cannot be drawn by observing time spent on merger enforcement during the initial years of the Reagan presidency. During 1982 and 1983, annual FTC professional time spent on merger investigations actually increased by five professional years per year from the late Carter years. Again, however, this observation must be tempered by the fact that the 1978-1981 data almost certainly under count time spent on investigations since data on matters concluded prior to 1980 are missing. In the case of merger enforcement in particular, antitrust inputs may have been supply-driven rather than the result of opposition to the anticompetitive consequences of mergers. That is, the existence of Hart-Scott-Rodino premerger notification requirements suggests that the large supply of mergers requiring review in the 1980s determined the level of enforcement activity. Of course, the FTC could choose to impose more stringent review, but notification requirements do place an effective floor on merger enforcement activity. Moreover, antitrust agency reviews became increasingly burdensome as the new "fix-it-first" merger policy began to be implemented (McGuckin). This internal enforcement standard was used as early as the 1981 Marathon Oil contested takeover (Davidson). Under this policy the FTC would agree not to oppose mergers in which the two entities would agree in advance to divest themselves of all clearly horizontal market overlaps. Considerable staff time was required to analyze and respond to these often complex divestiture proposals.

Time spent on merger enforcement dropped to a low of less than three professional years per annum during 1984-1985. Although the decline may be attributable to a fall in the "supply" of mergers to investigate, the four-fold drop in annual professional activity suggests that the FTC instituted a policy decision not to challenge mergers as vigorously as in previous years. Perhaps in response to public and Congressional calls for more active oversight of mergers among large food processors, FTC effort on merger enforcement rebounded to nearly 10 professional years per year during the final four years of the study period.

The relative importance of different types of violations in the FTC's enforcement portfolio can be determined by examining the proportion of the professional budget allocated to each alleged violation. Table 17.5 makes abundantly clear the overwhelming dominance of the Big Three cases during the first years of the study period. For the period 1978-1982, effort on alleged monopolization violations consumed more than half of the professional time expended on all food manufacturing matters. Together with price discrimination, the share jumps to more than 80 percent of all effort. Removing the effort expended on the Big Three cases paints quite a different picture (Figure 17.6). Without the Big Three cases, merger enforcement moves to the fore as the major domain for FTC activity among the food manufacturing industries during the 1978-1981 period. Even without the huge effort on the Big Three cases, monopolization matters still maintained a one-fifth share of professional activity during the Carter administration.

The Big Three cases all were initiated prior to 1978. Thus, the measures of professional effort excluding those cases portray a more accurate representation of discretionary decisions made during the study period about the allocation of FTC enforcement among food processors. When viewed as such, changes in emphasis over time on different types of alleged violations become apparent. On food industry matters other than the Big Three cases, merger enforcement consumed the lion's share of professional activity during the late Carter and early Reagan years. Merger enforcement's share of time fell by almost half during the lean enforcement years of 1984-1985, but still accounted for the largest share of effort. Merger investigations jumped to a commanding 81 percent share of enforcement activity during the resurgence of professional effort in the latter Reagan years.

Even without the addition of the large effort on the Big Three cases, monopolization matters still accounted for one-fifth of the professional time spent on food industry matters during the period 1978-1981. With the first budget under control by Reagan appointees, however, monopolization matters dropped to a scant three percent share of professional effort. Thereafter, this share rose slightly but remained small through the rest of the 1980s. Price discrimination matters never received much more than three percent of professional time when the Big Three cases are removed from the data.

Excluding effort on the Big Three cases, exclusive dealing and tying investigations took the third-largest share of professional effort under the Carter administration. This share rose to one-fourth of the professional budget during 1984-1985, but dropped to a scant three percent share during 1986-1989.

Price fixing investigations were a relatively unimportant component of the FTC's food manufacturing enforcement portfolio under the Carter administration

TABLE 17.5 Percentage Distribution of Federal Trade Commission Antitrust Effort Expended in the Food Manufacturing Industries, by Type of Violation Alleged, Fiscal Years 1978-1989

Type of Violation[a]	Carter Admin. 1978-81	Reagan Administration		
		1982-83	1984-85	1986-89
		Professional Years(%)		
All Effort				
Price Fixing (Sherman §1)	0.6	9.8	27.0	7.9
Monopolization (Sherman §2)	55.1	16.7	4.3	5.0
Price Discrimination (Clayton §2)	28.4	14.4	3.0	2.1
Exclusive Dealing and Tying (Clayton §3)	3.6	13.4	25.2	2.5
Mergers (Clayton §7)	12.2	45.5	33.8	80.8
Horizontal Interlocks	0.0	0.0	0.0	0.0
Other	0.2	0.1	6.8	11.8
total	100.0	100.0	100.0	100.0
Excluding "Big Three"[b]				
Price Fixing (Sherman §1)	2.6	13.4	27.4	7.9
Monopolization (Sherman §2)	20.0	3.2	4.4	5.0
Price Discrimination (Clayton §2)	2.5	2.8	3.1	2.1
Exclusive Dealing and Tying (Clayton §3)	17.1	18.2	25.6	2.5
Mergers (Clayton §7)	57.4	62.1	34.3	80.8
Horizontal Interlocks	0.0	0.1	0.0	0.0
Other	0.7	0.1	6.9	1.8
total	100.0	100.0	100.0	100.0

[a]If the same matter involves two violations, the effort is double counted.

[b]Numbers remove the effort expended on three huge monopolization cases begun before 1987, viz., Kellogg, *et al.*; ITT-Continental Baking; and General Foods.

budgets from 1978-1981. The share of professional time devoted to price fixing matters jumped by a magnitude of five under the early years of the Reagan administration, and doubled again to 27 percent during 1985-1985. During the final Reagan years, however, the share of effort on price fixing matters in the food manufacturing industries dropped back to eight percent.

Conclusions

The analysis presented in this study is an example of an application of positive economics to a public process, namely policy conduct. The period

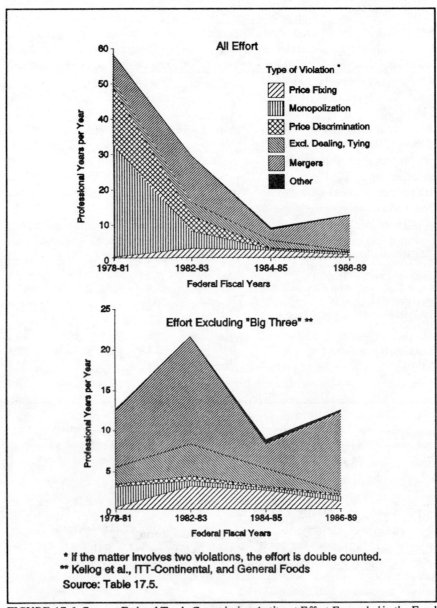

FIGURE 17.6 Current Federal Trade Commission Antitrust Effort Expended in the Food
Manufacturing Industries by Type of Violation Alleged, Matters Closed 1981-1989

examined is an exceptionally interesting watershed in antitrust enforcement. Economic theory became more important as a tool for selecting and developing cases, rather than serving solely in the position of litigation support after cases had already been selected. As the role of economic theory grew, a transition from reliance on mainstream to Chicago-school industrial organization theory occurred. Consequently, the antitrust policy stance moved from an essentially interventionist to a laissez-faire philosophy of government.

The description of antitrust activity provided herein demonstrates the advantages of moving from more aggregated to highly micro-level behavioral data. That is, much antitrust investigation and enforcement effort is concealed if the only measure of activity is tabulations of cases brought. The accounting of professional inputs shows more accurately how patterns of antitrust enforcement change over time. The analysis of internal FTC data obtained for the present study is analogous to observing individual sales transactions rather than industry aggregate shipments.

There are limitations to the use of professional inputs as measures of antitrust enforcement effort. One limitation of the current study is that data are absent from the Antitrust Division of the Department of Justice. Because of the joint enforcement responsibility of the FTC and the Antitrust Division, the actions of both agencies would need to be included if the data are to be used in economic models. Another limitation is that the study ignores antitrust effort involved in private suits brought under federal antitrust statutes, as well public and private litigation under state antitrust laws. During the early 1980s, there were more than 1,200 private cases filed per year under federal antitrust statutes. The number of private suits brought was about 12 times greater than the number of U.S. government cases (Salop and White). The number of state-level cases is doubtless much larger. Nevertheless, public federal effort is important because such cases set precedents and standards to which other cases must adhere.

Despite the limitations of professional time as an indicator of antitrust enforcement, the availability of the measure opens up some potentially fruitful directions for future research. In particular, one avenue would be to move from the essentially conduct-oriented analysis contained herein to a more performance-oriented one. Specifically, one could examine whether investigations (especially those that reach full phase) have had deterrence effects on industry merger patterns. That is, does the mere presence of antitrust oversight deter mergers? Conversely, have firms responded to the fix-it-first policy to the extent that it has stimulated merger activity among rivals once they observed how easy merger clearances had become?

Notes

1. Comparable data were not yet available from the Antitrust Division of the Department of Justice.

2. A caveat that must be reiterated is that the 1978-1981 figures shown in Table 17.4 account for matters closed during the period 1981 through 1989 only. Straightforward price fixing matters tend to be of relatively short duration. Thus, much of the effort on price fixing violations for the years prior to 1981 likely is absent from the data set.

3. Note that Tables 17.4 and 17.5 double count professional effort if the same matter involves two violations.

References

Baldwin, W. L. 1990. Efficiency and Competition: The Reagan Administration's Legacy in Merger Policy. *Review of Industrial Organization* 5(2): 159-174.

Connor, J. M., R. T. Rogers, B. W. Marion, and W. F. Mueller. 1985. *The Food Manufacturing Industries: Structure, Strategies, Performance, and Policies.* Lexington, Mass.: Lexington Books.

Connor, J. M., and F. E. Geithman. 1988. Mergers in the Food Industries: Trends, Motives, and Policies. *Agribusiness* 4(4): 331-346.

Davidson, K. M. 1985. *Megamergers: Corporate America's Billion-Dollar Takeovers.* Cambridge: Ballinger.

Gallo, J. C., J. L. Craycraft, and S. C. Bush. 1985. Guess Who Came to Dinner—An Empirical Study of Federal Antitrust Enforcement for the Period 1963-1984. *Review of Industrial Organization* 2: 106-131.

Kovacic, W. E. 1988. Public Choice and the Public Interest: Federal Trade Commission Antitrust Enforcement During the Reagan Administration. *Antitrust Bulletin* 33(3): 467-504.

Langenfeld, J., and D. T. Scheffman. 1990. The FTC in the 1980s. *Review of Industrial Organization* 5(2): 79-98.

McGuckin, R. H. 1990. Merger Enforcement: Out of the Courtroom After 75 Years. *Antitrust Bulletin* 35(3):677-694.

Posner, R. A. 1970. A Statistical Study of Antitrust Enforcement. *Journal of Law and Economics* 13: 365-420.

———. 1979. The Chicago School of Antitrust Enforcement. *University of Pennsylvania Law Review* 127: 925-948.

Salop, S. C., and L. J. White. 1988. Private Antitrust Litigation: An Introduction and Framework. In L. J. White ed. *Private Antitrust Litigation: New Evidence, New Learning.* Cambridge: MIT Press.

Subcommittee on Commerce, Consumer, and Monetary Affairs, U.S. House of Representatives. 1983. Subcommittee Oversight Hearing on Federal Trade Commission Operation. Staff memorandum. Washington, D.C.

Tollison, R. D. 1983. Antitrust in the Reagan Administration: A Report from the Belly of the Beast. *International Journal of Industrial Organization.* 1(2): 211-221.

U.S. Dept. of Agriculture. 1990. *Food Marketing Review, 1989-90.* Agricultural Economic Report No. 639. Washington, D.C.: Commodity Economics Division, Economic Research Service, USDA.

Walton, T. F., and J. Langenfeld. 1989. Regulatory Reform Under Reagan—The Right Way and the Wrong Way. In R. E. Meiners and B. Yandle, eds., *Regulation and the Reagan Era: Politics, Bureaucracy, and the Public Interest.* New York: Holmes and Meier.

About the Book

The economic performance of large firms in the food manufacturing and retail distribution industries are critical to the overall performance of the food sector. Contributors to this volume analyze the implications of recent events—such as the trend toward increased seller concentration due to mergers and leveraged buy-outs—that have helped increase food firm stock prices some 900 percent during the 1980s. Using diverse methods to analyze competitive strategy, the authors evaluate the contribution of many factors—including market share, concentration, cost conditions, relative efficiency of different-sized firms, pricing strategies, mobility barriers, and public policy—to increased profitability for leading food firms. The book is essential reading for industrial organization economists, food and agricultural marketing economists, strategic marketing analysts, and antitrust and regulatory policymakers.

About the Contributors

Keith B. Anderson is senior economic advisor to Commissioner Anne E. Brunsdale of the U.S. International Trade Commission and an economist in the Bureau of Economics at the Federal Trade Commission.

Wiltse Bailey is a project assistant in the Department of Agricultural Economics at the University of Wisconsin, Madison.

John M. Connor is a professor of Agricultural Economics at Purdue University, West Lafayette.

Ronald W. Cotterill is a professor of Agricultural and Resource Economics and director of the Food Marketing Policy Center at the University of Connecticut, Storrs.

Catherine A. Durham is an assistant professor of Agricultural Economics at Purdue University, West Lafayette.

Olan D. Forker is a professor of Agricultural Economics at Cornell University, Ithaca.

Stuart D. Frank is an agricultural economist at Packers and Stockyards Administration, U.S. Department of Agriculture.

Frederick E. Geithman is an associate research scientist in the Agricultural Economics Department at the University of Wisconsin, Madison.

Lawrence E. Haller is a research assistant at the Food Marketing Policy Center, Department of Agricultural and Resource Economics, University of Connecticut, Storrs.

Charles R. Handy is an agricultural economist with the U.S. Department of Agriculture and the Economic Research Service.

Keith Heimforth is a project assistant in the Agricultural Economics Department at the University of Wisconsin, Madison.

Dennis R. Henderson is a professor of Agricultural Economics at Ohio State University, Columbus.

Clement W. Iton is a post-doctoral fellow at the Food Marketing Policy Center, Department of Agricultural and Resource Economics, University of Connecticut, Storrs.

Larry S. Karp is an associate professor of Agricultural and Resource Economics at the University of California, Berkeley.

Phil R. Kaufman is an agricultural economist with the U.S. Department of Agriculture and the Economic Research Service.

John E. Lenz is a research associate in the Agricultural Economics Department at Cornell University, Ithaca.

Rigoberto A. Lopez is an associate professor of Agricultural and Resource Economics at the University of Connecticut, Storrs.

Bruce W. Marion is a professor of Agricultural Economics at the University of Wisconsin, Madison.

Emilio Pagoulatos is professor and department head of Agricultural and Resource Economics at the University of Connecticut, Storrs.

Jeffrey M. Perloff is a professor of Agricultural and Resource Economics at the University of California, Berkeley.

Warren P. Preston is an assistant professor of Agricultural Economics at Virginia Polytechnic Institute and State University, Blacksburg.

Richard J. Sexton is an associate professor of Agricultural Economics at the University of California, Davis.

Ian M. Sheldon is an associate professor of Agricultural Economics at Ohio State University, Columbus.

Steven T. Sonka is a professor of Agricultural Management and Director of the Food and Agribusiness Management Program at the University of Illinois, Urbana.

Gunta S. Vitins is a market development officer in the Food Industry Branch at the B.C. Ministry of Agriculture, Fisheries & Food, Canada.

Joyce Wann is an associate professor at the Research Institute of Agricultural Economics, National Chung-Hsing University, Taiwan, Republic of China.

Randall E. Westgren is an associate professor and department chair of Agricultural Economics at McGill University, Sainte Anne De Bellevue, Quebec, Canada.